DEMOGRAPHIC
TRENDS AND PATTERNS
IN THE SOVIET UNION
BEFORE 1991

The International Institute for Applied Systems Analysis

is an interdisciplinary, nongovernmental research institution founded in 1972 by leading scientific organizations in 12 countries. Situated near Vienna, in the center of Europe, IIASA has been for more than two decades producing valuable scientific research on economic, technological, and environmental issues.

IIASA was one of the first international institutes to systematically study global issues of environment, technology, and development. IIASA's Governing Council states that the Institute's goal is: *to conduct international and interdisciplinary scientific studies to provide timely and relevant information and options, addressing critical issues of global environmental, economic, and social change, for the benefit of the public, the scientific community, and national and international institutions.* Research is organized around three central themes:

- Global Environmental Change;
- Global Economic and Technological Change;
- Systems Methods for the Analysis of Global Issues.

The Institute now has national member organizations in the following countries:

Austria
The Austrian Academy of Sciences

Bulgaria
The National Committee for Applied
Systems Analysis and Management

Canada
The Canadian Committee for IIASA

Czech and Slovak Republics
The Committees for IIASA of the
Czech and Slovak Republics

Finland
The Finnish Committee for IIASA

Germany
The Association for the Advancement
of IIASA

Hungary
The Hungarian Committee for Applied
Systems Analysis, Hungarian
Academy of Sciences

Italy
The National Research Council (CNR)
and the National Commission for Nuclear
and Alternative Energy Sources (ENEA)

Japan
The Japan Committee for IIASA

Netherlands
The Netherlands Organization for
Scientific Research (NWO)

Poland
The Polish Academy of Sciences

Russia
The Russian Academy of Sciences

Sweden
The Swedish Council for Planning and
Coordination of Research (FRN)

United States of America
The American Academy of Arts and
Sciences

DEMOGRAPHIC TRENDS AND PATTERNS IN THE SOVIET UNION BEFORE 1991

Edited by
Wolfgang Lutz,
Sergei Scherbov and
Andrei Volkov

London and New York

International Institute for Applied
Systems Analysis
Laxenburg, Austria

First published 1994
by Routledge
2 Park Square, Milton Park, Abingdon, Oxon, OX14 4RN

Simultaneously published in the USA and Canada
by Routledge
270 Madison Ave, New York NY 10016

Transferred to Digital Printing 2011

Typeset by International Institute for Applied Systems Analysis

British Library Cataloguing in Publication Data
A catalogue reference for this book is available from the British
Library

Library of Congress Cataloging in Publication Data
has been applied for

ISBN 0–415–10194–8

Publisher's Note
The publisher has gone to great lengths to ensure the quality of this reprint
but points out that some imperfections in the original may be apparent.

Contents

III Components of Mortality Trends

IV Changes in Population Structure

Foreword

Each day brings new examples of collaboration between East and West in different spheres of science. The book that you hold in your hands is an example; it is no ordinary book.

This book is devoted to demographic processes in the USSR. Demography is a field that involves observation of some general features of human development, and that could explain why, in the former USSR, this field of science was partly closed to researchers: the similarity in demographic processes in different social systems could have disturbed some of the sociological myths of socialist development.

Nevertheless, the search for demographic patterns took place on both sides of the Iron Curtain. It was not scientists' fault that linguistic and political barriers prevented them from knowing the work of their foreign colleagues. The seminar held in 1990 in Tbilisi, Georgia, which provided material that serves as the basis of this book, was virtually the first time that demographic scientists from East and West could work together. The success of the seminar and this book underscore the simple fact that, with joint effort, we can achieve more.

For Western readers it may mark their first opportunity for learning about demographic processes in the former USSR, as evaluated by experienced scientists after they finally got access to data. I hope that the authors of this book will broaden and deepen their joint research. I know that after such a successful beginning readers will join me in wishing them no less success in the continuation of their work.

Stanislav S. Shatalin
Moscow
September 1992

List of Figures

List of Tables

Contributors

Barbara A. Anderson is a professor of sociology at the University of Michigan, Population Studies Center. She received a Ph.D. in 1974 from Princeton University. Her research interests include mortality patterns in the republics of the former USSR.

Evgeny M. Andreev is head of the demography information system laboratory at the Institute of Statistics and Economic Research, State Committee of the Russian Federation on Statistics, Moscow. He received a Ph.D. from Moscow University. His main research interests include analysis of dynamics and differentiation of mortality and mathematics modeling of population trends.

Alexandr Anichkin is a senior researcher at the Institute for Employment Studies Center of Demography and Human Ecology, Russian Academy of Sciences and Ministry of Labor and Employment, Moscow. He received his Ph.D. in 1988. His main field of interest is demographic analysis of population reproduction.

Alexandr Avdeev is chief of the population economics and socio-demographic development division at the Center for Population Problems Studies, Moscow State University. He received his Ph.D. from the same university in 1984. His research areas include fertility, family planning, and population economics.

Valentina Bodrova is an adviser on women and family programs at the All-Russian Center for Public Opinion and Market Research in Moscow. She received her Ph.D. in 1965. She has specialized in studies on population development, population policy, and the status of women and the family.

Galina Bondarskaya is a senior researcher in the department of demography at the Institute of Statistics and Economic Research, State Committee of the Russian Federation on Statistics, Moscow. In 1974 she received a Ph.D. from the Research Institute of Statistics. Her main research interests are analyses and projections of fertility.

Ansley J. Coale is Professor Emeritus of Economics and Public Affairs at Princeton University and the senior demographer at the Office of Population Research. He received a Ph.D. from Princeton University in 1947. His research interests include mathematical demography, historical demography, mortality, fertility, nuptiality, and methods of estimation.

Leonid E. Darsky is head of the population reproduction analysis laboratory at the Institute of Statistics and Economic Research, State Committee of the Russian Federation on Statistics, Moscow. He received his Ph.D. from the Research Institute of Statistics in 1968. His main scientific interests include analyses and projections of dynamics and differentiation of fertility, nuptiality, and marital status.

Irina Ilyina is a senior researcher in the department of demography at the Institute of Statistics and Economic Research, State Committee of the Russian Federation on Statistics, Moscow. In 1976 she received her Ph.D. from the Research Institute of Statistics. Nuptiality is her main area of interest.

Vida Kanopiene is a senior researcher at the Institute of Economics, Lithuanian Academy of Sciences, Vilnius. She received her D.Sc. from Vilnius University in 1983. Her main research interest includes women's employment.

Kalev Katus is the director of the Estonian Interuniversity Population Research Center. He received his Ph.D. in 1982. His main areas of research are long-term trends of population development and fertility analysis.

Nathan Keyfitz is a member of the US National Academy of Science and Professor Emeritus of Harvard University. He is currently Institute Scholar at IIASA. His main interests are aging and population and environment.

Tatiana L. Kharkova is a senior researcher in the department of demography at the Institute of Statistics and Economic Research, State Committee of the Russian Federation on Statistics, Moscow. She received her Ph.D. from the Research Institute of Statistics in 1983. Her research interests include analyses of demographic behavior and studies of demographic history.

W. Ward Kingkade is a senior demographer in the Europe/Commonwealth of Independent States Branch of the Center for International Research of the US Bureau of the Census. He received a Ph.D. in 1983 from Brown University. His research focuses on the demography of the newly independent states and includes the US government's official population projections for these countries.

Juris Kruminš is an associate professor in the department of statistics and demography at Latvian State University in Riga.

Natalia Ksenofontova is a researcher in the department of demography at the Institute of Statistics and Economic Research, State Committee of the Russian Federation on Statistics, Moscow. She has studied at the Moscow Institute of Economics and Statistics. Her main research area is infant mortality.

Andis Lapinš is a professor at Latvian State University in Riga.

Wolfgang Lutz is leader of the Population Project at IIASA. He received his Ph.D. from the University of Pennsylvania in 1983. His main interests include European population and environmental effects of population growth.

Sergei Pirozhkov is director of the National Institute for Strategic Research in Kiev. He studied statistics at the Kiev Institute of National Economics. Population reproduction modeling is his current area of research.

Lidia Prokophieva is a senior researcher at the Institute for Socioeconomic Studies of Population, Russian Academy of Sciences, Moscow. She received a Ph.D. from the Moscow Institute of Economics and Statistics in 1984. The family life cycle and the problems of poverty are her main research interests.

Natalia Rimashevskaya is the director of the Institute for Socioeconomic Studies of Population, Russian Academy of Sciences and Ministry of Labor and Employment, Moscow. She received a Ph.D. in 1974 from Moscow State University. Her main area of interest is social demography.

Gaiané Safarova is a researcher at the St. Petersburg Institute for Economics and Mathematics. She received a Ph.D. from Leningrad (St. Petersburg) State University in 1987. She has specialized in the study of mathematical models of demographic processes, population projections, and the labor market.

Sergei Scherbov is a senior lecturer at the Population Research Centre, University of Groningen, the Netherlands. He received a Ph.D. in 1983 from the All-Union Research Institute for Systems Studies of the USSR Academy of Sciences. His research interests include demographic modeling and software development, applications of multistate demography, and population projections.

Vladimir M. Shkolnikov is head of the laboratory of the Institute for Employment Studies Center of Demography and Human Ecology at the Russian Academy of Sciences and Ministry of Labor and Employment in Moscow. He received his Ph.D. from the Institute of Geography in 1987. His main research interests include population geography.

Brian D. Silver is the chairman of the department of political science at the University of Michigan. He received a Ph.D. in 1972 from the University of Wisconsin. His research interests include mortality patterns in the republics of the former USSR.

Vlada Stankuniene is leader of the demographic department at the Institute of Economics, Lithuanian Academy of Sciences in Vilnius. He received a D.Sc. in 1978 from Vilnius University. His main research interests are demographic evolution, population projections, family policy, and reproduction behavior.

Sergei A. Vassin is a senior researcher at the Institute for Employment Studies Center of Demography and Human Ecology at the Russian Academy of Sciences and Ministry of Labor and Employment in Moscow. He studied at the Moscow Institute of Economics and Statistics. His main scientific interests include mortality analysis and demographic data accuracy.

Anatoli Vishnevsky is head of the Center of Demography and Human Ecology and deputy director of the Institute for Employment Studies, Russian Academy of Sciences and Ministry of Labor and Employment in Moscow. He has conducted research at the Kharjkov University and the Russian Academy of Natural Science. His main research interests include social demography and demographic analysis.

Andrei Volkov is head of the department of demography at the Institute of Statistics and Economic Research, State Committee of the Russian Federation on Statistics, Moscow. In 1971 he received a Ph.D. from the Moscow Institute of Economics and Statistics. His main research interests are family demography and population statistics.

Frans Willekens is a professor of demography and head of the Population Research Centre, University of Groningen, and deputy director of the Netherlands Interdisciplinary Demographic Institute (NIDI), The Hague. He received a Ph.D. from Northwestern University. His research interests include demographic theory and models of change.

Sergei V. Zakharov is the head of laboratory at the Institute for Employment Studies Center of Demography and Human Ecology, Russian Academy of Sciences and Ministry of Labor and Employment, Moscow. In 1991 he received his Ph.D. from the Moscow Institute of Economics and Statistics. His research interests include population modeling and regional demography.

Introduction: Past and Present Studies of the Soviet Population

The set of papers in this volume originates from an international conference held in Tbilisi (Georgia) 8–12 October 1990 which was entitled "Demographic Processes in the USSR in the Context of the European Experience" and organized by IIASA together with the Institute for Socioeconomic Studies of Population, USSR Academy of Sciences, and the Educational and Scientific Center (Georgia) of the Soviet branch of the World Laboratory. It was the first large-scale international demographic conference in the Soviet Union for many decades.

The purpose of the conference was twofold: first to improve scientific contacts between Soviet and Western demographers; and second to provide an overview of demographic trends and patterns in the republics of the Soviet Union. While the first goal was already achieved in and around Tbilisi, a process which was greatly enhanced by the famous Georgian hospitality, the second goal was much more difficult and time-consuming. A comprehensive survey of Soviet demographic trends and patterns written by 31 authors of many different linguistic, ethnic, and scientific backgrounds required extensive editing and revisions that went far beyond usual text correction. Nevertheless, we are happy to have the book published two years after the conference and hope that it will prove to be a useful compendium of demographic research on the Soviet Union.

The selection of papers was guided by several principles. First the papers had to focus on the Soviet Union. For this reason some interesting papers presented at the conference that primarily looked at other European countries were not included. Second we wanted to make sure that all important demographic aspects received proper and fair treatment. Finally, in addition to quality and originality, emphasis was put on presenting a broad view of Soviet demography and making relevant information accessible to the Western reader who does not know the Russian language. Nevertheless, despite all the editing effort the book is still somewhat uneven and heterogeneous in some respects. But we hope that the reader will appreciate the intrinsic difficulties in producing such a volume which is the first of its kind.

Recent changes in the former Soviet Union caused many uncertainties and problems. One of the least important but still difficult questions was finding an appropriate title for the book. Reference to the Commonwealth of Independent States was considered inappropriate because it excludes the Baltic states which have a prominent position in this volume. Since none of the analyses go beyond 1990, a year when the Soviet Union still existed, it was decided to take a strictly historical approach and refer to the political conditions of the pre-1991 period. Following the same logic, if the name of a region or city has changed since 1990, we use the former name irrespective of recent changes.

The republics that were formerly part of the Soviet Union have a total population of about 290 million. This is more than 5 percent of the world population and more than the population of the USA and Canada. In terms of cultural, political, economic, military, and environmental impacts on the world this 5 percent is very significant notwithstanding the collapse of the Soviet Empire. In demographic terms most of the populations described in this volume show great structural similarities to the rest of Europe. This is why we chose to consider the Soviet population trends and patterns in the context of the European experience.

History of Demography in the USSR

Today's demographic patterns are to a large extent determined by the past. For three-quarters of this century the Soviet demographic history was affected by dramatic social changes and abrupt breaks in social traditions. These changes and breaks severely affected the population's living conditions, social norms, and demographic behavior.

The country entered the 20th century as a conglomerate of lifestyles, religions, and cultures. During the course of the 20th century, socioeconomic development – which determined the demographic transition processes – started at different times and proceeded differently in different regions and in various sociodemographic groups, resulting in increasing heterogeneity. Today the western and northern parts of the former Soviet Union have the demographic characteristics of a developed country while the demographic trends in the Central Asian republics resemble the demographic patterns of developing countries.

In the course of this century the country experienced such social shocks as World War I and the civil war, collectivization and the catastrophic famine of 1933, mass repressions and genocide of entire social and ethnic subgroups of the population, the bloody war of 1939–1940 (with Finland), and World War II and the postwar famine years. All these events had their repercussions on demographic patterns (mortality, fertility, and migration); their demographic consequences still have some effect on the population and will not disappear very soon.

All these discontinuities complicate the study of demographic trends and patterns in the republics of the Soviet Union. Even worse is the lack of reliable statistical information for extended periods. At the beginning of the 20th century Russia had incomplete demographic statistics. Before the 1917 revolution there was only one population census of the modern type, in 1897. During the civil war and the first few years after it, registration of demographic events was not organized. Handing over the registration of these events from the church to the civil administration demanded a new way of organizing registration, and it took many years. Evidence of the intensive attempts at obtaining statistical information about the population is the fact that in less than 10 years there were three population censuses in the country [1920, 1923 (of urban population only), and 1926].

The 1920s were quite favorable for the development of demography as a science. These were the years of fruitful work by Soviet demographers among whom Kvitkin, Novosel'skiy, Payevskiy, and Korchak-Chepurkovskiy stand out. In 1919, on the initiative of a well-known Russian economist, Tugan-Baranovskiy, the Institute of Demography of the Ukrainian Academy of Sciences started its work, with the prominent scientist Ptoukha as its director. During the 19 years of its operation this institute did much work, which is still available, on the history of Soviet demography and statistics. At the end of 1930 in Leningrad the Demographic Institute of the USSR Academy of Sciences was established and its activity was directed by the outstanding demographers Novosel'skiy and Payevskiy. Its brief operation enriched demographic science with many important and interesting studies.

Beginning in the early 1930s, the publication of demographic statistical data was stopped to conceal the negative demographic consequences of collectivization and famine. The estimates which appeared in the press deliberately overstated the population size of the country. The 1937 population census, with its more realistic but lower estimate of the population, was announced as being imperfect; many statisticians who worked on this census were later repressed. The elimination of experienced statistical personnel, the fear of further repression, and World War II impeded improvements in the quality and reliability of demographic data. All these things had a severe influence on the development of scientific population studies.

The repressions of the 1930s affected many demographers, too. The closure of the Leningrad and Ukrainian institutes in 1934 and 1938, respectively, the discontinuation of demographic data publications, and the substitution of the results of concrete analysis for ideological dogmas in demographic theory slowed down the development of demographic science in the USSR. To a great extent the continuity in the development of methods of study and demographic theory was lost.

Statistical analysis of demographic processes in those years was almost completely conducted in the Central Statistical Office and its local branches. However, in many respects, it was limited to a description of the demographic situation and processes on the basis of crude vital rates, which only provided a rather superficial picture. But even those results remained mostly unknown to the public.

The revival of demographic studies began in the 1960s and was induced by the desire of the administration and economic-planning institutions to have information on population growth. It was made possible by the general liberalization of social and political life during this period. Attention was being paid to the determinants of fertility decline and the still relatively high mortality level. At this time the details of the 1959 census became available to scholars, and the very general information given in the official statistics gave rise to interest in conducting a series of local sample surveys. The official statistics became more widely used in the 1960s, though much of the data have not been published yet. Several demographic research groups were organized.

During this period methods of demographic analysis were being further developed. Emphasis was put on cohort analysis, and the study of attitudes and demographic behavior began. Also the first study by Soviet demographers on the demographic transition in various regions of the country occurred in this period. Although much of the work is still of a descriptive nature, and the interpretation of trends is not free from dogmatism, the scientific analysis of population continues on a larger and more profound scale. Training in demography was improved, new demographic research teams in the large cities of various republics were established, and publications on demographic issues became numerous. The results of these studies conducted in the 1960s were later (in the 1980s) summarized in several comprehensive monographs but remained almost unknown to Western readers.

However, the demography of the Soviet Union is far from being *terra incognita* for Western readers. In addition to the fundamental work by Frank Lorimer there has been important work on the Soviet Union by the American demographers Ansley Coale, Barbara Anderson, and Brian Silver, and Murray Feshbach, by the French scholars Jean-Noel Biraben, Roland Pressat, and Alain Blum, and by several other Western scientists. Demographic studies published in Russian, however, are not as well known in the West.

There is no doubt that the demographic analysis of the populations of the Soviet Union, within the USSR and in the West, suffered considerably from the disconnection of Soviet demographers and their foreign colleagues. The lack of possibilities for exchanging data and discussing methods and substantive findings was a serious impediment to state-of-the-art, comprehensive analysis of Soviet population trends. This separation of scientists partly led to mutual misunderstandings and in consequence the impoverishment of demographic analysis in its aim to comprehend the complicated processes of demographic development.

Fortunately, now, the end of the Cold War has reached the sphere of demography as well. More data have begun to be published; contacts between Soviet scholars and their foreign colleagues are more frequent; joint surveys (though still rare) are being carried out; and Russian,

Ukrainian, and Baltic research teams are taking part in international demographic projects. The IIASA Population Program over the past 15 years has played a key role in facilitating the collaboration of demographers and establishing scientific ties that now, under less restricted circumstances, produce visible fruits. One of them is this volume.

Structure of the Volume

The 25 chapters of this book are grouped into four parts. The first eight chapters in Part I discuss trends and patterns of fertility. Different fertility indicators are studied in their relationship to regional, ethnic, sociocultural, and other variables. Many of the papers are structured around the concept of fertility transition with emphasis both on the description of the transition over time and on the cross-sectional analysis of different parts of the USSR that are currently at different stages of demographic transition. Part II assembles a set of eight papers on family and household dynamics as well as on public opinion and attitudes toward the family. Part III includes six contributions describing different aspects of changing mortality patterns in the regions of the USSR with special emphasis on the cause-of-death structures that underlie the unfavorable mortality trends of the 1970s and early 1980s. Part IV finally presents three papers that look at the interplay of the various components of population change and their impacts on age distributional aspects. These four parts cover all important aspects of population dynamics in the Soviet Union except for migration on which not enough information was available.

Part I: Fertility Trends and Patterns

The first chapter in Part I gives a long-term perspective of the trends in fertility and marriage in some Soviet republics in comparison to neighboring countries. Written by Coale it is based on the rich materials of the Princeton European Fertility Project but extends the analysis to the year 1985. One interesting pattern revealed in this paper is that the Baltic states show great demographic similarities with Scandinavia up to the 1960s; thereafter patterns diverge.

In Chapter 2 Lutz and Scherbov give a comprehensive survey of fertility trends in all republics of the Soviet Union between 1959, when modern fertility statistics were first recorded, and 1990. Using fertility indicators of increasing sophistication – from crude birth rates to cohort analysis and parity-specific information – this paper indicates that the European republics recently showed a movement toward the two-child family, following the trend of many European countries during the 1960s and early 1970s. Anichkin and Vishnevsky, in Chapter 3, provide a closer look at the stages of the fertility transition in selected republics which are typical for three groups of republics: Estonia, a post-transitional republic; Azerbaijan, a well-advanced republic; and Tajikistan, a republic at the beginning of the fertility transition.

Using the data from a 1985 sociodemographic survey (covering 5 percent of the Soviet population), Darsky presents in Chapter 4 a more detailed cohort analysis of birth intervals and parity progression ratios among various subpopulations of the Soviet Union. Using the same survey Bondarskaya, in Chapter 5, studies ethnic and territorial differentials in marital fertility. Her paper clearly indicates the important role of cultural traditions and also demonstrates the concentration on the two-child family in European republics.

In Chapter 6 Katus uses long historical time series (for Estonia since the late 18th century) to study the fertility transition in the three Baltic states. He finds that Estonia and Latvia experienced an early fertility transition (starting already in the late 19th century), whereas Lithuania was later and more closely related to the Eastern European pattern. Zakharov in Chapter 7 takes a closer look at the fertility transition in the regions of the Russian Federation. He finds that the provinces of Novgorod, Yaroslavl, Leningrad, and Moscow were part of the vanguard group among all the provinces.

Finally, in Chapter 8 Avdeev treats the issue of contraception and abortion in the USSR. He explains why induced abortion has become the main method of family planning in the USSR, which has one of the highest abortion rates in the world. Frequencies of abortion and contraceptive prevalence are estimated using the Bongaarts model. He also shows that recent government efforts to increase contraceptive use have not been successful.

Part II: Family Dynamics and Changing Attitudes

Chapter 9 by Volkov studies the demographic interactions of trends in nuptiality, marital dissolution, fertility, and the separation of young couples from their parents. He uses both aggregate statistics and data from two large surveys on family formation in the USSR in 1984 and 1989. The 1984 survey showed that more than half of all newly married couples in the USSR lived with their parents. Next, in Chapter 10, Ilyina describes trends in the marital status composition of the Soviet population by analyzing available historical time series on the changing prevalence of never married, married, divorced, and widowed men and women by ethnicity. In Chapter 11 Willekens and Scherbov build a multistate marital life table expanded by a staging process approach for further analysis of nuptiality characteristics in the USSR. Based on data on the number of children born they also estimate parity-specific fertility rates and construct cohort fertility careers of Soviet women.

Chapter 12 by Lapinš deals with the special question of teenage marriage using Latvian data. The paper studies demographic, social, and economic questions of married couples under age 18. Another special group is studied in Chapter 13 by Prokophieva, namely, the group of large families in low fertility regions. Using data from the Taganrog socioeconomic survey, Prokophieva studies the motivation for having more than four children and presents a social portrait of such families.

Chapters 14 (by Bodrova) and 15 (by Stankuniene and Kanopiene) deal with public opinion on family, fertility, and population issues. Bodrova describes the results of a poll carried out in 28 towns and 13 villages in 7 republics and gives information on desired family sizes, the demand for family support, and various family policies by selected socioeconomic background variables. In Chapter 15, Stankuniene and Kanopiene use data from two surveys carried out in 1988 and 1990 on Lithuanian family policies and related issues. One interesting finding was the clear majority view that the best family policy would be to give men the opportunity for earning as much as possible to support their families. It was also clear that the preferred way of caring for preschool children was at home by the mother. However, those with higher education were in favor of kindergartens.

In Chapter 16, Rimashevskaya gives a comprehensive description of the social roles and the status of women in the USSR. Her considerations range between the social and professional mobility of women and a comparison of employment careers of both spouses. She finds that despite the decades of very high female labor force participation, the patriarchal structure in society has hardly changed and is even being strengthened during the present period of economic and political restructuring.

Part III: Components of Mortality Trends

Chapter 17 by Andreev studies the differences in causes of death between the USSR and a group of five large Western countries. He shows that higher mortality in the USSR is mostly attributable to circulatory and respiratory diseases.

Anderson and Silver, in Chapter 18, take a close look at mortality trends in the working ages for the period between 1959 and 1988. After comparing the Soviet data to other industrialized countries, they conclude that mortality in the working ages, especially for men in Russia, is sufficiently high to warrant concern about the number of years of life lost. In Chapter 19 Kingkade focuses on regional variations of mortality by cause of death. Using years of potential life lost as an indicator, Kingkade concludes that substantial saving of human potential is possible through mortality reduction, which would be tied to an effective health-care system.

One of the important obstacles for the adequate analysis of demographic trends in the USSR is the unreliability of data and the problem of separating real trends from gradual improvements in the completeness of vital registration. In Chapter 20, Ksenofontova estimates the reliability of data on infant mortality and adjusts the series of infant mortality rates.

Shkolnikov and Vassin, in Chapter 21, analyze the changes in spatial differentials of life expectancy for urban and rural populations of Russia in 1979 and 1988. Studying mortality trends for 53 territories of European Russia, they show that between 1979 and 1988 life expectancy was increasing in the direction from northeast to southwest.

Chapter 22 by Kruminš is also a comparative analysis of mortality trends with a view to the Baltic states: Latvia, Lithuania, and Estonia. The author shows that before 1940 the Baltic states were more advanced in socioeconomic terms which was also reflected in higher life expectancy in comparison with the USSR and that after becoming part of the USSR the Baltic states started losing their good position in life expectancy.

Part IV: Changes in Population Structure

How the age structure of the Soviet population has been influenced by regular and irregular changes in fertility and mortality from the 1920s to the 1940s is described in Chapter 23 by Andreev, Darsky, and Kharkova.

The paper is an attempt at reconstructing the population dynamics during this period, which was characterized by several cataclysms and only very fragmentary statistics. The authors also estimate the human losses in the Soviet Union during World War II.

A similar effort using a different method is made by Pirozhkov and Safarova in Chapter 24. They apply the stable population model to estimate population losses during World War II and during the postwar period. They distinguish between direct losses by mortality and indirect losses of potential by exceptionally depressed fertility as compared with the fertility level that could be expected from an uninterrupted demographic transition process.

In Chapter 25, Keyfitz goes beyond stable population theory by looking at the trends in intercohort increases or decreases in the USSR, the USA, and Europe. He points out that the size of the upcoming generation is the most important single question in demography and hints at several political consequences that discontinuities in cohort sizes may have on the countries concerned.

This book is certainly not an exhaustive treatment of demographic trends and patterns in the populations that used to form the Soviet Union. It rather is a beginning for more intensive scientific collaboration between demographers from the West and those from the Commonwealth of Independent States and the Baltic countries. We hope that this volume will result in stimulating further critical analysis of demographic trends in the former Soviet Union and a strengthening of scientific cooperation in studying its demographic future.

The editors gratefully acknowledge the assistance of Eva Toth and Marilyn Brandl in preparing the manuscript.

Wolfgang Lutz
Sergei Scherbov
Andrei Volkov

Part I

Fertility Trends and Patterns

Chapter 1
Nuptiality and Fertility in USSR Republics and Neighboring Populations

Ansley J. Coale

In the late 1970s Coale, Anderson, and Härm (1979) wrote a book on the decline in fertility in Russia since the 19th century, and in the mid-1980s Coale and Watkins (1986) summarized some of the principal findings of the Princeton European Fertility Project, a project that assembled data on overall fertility, marital fertility, and proportions married in more than 600 European provinces. The Princeton European Fertility Project (of which the study of Russia was a part) was intended to document the major decline in marital fertility in all of the provinces, and the associated changes in nuptiality, and to explore the socioeconomic circumstances under which the changes occurred. The results showed that some of the best-known themes of the so-called demographic transition are far from universally valid. For example, a sustained decline in fertility often preceded any major reduction in infant mortality; rural areas in France inhabited by peasants at moderate levels of literacy and rather high mortality began a major reduction in the frequency of births many decades before there was any decline in urbanized and industrialized England. There were also several positive findings.

A particular result that emerged from the study of Russia, and also from the broader study of Europe, is the focus of this chapter. In some West European countries it was found that levels and changes in fertility tended to cluster geographically, even though within each cluster of provinces there were large differences in the mortality, literacy, and occupational structures. The clusters seem to share a common language, or dialect, and cultural history. When Leasure showed a coded map indicating the level of marital fertility in the provinces of Spain in 1910 to a professor of Spanish literature, the professor said that it looked like a linguistic map of Spain. In exploring the clear distinction in the fertility of French-speaking and Flemish-speaking populations in Belgium, Lesthaeghe identified pairs of villages on opposite sides of the language border, separated by a short distance, but matched in size and other characteristics, and found systematically lower fertility in the French-speaking villages.

In Russia, there were also relations between fertility and ethnicity or nationality, as this factor is labeled in Soviet statistics. Within European Russia in 1926 there was a strong association between the proportion of a province that was "Western" non–Great Russian nationality and the level of marital fertility (the higher the proportion Western, the lower the fertility level). Western nationalities included Poles, Ukrainians, Byelorussians, Jews, Lithuanians, and Western Finnic peoples. The association was not much weakened when the two very low-fertility provinces with very high proportion Western (Estonia and Latvia) were omitted from the analysis. In 1959 there was little remaining negative correlation between proportion Western and fertility, but a strong positive relation between the proportion of "Eastern" non–Great Russian nationalities and fertility. Eastern nationalities included Turkic peoples, a Mongolian group, and several Eastern Finnic peoples.

The most distinctive differences in fertility trends in Russia are not within European Russia, but between the European area, on the one hand, and the Transcaucasian and the Central Asian regions, on the other.

This paper is a comparison of the evolution of nuptiality and fertility in the republics of the Soviet Union with selected neighboring populations in Europe and in Asia, especially in the Chinese province that borders on the Central Asian republics. Changes in the mean age at first marriage and in the total fertility rate in the European republics of the

USSR are compared with the changes in selected European countries. Analogous comparisons are made between the non-European republics and China and other populations that until recently were characterized by very early marriage.

The theme is the different patterns of change in nuptiality in Europe, the similarities and differences compared to the European republics, and analogous comparisons between the non-European republics and the non-European populations selected. The paper shows that different patterns of age at marriage have been associated with differences in marital fertility, and that the modern transition in fertility has been different in populations with different mean ages when first married.

1.1   Three Distinctive Patterns of Marriage

Within Europe in the mid-19th century, there were two major patterns of marriage: the West European pattern of very late marriage, with mean age for women from 23 to 28 years and high proportions permanently single, 10 percent to 15 percent still single at age 50; and the Eastern European pattern of marriage, with a mean age around 20 and 5 percent or less single at age 50. Outside of Europe (and areas settled by Europeans) a third very different pattern of marriage was traditional. In most of Asia and Africa marriage occurred at much earlier ages (mean age 16 to 18 years, except in South Asia where the custom of child marriage led to mean ages as low as 11 years); the proportion single at age 50 and above was very low indeed (generally less than 1 percent).

Before the transition in marital fertility (the large sustained decline in fertility involving the introduction and diffusion of contraception and abortion to prevent higher-order births) began, the fertility of married women was lower in the populations in Asia and Africa, where as many as 90 percent of women between 15 and 49 were currently married, than in the populations of Europe, where as few as 50 percent of women between 15 and 49 were currently married. As a result, the overall level of fertility (as represented by the total fertility rate, TFR) was about the same, or a little higher in the earlier-marrying populations, before the modern sustained fall in marital fertility began. The large decline in marital fertility in Western Europe was typically accompanied by a fall in the mean age at first marriage. The fall in marital fertility began in France probably before 1800, and continued during the 19th century;

age when married also fell during these years. The median date for a fall of 10 percent in marital fertility (on the way to a large decline) in the provinces of Europe was 1903; in most countries age at marriage changed only slightly before 1930, when in the late-marrying Western European populations it began to decline. In the earlier-marrying Eastern European populations, there was less change in age at marriage even during the years when marital fertility fell.

In Asia and Africa, a major fall in marital fertility is not yet universal. With or without a reduction in the fertility of married women, there has been an increase in the mean age of entry into first marriage in Asia and North Africa.

1.2 Nuptiality and Fertility Change in Scandinavia and the Baltic Republics

The Scandinavian and Baltic fertility changes are compared by tracing the mean age at first marriage and the total fertility rate from 1900 until 1985 (*Figures 1.1* and *1.2*). In all five areas, mean age at marriage was 26 years or older in 1900, was still very high in 1930, and fell below 24 years by the 1960s. The especially high mean age in the Baltic republics around 1930 may have been the result of a shortage of eligible males after the First World War.

A larger divergence is the rise in age at marriage in Scandinavia after 1970, which was quite steep and not matched in the Baltic republics except by a modest increase in Estonia. The upturn in age at marriage in Sweden, Denmark, and Norway was part of a profound change in marriage customs in some Western populations, a change that was especially large in the Scandinavian countries. The change included large increases in the cohabitation of couples, in divorce rates, and in the proportion of all births to unmarried women, and a fall in remarriage rates of the divorced.

The overall level of fertility as measured by the total fertility rate was very similar in the Baltic and Scandinavian populations (*Figure 1.2*) from 1900 to 1930. The TFR in 1960 was higher in the Scandinavian countries because of a mild postwar baby boom, which did not occur in the Baltic republics. The Scandinavian decline from this 1960 high level also did not occur in the Baltic populations.

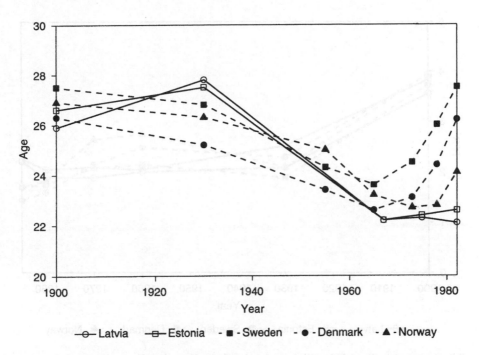

Figure 1.1. Average age of females at first marriage in Scandinavia, Estonia, and Latvia, from 1900 to 1985.

Because the link between marriage and childbearing was weakened in Scandinavia, the decline in the total fertility rate after 1970 was less conspicuous in Scandinavia than the decline in the proportion married among women of childbearing age; the reduced number of births to married women, because the proportion married was reduced, was partially offset by a large increase in the number of births to the unmarried.

Religion, language, level of education, and other aspects of the Baltic provinces in Russia were more Scandinavian or German in character than Russian in the 19th century; these common cultural traits underlie demographic similarities with Scandinavia, such as late age when married and a relatively early reduction in fertility. The index of proportion married employed in the Princeton European Fertility Project (I_m) for the Scandinavian countries and Estonia and Latvia (from 1900 to 1985) is shown in *Figure 1.3*. The similarity in nuptiality in these populations seen in *Figure 1.2* reappears. An index calculated for individual years would show greater differences; for example, war casualties reduced the

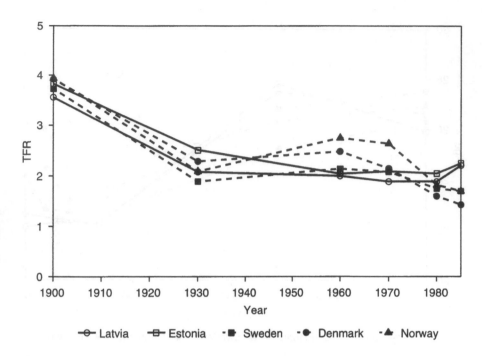

Figure 1.2. Total fertility rates in Scandinavia, Estonia, and Latvia, from 1900 to 1985.

proportion married in Russia after World War II substantially, and the high I_m in Scandinavia in 1960 reflects the postwar baby boom.

In the 1970s and 1980s the profound changes in marriage that occurred in Sweden and Denmark (and many other Western populations) were not duplicated in Latvia and Estonia. In these recent years, the cultural heritage these populations shared no longer generated as similar a movement in nuptiality as in the earlier part of this century.

1.3   Nuptiality and Fertility Changes in the Non–Baltic European Republics and East European Countries

This section examines the change in overall fertility and age at first marriage in the non–Baltic European republics of the Soviet Union and in selected Eastern European countries. In *Figures 1.4* and *1.5* mean age

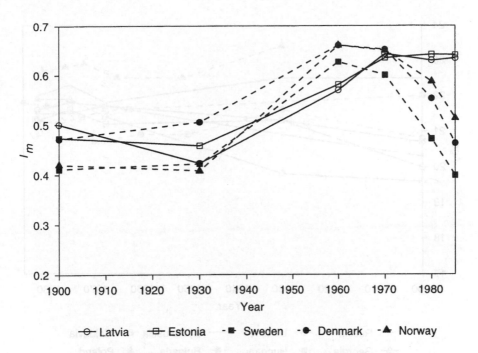

Figure 1.3. Index of proportion married (I_m) in Scandinavia, Estonia, and Latvia, from 1900 to 1985.

at first marriage and the total fertility rate are compared in Russia, the Ukraine, Byelorussia, Moldavia, Georgia, and selected Eastern European countries. The major changes in age at marriage seen in Scandinavia and the Baltic republics did not occur in these populations. The mean age at marriage in the European republics lies between 20 and 22 years with the exception of Georgia in 1900 (19.6) and Byelorussia, where there is a flat trend at a mean age a little above 22 years. Age at first marriage is nearly constant in the Eastern European populations. The average age at first marriage in Poland is slightly higher than in the other areas, but similarly free of a significant trend up or down. Evidently when the traditional age at marriage was a little above 20 years, there was little tendency for the mean age to rise or fall during the period of major decline in overall fertility.

Two time patterns of age at first marriage are evident. The first is a large reduction from mean ages above 25 in 1900 to between 22 and 24 in the 1960s, shared by the Scandinavian countries and Latvia

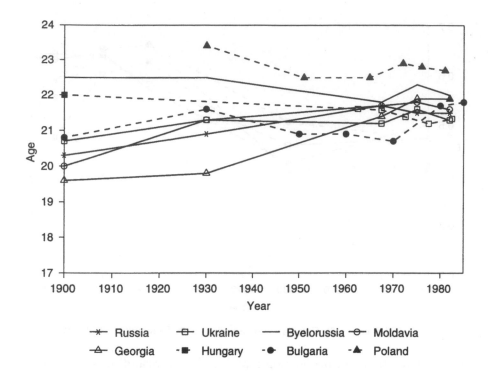

Figure 1.4. Average age of females at first marriage in European republics and Eastern European countries, from 1900 to 1985.

and Estonia; in the 1970s and 1980s the mean age rose extensively in Scandinavia but only slightly in Estonia and not at all in Latvia. The second pattern is a nearly unchanging mean age at first marriage, at a level near 21 years, shared by the non–Baltic republics of the European USSR and the neighboring countries in Eastern Europe.

The time pattern of overall fertility in the past 50 years was similar in the two areas that experienced a big reduction in age at marriage from 1900 to the 1960s, except that a mild increase in TFR from 1930 to 1960 in Scandinavia and a subsequent down turn did not occur in the Baltic republics. In 1900, the non–Baltic European republics had TFRs at about the same level as the Eastern European populations with which they are compared, except for Hungary where fertility was lower than in the other populations (*Figure 1.5*). The decline in overall fertility in the European republics was later than in the Eastern European countries,

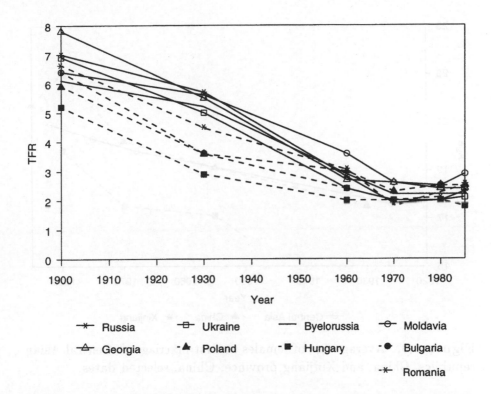

Figure 1.5. Total fertility rates in European republics and Eastern European countries, from 1900 to 1985.

but the two sets of TFRs converge by 1970, and remain steady after that date.

1.4   Nuptiality and Fertility Changes in the Central Asian Republics and China

In *Figures 1.6* and *1.7* the mean age at first marriage and the total fertility rate in the Central Asian republics are compared with China and with the Chinese province that directly borders on the Central Asian republics for selected years. The similarity in the rise in the mean age at first marriage in China and Soviet Central Asia is remarkable until 1970. After that date marriage age in China rose steeply and then fell, while age at marriage in the Central Asian republics continued its

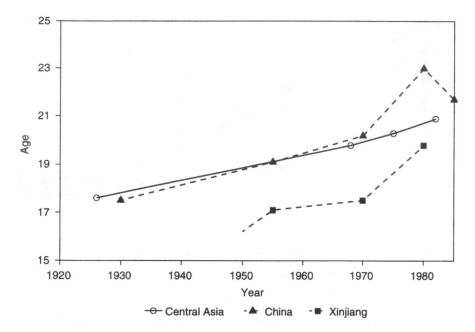

Figure 1.6. Average age of females at first marriage in Central Asian republics, China, and Xinjiang province, China, selected dates.

steady increase. Both populations exemplify the modern increase in age at marriage shared by almost all populations that approached the middle of this century still adhering to a traditional early female marriage age (Coale, 1983). The mean ages of 18 and under that were the rule in these populations connoted marriages that were arranged by families, rather than marriages that followed a period of courtship. Female education and employment opportunities outside the family caused the erosion of the custom of arranging marriages, offered an alternative to early marriage, and hence increased the mean age to 20 years or more.

In China, a strong government effort to reduce fertility during the 1970s was the major source of the steepened increase in mean age at marriage from 1970 to 1980. One aim of the program was to postpone marriage until age 23 in rural areas and 25 in the cities. To marry, permission from local officials was required; permission was withheld until the woman was 23 (or 25 in the cities). In 1980 refusal of permission to brides below 23 (or 25) was relaxed, and the mean age at marriage fell. Perhaps without this strong intervention in regulating age at marriage,

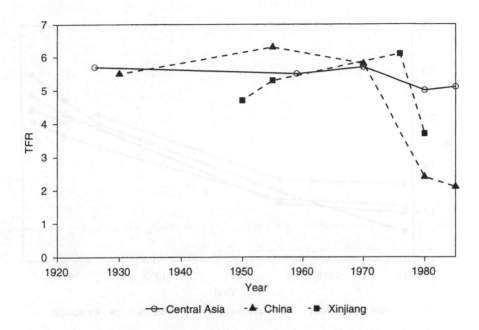

Figure 1.7. Total fertility rates in Central Asian republics, China, and Xinjiang province, China, selected dates.

age of entry into marriage in China would have continued to parallel the mean age at marriage in the Central Asian republics.

Just as the Central Asian republics that border China have retained relatively early marriage, the westernmost province of China, Xinjiang (or more precisely, Xinjiang Uygur autonomous region), has retained the lowest mean age at marriage, nearly 2.5 years below the national mean. The very low age at marriage of the Wei (as most sources call the Uygur, who constituted nearly 50 percent of the population of Xinjiang province in 1982) accounts for the low mean age at marriage in the province (see *Figure 1.6*). In fact the mean age at first marriage of the Wei around 1980 was only 18.7 years, while in the urban population of Xinjiang, practically devoid of Wei, the mean age of 24.9 at marriage was the same as among the urban population of all China. In administering their antinatalist policy (including the "one-child" norm introduced in 1979), the Chinese have explicitly exempted certain minorities, and also populations in sparsely settled areas. Any attempt to bring large changes

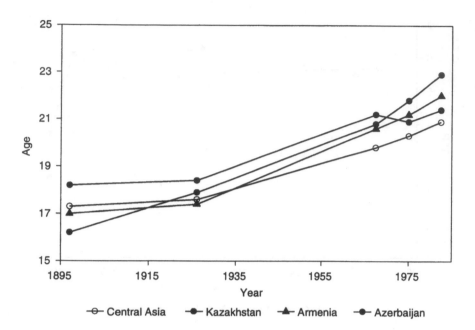

Figure 1.8. Average age of females at first marriage in Central Asian republics and other non-European republics, from 1895 to 1980.

in age at marriage and in marital fertility among the Wei might have been unsuccessful.

The changes in the total fertility rate from 1930 until 1970 were broadly similar in China and the Central Asian republics (*Figure 1.7*). Then there was a marked decline in China, and a much smaller reduction in the Soviet area. The fall in TFR in Xinjiang after 1970 was mostly a response (as was the fall in China) to the government family-planning program. Lacking data on fertility by nationality, one may conjecture that most of the decline occurred among the non-Wei population, which is more than 50 percent urban.

The non-European republics other than those included in Central Asia are Kazakhstan (which neighbors the Central Asian republics and like Tajikistan and Kirghizia extends east to the border with China) and the Transcaucasian republics of Armenia and Azerbaijan. The evolution of mean age at first marriage and the total fertility rate in these republics is shown in *Figures 1.8* and *1.9*. These republics are the home of non-European nationalities with distinct languages and cultural traditions.

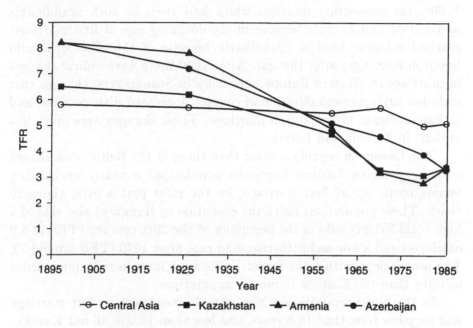

Figure 1.9. Total fertility rates in Central Asian republics and other non-European republics, from 1895 to 1985.

They shared with the Central Asian republics a very low age at first marriage around 1930, but experienced a somewhat greater increase after 1960. The total fertility rate fell markedly by 1960, but was still well above the fertility level in the European republics in 1980.

1.5 Discussion

This chapter has presented information on the changing age at marriage and overall fertility in four sets of republics in the Soviet Union: the two Baltic republics of Estonia and Latvia, the non–Baltic European republics, the Central Asian republics, and three other non-European republics. Three sets of republics shared similar patterns of change in age at marriage or overall fertility or both with different neighboring populations outside the Soviet Union. Latvia and Estonia had mean ages at first marriage and overall levels of fertility that were very similar to the Scandinavian countries in 1900, and shared similar large reductions in age at marriage and overall fertility before World War II. In the

1960s, the proportion married, which had risen in both Scandinavia and Latvia and Estonia because of the declining age at first marriage, reached a higher level in Scandinavia because of the effect of a mild boom in marriages after the war. After 1960 there were radical changes in marriage in Western Europe, especially in Scandinavia, changes that included extensive cohabitation of couples, increased divorce rates, and a sharp increase in age at first marriage. These changes were much less evident in Estonia and Latvia.

The European republics, other than those in the Baltic area, shared with neighboring Eastern European populations a nearly unchanging female mean age at first marriage, for the most part a little above 20 years. These populations (with the exception of Hungary) also shared a high total fertility rate in the beginning of the 20th century (TFR of 5.9 or above) and a low and little-changing rate after 1970 (TFR around 2). The populations within the Soviet Union were later in reducing marital fertility than the Eastern European populations.

In the non-European republics, female mean age at first marriage was very low (less than 18.5 years, and less than 18.0 in all but Kazakhstan) at the turn of the century and even until 1930. Early marriage was common in diverse populations in both Eastern and Western Asia; in North Africa; in Taiwan, Korea, Malaysia, and China; in Afghanistan, Iran, and Kuwait; and in Morocco and Libya, to cite a few examples. In both the Soviet populations and the others, age at marriage has risen to a mean ranging between just over 20 and 23 years. In these traditionally early-marrying populations, the rise in age at marriage has preceded the beginning of a sustained reduction of marital fertility. In fact, the index of marital fertility (I_g) has often risen in these populations before beginning to fall. Such was the case in Taiwan, Korea, and China where marital fertility rose from 1930 to the late 1950s.

In all but Central Asia, the rise in age at marriage since 1930 has been accompanied by a decrease in marital fertility; in these republics, I_g rose from 1926 to 1959, and from 1959 to 1970, analogous to the increases from 1930 to the late 1960s in East Asia, but in a later period. Such increases are probably the result of the weakening of traditional practices, including prolonged breast-feeding, that lower marital fertility. A significant instance of cultural influence upon demographic trends in contiguous areas inside and outside the Soviet Union is the very early marriage and high fertility in both the non-Han (Wei) nationality in Xinjiang province and the non-Russian nationalities in the Central Asian republics. These inhabitants of adjacent areas on opposite sides of a national boundary have closely related languages, religions, and traditions.

These examples show that despite the different social and economic situations within the Soviet Union and in diverse neighboring territories, there have been common features between the Soviet republics and nearby countries in the evolution of nuptiality and marital fertility during an era of major change. The most striking similarities are the change from very late to much earlier marriage in Scandinavia and Latvia and Estonia and the change from very early to later marriage in the Central Asian republics and neighboring China. In both instances, the changes in fertility inside and outside the Soviet Union had common features.

There is little doubt that Livi-Bacci was correct in saying that couples living in urban apartments and owning automobiles and television no longer have eight children; in the extreme the proposition that modern social change is accompanied by lower fertility seems valid. But culture, religion, and traditions complicate any simple statement of the transition in fertility. In Kuwait in 1975, per capita income was the highest in the world, and female life expectancy at birth was about 70 years. Yet the total fertility rate was 6.2, about the same as in Pakistan with an expectation of life of less than 50 years and much lower per capita income.

References

Coale, A.J., 1983. Recent Trends in Fertility in Less Developed Countries. *Science* **221**:828–832.

Coale, A.J., and S.C. Watkins, 1986. *The Decline of Fertility in Europe.* Princeton University Press, Princeton, NJ.

Coale, A.J., B. Anderson, and E. Härm, 1979. *Human Fertility in Russia Since the Nineteenth Century.* Princeton University Press, Princeton, NJ.

Chapter 2

Survey of Fertility Trends in the Republics of the Soviet Union: 1959–1990

Wolfgang Lutz and Sergei Scherbov

The demographic history of the USSR during the 20th century may be divided into two periods: one period with only fragmentary information on vital statistics, which lasted until the late 1950s, and the following 30-year period with complete statistical coverage. Although fertility in the first period has been comprehensively studied in *Human Fertility in Russia Since the Nineteenth Century* (Coale *et al.*, 1979), less systematic work has been done for the more recent period. This chapter looks exclusively at the period from 1959 to 1990 and systematically studies changes in age-specific fertility behavior in the 15 republics of the USSR during this time.

One might argue that fertility behavior is more closely related to ethnic and linguistic groups than to administrative boundaries – such as republics – and therefore should be studied according to these characteristics (Chapter 1). However, such studies of ethnic and linguistic fertility differentials can only be done for census years or based on special surveys. For information on single years one has to suffice with the republics. Despite all ethnic enclaves and mixing of different nationalities, in the same republic, there was a strong connection between the structure of the republics and the national composition of the Soviet Union. Furthermore the recent breakup of the USSR has made the republics by far the most important political units.

2.1   Trends in Absolute Numbers of Births

Absolute numbers of births are not indicators of demographic behavior but they are relevant to the future size and age structure of the population. During the early 1980s, on average, 5.1 million babies were born annually in the USSR. These births represented about 4 percent of all children born in the world per year. The number was higher than the number of babies born in the USA (3.7 million), but lower than the number born in Europe, excluding the USSR (6.6 million). Among all children born in the more developed world (as defined by the UN) one out of three was born in the Soviet Union.

Within the Soviet Union 45 percent of all births occurred in Russia, which held more than half of the total population; 19 percent of all births were in the Ukraine, Byelorussia, and Moldavia. The three Baltic states together accounted for only a little more than 2 percent, while 28 percent of all children were born in Uzbekistan, Kazakhstan, Kirghizia, Tajikistan, and Turkmenistan. Over the past 30 years, these proportions have been changing significantly with the European republics losing and the Central Asian republics gaining. In the early 1960s more than half of all Soviet babies were born in Russia while Central Asia accounted for only 19 percent (see *Table 2.1*).

Crude birth rates (CBRs) take into account the total size of the population but are still influenced by the population's age composition. The number of births per 1,000 inhabitants has been decreasing in the Soviet Union from 26.3 in the late 1950s to 17.9 in the late 1960s. It started to increase slightly to 19.1 in the early 1980s. The value of the 1990 CBR was higher in the USSR than in all European countries except Ireland. This value for the USSR, however, hides the tremendous heterogeneity within the USSR. Some European republics have rates around 16 per 1,000 inhabitants, while Uzbekistan, Turkmenistan, and Tajikistan have values of up to 44 per 1,000 inhabitants.

Table 2.2 shows that over the past 30 years the crude birth rate decreased significantly in all republics except Tajikistan. The period of

Table 2.1. Distributions of births over time by republics.

Years	No. of births (in 1,000s)	Share of regions in %				
		Russia	Non-Baltic European republics	Baltic republics	Trans-caucasian republics	Central Asian republics
1960–1964	24,706	50.1	21.9	2.3	6.9	18.8
1965–1969	20,763	45.6	22.0	2.5	7.4	22.4
1970–1974	21,934	45.4	21.7	2.5	6.6	23.7
1975–1979	23,595	45.6	20.4	2.3	6.4	25.2
1980–1984	25,692	45.4	19.7	2.3	6.5	26.2
1985–1989	27,027	43.9	18.5	2.3	6.6	28.7

Non-Baltic republics: Ukraine, Byelorussia, and Moldavia.
Baltic republics: Latvia, Lithuania, and Estonia.
Transcaucasian republics: Georgia, Azerbaijan, and Armenia.
Central Asian republics: Uzbekistan, Kazakhstan, Kirghizia, Tajikistan, and Turkmenistan.

fastest decline in most republics was between 1960 and 1965. Within this five-year period, the CBR in Armenia declined from 40.1 to 28.6, in Kazakhstan it declined from 37.2 to 26.9, and the decline was from 23.2 to 15.7 in Russia. On average, this decline was 26 percent between 1960 and 1965, and 32 percent over the 1959 to 1970 period. *Table 2.3* gives a decomposition of the CBR decline into the fertility component and the age-structural component. This is done according to the formula

$$b^1 b^0 = \underbrace{\frac{\sum w_i^1 f_i^1}{\sum w_i^1 f_i^0}}_{F1} \times \underbrace{\frac{\sum w_i^1 f_i^0}{\sum w_i^0 f_i^0}}_{F2} ,$$

where b^0 and b^1 are the crude birth rates at the beginning and the end of the period considered, w_i^0 and w_i^1 are the proportions of women in age group i of the total population (at time 0 or 1), and f_i^0 and f_i^1 are the age-specific fertility rates. The first term, F1, gives the effect of differential fertility patterns, whereas the second one, F2, gives the effect of relevant changes in the age and sex composition of the population.

In *Table 2.3* we can see that the 32 percent decline in the period 1959–1970 in the USSR is the combined effect of a 15 percent decrease in fertility and a 21 percent decrease caused by the change in the age structure – that is, less women in the prime childbearing ages. This trend is rather similar to most of the western republics of the USSR, except for Estonia where fertility slightly increased. In Georgia the level

Table 2.2. Crude birth rates in the republics of the Soviet Union, 1959–1989.

Year	RSFSR	UKRSSR	BYELSSR	UZBSSR	KAZSSR	GRSSR	AZSSR	LITSSR
1960	23.2	20.5	24.4	39.8	37.2	24.7	42.6	22.5
1961	21.9	19.5	23.4	38.2	36.0	24.7	42.1	22.2
1962	20.2	18.8	22.1	37.0	33.6	23.7	40.4	20.9
1963	18.7	17.9	20.6	35.8	31.1	23.0	40.8	19.7
1964	16.9	16.5	19.0	35.0	28.4	22.0	39.8	19.0
1965	15.7	15.3	17.9	34.7	26.9	21.2	36.6	18.1
1966	15.3	15.6	17.6	34.1	25.7	20.3	35.4	18.0
1967	14.4	15.1	16.8	33.0	24.7	19.5	32.5	17.7
1968	14.1	14.9	16.4	34.3	23.8	19.4	32.1	17.6
1969	14.2	14.7	15.9	32.8	23.4	18.7	29.3	17.4
1970	14.6	15.2	16.2	33.6	23.4	19.2	29.2	17.6
1971	15.1	15.4	16.4	34.4	23.8	19.0	27.7	17.6
1972	15.3	15.5	16.1	33.0	23.5	18.0	25.6	17.0
1973	15.1	14.9	15.7	33.5	23.4	18.3	25.2	16.0
1974	15.6	15.1	15.8	34.1	24.2	18.4	24.9	15.8
1975	15.7	15.1	16.6	34.2	24.3	18.3	24.9	15.7
1976	15.9	15.2	15.7	35.0	24.5	18.3	25.5	15.7
1977	15.8	14.7	15.7	33.4	24.2	17.9	25.0	15.5
1978	15.9	14.7	15.9	33.9	24.4	17.8	24.9	15.3
1979	15.8	14.7	15.8	34.4	24.0	17.9	25.2	15.2
1980	15.9	14.8	16.0	33.8	23.8	17.7	25.2	15.1
1981	16.0	14.6	16.3	34.9	24.3	18.2	26.3	15.1
1982	16.6	14.8	16.3	35.0	24.3	17.9	25.3	15.2
1983	17.5	16.0	17.6	35.3	24.3	18.0	26.1	16.3
1984	16.9	15.6	17.0	36.2	25.4	18.5	26.6	16.2
1985	16.5	15.0	16.5	37.2	24.9	18.7	26.7	16.3
1986	17.2	15.5	17.1	37.8	25.5	18.7	27.6	16.5
1987	17.1	14.8	16.1	37.0	25.5	17.9	26.9	16.2
1988	16.2	14.0	15.9	37.0	24.5	17.5	27.9	15.4
1989	14.6	13.3	15.0	33.3	23.0	16.7	26.4	15.1

Table 2.2.   Continued.

Year	MOLSSR	LATSSR	KIRSSR	TAJSSR	ARMSSR	TURKSSR	ESTSSR	USSR
1959								25.0
1960	29.3	16.7	36.9	33.5	40.1	42.4	16.6	24.9
1961	28.2	16.7	35.8	34.0	37.2	41.0	16.5	23.8
1962	25.6	16.0	33.9	33.6	34.7	40.1	16.1	22.4
1963	24.4	15.3	33.3	34.5	32.6	39.5	15.3	21.1
1964	22.5	14.7	31.8	34.7	30.2	38.1	15.4	19.5
1965	20.4	13.8	31.4	36.8	28.6	37.2	14.6	18.4
1966	21.0	14.0	30.8	35.4	27.1	37.6	14.3	18.2
1967	20.7	13.9	30.5	35.2	24.4	35.5	14.2	17.3
1968	20.0	14.0	30.8	36.7	23.9	35.6	14.9	17.2
1969	19.0	14.0	30.1	34.7	22.8	34.3	15.5	17.0
1970	19.4	14.5	30.5	34.8	22.1	35.2	15.8	17.4
1971	20.2	14.7	31.7	36.8	22.6	34.8	16.0	17.8
1972	20.6	14.5	30.7	35.3	22.4	34.0	15.6	17.8
1973	20.4	14.0	30.8	35.5	22.0	34.5	15.0	17.6
1974	20.4	14.2	30.8	36.9	21.8	34.5	15.1	18.0
1975	20.6	14.1	30.7	37.0	22.2	34.7	14.9	18.1
1976	20.6	13.9	31.7	38.2	22.6	35.1	15.1	18.4
1977	20.2	13.7	30.6	36.4	22.4	34.6	15.1	18.1
1978	20.1	13.6	30.4	37.5	22.2	34.4	14.9	18.2
1979	20.3	13.7	30.1	37.8	22.9	34.9	14.9	18.2
1980	20.0	14.0	29.6	37.0	22.7	34.3	15.0	18.3
1981	20.5	14.0	30.8	38.3	23.4	34.3	15.4	18.5
1982	20.6	14.6	31.2	38.2	23.2	34.7	15.4	18.9
1983	22.5	15.7	31.4	38.3	23.6	35.1	16.0	19.8
1984	21.9	15.7	32.1	39.8	24.2	35.2	15.9	19.6
1985	21.9	15.2	32.0	39.9	24.1	36.0	15.4	19.4
1986	22.7	15.9	32.6	42.0	24.0	36.9	15.6	20.0
1987	21.8	15.8	32.6	41.8	22.9	37.2	16.0	19.8
1988	20.7	15.1	32.3	44.0	22.2	38.1	15.9	19.1
1989	18.9	14.5	30.4	38.7	21.6	35.0	15.4	17.6

Table 2.3. Relative impact of age-specific fertility changes and changes in age composition on the CBR in the USSR and republics.

	1959–1970			1970–1979			1979–1987		
	Total change in CBR	Due to changes in F1	F2	Total change in CBR	Due to changes in F1	F2	Total change in CBR	Due to changes in F1	F2
USSR	0.68	0.85	0.79	1.06	0.98	1.08	1.09	1.11	0.98
RSFSR	0.60	0.75	0.79	1.10	0.98	1.11	1.08	1.15	0.93
UKRSSR	0.70	0.88	0.79	0.99	0.97	1.02	1.02	1.06	0.96
BYELSSR	0.62	0.81	0.76	0.99	0.92	1.07	1.04	1.01	1.03
UZBSSR	0.88	1.12	0.78	1.03	0.94	1.09	1.09	0.96	1.14
KAZSSR	0.63	0.74	0.84	1.04	0.94	1.10	1.05	1.06	0.98
GRSSR	0.78	1.00	0.78	0.95	0.89	1.06	1.02	1.01	1.01
AZSSR	0.71	0.93	0.75	0.86	0.77	1.12	1.08	0.88	1.22
LITSSR	0.76	0.88	0.86	0.87	0.88	0.98	1.06	1.05	1.01
MOLSSR	0.61	0.71	0.85	1.05	0.96	1.09	1.10	1.17	0.94
LATSSR	0.84	0.97	0.86	0.96	0.96	0.99	1.16	1.15	1.00
KIRSSR	0.88	1.13	0.78	1.01	0.93	1.08	1.08	1.00	1.08
TAJSSR	1.16	1.51	0.76	1.09	1.05	1.03	1.11	0.98	1.12
ARMSSR	0.54	0.67	0.80	1.02	0.84	1.21	1.03	1.04	0.99
TURKSSR	0.88	1.15	0.76	1.00	0.90	1.10	1.06	0.94	1.13
ESTSSR	0.94	1.09	0.86	0.95	0.94	1.01	1.06	1.08	0.98

of fertility stayed constant over that time and the 22 percent decrease in the crude birth rate was due to the changing age structure. In Uzbekistan, Kirghizia, and Turkmenistan the level of fertility even increased between 1959 and 1970 despite declining crude birth rates. In other words, a strong negative age-structural effect turned an increase in fertility levels into a decline in CBR. In Tajikistan, fertility even increased by 50 percent and, due to the negative age-structural influence, the CBR increased only by 16 percent.

For the periods 1970–1979 and 1979–1987, the observed changes are less dramatic. The CBR, on average, increased by 6 percent during the 1970–1979 period and 9 percent during the 1979–1987 period. In contrast with the first period, the age structure in the second period generally exerted a positive influence on the CBR seen by index numbers larger than 1.0. Over the 1979–1987 period, fertility declined somewhat in all republics except Tajikistan but, because of the positive age-structural effects, the CBR increased in most republics. Between

1979 and 1987 the trend reverses with CBR on the average increasing less than fertility levels. This exercise shows us that age-structural changes in the USSR played an important role in influencing the CBR trends. For the early period, 1959–1970 especially, CBR greatly exaggerates the impression of fertility decline.

2.2   Princeton Indexes

In *Human Fertility in Russia Since the Nineteenth Century*, Coale *et al.* (1979) give indexes of overall fertility, marital fertility, and proportions married for all republics of the Soviet Union and many smaller regions for 1926, 1959, and 1970. In several cases they even go back to 1897. This was part of a much larger project called the Princeton European Fertility Project, which calculated and analyzed these indexes for almost all provinces of Europe. These indexes are based on indirect standardization using the Hutterite fertility schedule as weights. They have been studied in other publications (Coale *et al.*, 1979; Coale and Watkins, 1986). In this chapter we have updated the time series of indexes by including information from the 1979 census.

Table 2.4 shows that the 1960s were characterized by a strong decline in the index of overall fertility (I_f) in the European parts of the Soviet Union, whereas in many Central Asian republics the level of fertility increased. During the 1970s the pattern is reversed with fertility indexes slightly increasing in the European part and decreasing in most Central Asian republics. On the aggregate this leads to a decline in I_f for the USSR between 1959 and 1979 and a slight increase from 1970 to 1979.

When looking at the decomposition of the index of total fertility (I_f) in the marital fertility component (I_g) and the proportions married (I_m), we see some trends in the components that partly compensate for each other. In the whole of the USSR between 1959 and 1970, the proportion married increased while marital fertility decreased. The net was still a decrease in fertility, but this decrease would have been much faster if the proportion of those married had not increased. Between 1970 and 1979 the reverse movement is evident again, with proportions of those married decreasing and marital fertility increasing.

As for individual republics between 1959 and 1979, we find an increase in the proportions married in all non–Central Asian republics.

Table 2.4. Princeton indexes: total fertility (I_f), martial fertility (I_g), and proportion married (I_m).

	I_f			I_g			I_m		
	1959	1970	1979	1959	1970	1979	1959	1970	1979
USSR	0.236	0.185	0.194	0.392	0.279	0.297	0.604	0.665	0.653
RSFSR	0.222	0.151	0.164	0.374	0.232	0.252	0.594	0.653	0.650
UKRSSR	0.192	0.158	0.161	0.329	0.233	0.237	0.583	0.678	0.682
BYELSSR	0.238	0.174	0.173	0.422	0.264	0.265	0.564	0.659	0.653
UZBSSR	0.403	0.432	0.397	0.545	0.599	0.610	0.738	0.721	0.651
KAZSSR	0.370	0.261	0.252	0.556	0.379	0.389	0.665	0.688	0.647
GRSSR	0.219	0.203	0.191	0.366	0.306	0.308	0.598	0.665	0.621
AZSSR	0.407	0.364	0.270	0.603	0.534	0.491	0.676	0.681	0.549
LITSSR	0.218	0.187	0.167	0.396	0.290	0.283	0.551	0.645	0.589
MOLSSR	0.293	0.201	0.204	0.450	0.306	0.305	0.650	0.656	0.668
LATSSR	0.161	0.151	0.150	0.283	0.236	0.238	0.569	0.642	0.629
KIRSSR	0.353	0.369	0.347	0.491	0.526	0.536	0.718	0.701	0.648
TAJSSR	0.307	0.451	0.453	0.405	0.610	0.660	0.757	0.739	0.687
ARMSSR	0.393	0.247	0.223	0.595	0.366	0.371	0.660	0.676	0.602
TURKSSR	0.414	0.451	0.399	0.552	0.624	0.614	0.750	0.724	0.651
ESTSSR	0.163	0.168	0.165	0.280	0.264	0.257	0.580	0.634	0.641

Only in Uzbekistan, Kirghizia, Tajikistan, and Turkmenistan does marital fertility increase over that period. After 1970 the trend in Central Asia is not reversed but marital fertility continues to increase on the whole, while the proportion married further decreases. Judging from the level and trend of the marital fertility index, most of Central Asia in 1979 can still be considered to be in a pre-transitional state while the Transcaucasian republics had clearly entered their marital fertility transition.

2.3   Period Total Fertility Rates

A look at period total fertility rates (TFR) in selected Soviet republics since 1959 (*Figure 2.1*) shows an interesting divergence of trends between Central Asia and Transcaucasia. In Uzbekistan and Tajikistan TFRs increase up to the mid-1970s. In Tajikistan, this increase is greater than 50 percent from four to more than six children per woman. In Azerbaijan fertility rates also increased somewhat up to 1964 but then began a steep decline. In Armenia fertility declined over the whole period of

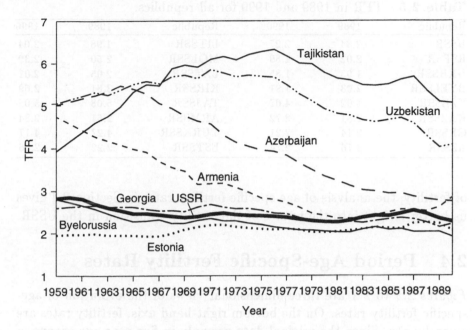

Figure 2.1. Trends in total fertility rate in selected Soviet republics, 1959–1989.

observation. In Georgia, Byelorussia, and Estonia fertility stayed at a constant low level.

The picture of different fertility trends in *Figure 2.1* clearly shows that these are populations at different stages in their secular fertility transitions. It has often been observed in the history of fertility declines, especially in the Third World, that before decline occurs, fertility can even increase. This increase may be due to changes in the marriage pattern, changes in the traditional patterns of breast-feeding, improved health, and so on. These fertility-increasing aspects of modernization are then soon outweighed by decreases in desired family size and the spread of family-size limitation.

Figure 2.1 shows that for Uzbekistan and Tajikistan the turning point toward family limitation was in the mid-1970s, for Azerbaijan it was in the early 1960s, and it was the 1950s for Armenia. *Table 2.5* lists the TFR for all republics in 1989 and 1990. Since the move to family limitation is typically associated with changes in the age pattern

Table 2.5. TFR in 1989 and 1990 for all republics.

Republic	1989	1990	Republic	1989	1990
USSR	2.34	2.27	LITSSR	1.98	2.04
RSFSR	2.02	1.89	MOLSSR	2.50	2.39
UKRSSR	1.93	1.85	LATSSR	2.05	2.01
BYELSSR	2.03	1.91	KIRSSR	3.81	2.69
UZBSSR	4.02	4.07	TAJSSR	5.08	5.05
KAZSSR	2.81	2.72	ARMSSR	2.61	2.84
GRSSR	2.14	2.21	TURKSSR	4.27	4.17
AZSSR	2.76	2.74	ESTSSR	2.22	2.06

of fertility, the analysis of age-specific fertility rates in Section 2.4 gives us better insight into the nature of the fertility transition in the USSR.

2.4 Period Age-Specific Fertility Rates

Figures 2.2 to *2.4* are three-dimensional views of the evolution of age-specific fertility rates. On the bottom right-hand axis, fertility rates are given by age. Since the original data are only in five-year age groups, a four-parameter model by Coale and Trussell (1974), which is a combination of their marriage and marital fertility models, was used to fit a fertility trajectory to the grouped data. On the bottom left-hand axis, time is in single years from 1959 to 1988. Hence age-specific fertility schedules of any year may be produced by diagonally cutting the mountain from left to right along the lines indicating calendar years.

For all populations of the Soviet Union, *Figure 2.2* indicates little change in age-specific fertility schedules over time. This is due to the fact that the majority of the population lives in the European republics, which are well beyond their demographic transition and experienced some fluctuations only at a low level of fertility. Only between 1959 and 1969 do we see a slight decrease followed by a weak increase. This on the aggregate seemingly stable condition, however, hides diverging and partly reversed trends in individual republics.

For Armenia (*Figure 2.3*) we see that the age-specific fertility schedule of 1988 has the typical shape of a moderate- to low-fertility population: a fast increase from age 15 to a peak at ages 22–23 followed by an initially fast and then slower decline in fertility rates. This pattern in Armenia has prevailed only over the past 10 years. In the early 1970s we find that fertility rates were somewhat higher and the decline after the

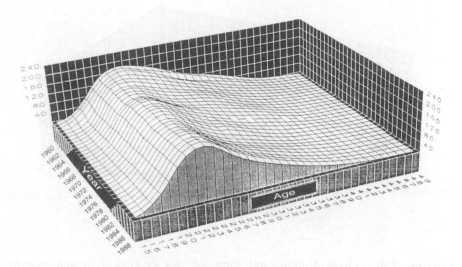

Figure 2.2. Three-dimensional view of the evolution of age-specific fertility rates in the USSR, 1959–1988.

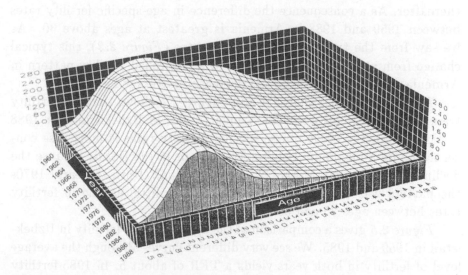

Figure 2.3. Three-dimensional view of the evolution of age-specific fertility rates in Armenia, 1959–1988.

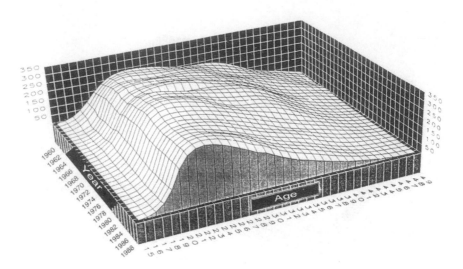

Figure 2.4. Three-dimensional view of the evolution of age-specific fertility rates in Uzbekistan, 1959–1988.

peak a bit slower. Going back to 1959 we find a very different pattern: a slower increase peaking at a later age (25–26) and a much slower decline thereafter. As a consequence the difference in age-specific fertility rates between 1959 and 1988 in Armenia is greatest at ages above 30. As we saw from the trend in total fertility rates (*Figure 2.2*), this typical change from a moderately controlled to a controlled fertility pattern in Armenia resulted in a substantial decrease in average fertility.

A comparison of *Figure 2.3* and *Figure 2.4* shows that the fertility transition is more advanced in Armenia than in Uzbekistan. In 1988 in Uzbekistan the curve of age-specific fertility rates is much less concentrated on a maximum than in Armenia or in the USSR, because the decline from the peak at age 22 is very gradual. Until the early 1970s the pattern was even more extreme with very little decline in fertility rates between ages 25 and 38.

Figure 2.5 gives a comparison of the age profiles of fertility in Uzbekistan in 1959 and 1985. We see very different curves although the average level of fertility in both years yields a TFR of about 5. In 1985 fertility rates are higher in the 20s and lower in the 30s than in 1959. The degree of transition from a concave to a more convex curve beyond age 30 has been suggested as a measure of deviation from natural fertility.

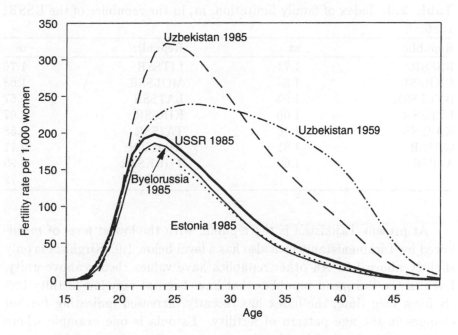

Figure 2.5. Age patterns of fertility in selected republics and years.

2.5   Index of Family Limitation

One of the parameters in the Coale-Trussell model of marital fertility as well as in the model combining marriage and marital fertility – denoted as m (see Coale and Trussell, 1974) – is called the index of family limitation. It indicates the degree to which the observed age pattern of fertility deviates from a standard pattern of natural fertility – namely, without evidence of family limitation – as described by Henry (1961). The concept of natural fertility and the parameters of the model have been described in Coale and Trussell (1974) and United Nations (1981).

Table 2.6 gives the estimates for m in all republics of the Soviet Union in 1986. (For detailed analysis of m in the republics of the USSR, see Lutz and Scherbov, 1989.) A high level of m indicates a significant departure from natural fertility. The value 1.0 is often taken as a threshold value from natural to controlled fertility. For Uzbekistan we see that in 1986 the level reached exactly 1.0, whereas over the period 1959–1975 m was still around 0.1 – that is, fertility was virtually *natural*.

Table 2.6. Index of family limitation, m, in the republics of the USSR, 1986.

Republic	m	Republic	m
RSFSR	1.71	LITSSR	1.76
UKRSSR	1.84	MOLSSR	1.68
BYELSSR	1.93	LATSSR	1.67
UZBSSR	1.00	KIRSSR	1.07
KAZSSR	1.40	TAJSSR	0.58
GRSSR	1.81	ARMSSR	2.11
AZSSR	1.69	TURKSSR	0.95
		ESTSSR	1.72

At present Tajikistan is the republic with the lowest level of m followed by Turkmenistan which also has a level below 1.0. Kirghizia is only slightly above 1.0. All other republics have values clearly above unity. Even in the European republics that have been at very low-fertility levels for a long time, the index has recently increased indicating further changes in the age pattern of fertility. Estonia is one example where changes in the level of fertility do not necessarily correspond to changes in the shape of the age pattern of fertility. From the mid-1960s to the mid-1970s, m increased sharply in Estonia from around 1.0 to 1.5; the TFR also increased over that period.

This index of family limitation gives us a better indication of the process of fertility transition than the average level of period fertility.

2.6   Cumulative Cohort Fertility

All the comparisons so far have looked at the variations of births by age and year. The analysis of fertility for real cohorts of women is made difficult by the fact that no systematic information is given for the pre-1959 period. Nevertheless, it seems worthwhile to compute all information that can be extracted from the 30 years for which we have age-specific fertility rates. One interesting summary measure is the cumulative fertility of birth cohorts up to a certain age. We are in a position to do this for 10 single-year birth cohorts at least up to age 40. If we go down to age 35 or 30, the number of cohorts increases to 15 or 20, correspondingly.

These data can be compared with the findings from sample surveys. Although the findings from both sources correspond quite well, the data

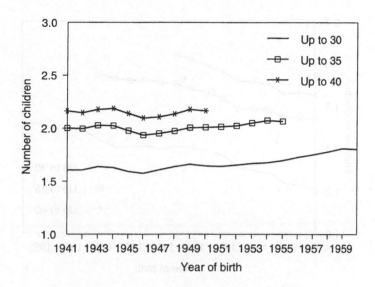

Figure 2.6. Cumulative cohort fertility in the USSR.

based on vital registration give slightly lower estimates. This might be taken as an indication of the incompleteness of birth registration.

Figures 2.6 to *2.9* give the trends in cumulative cohort fertility up to ages 30, 35, and 40 in the USSR and three selected republics. For the total of the USSR (*Figure 2.6*) the stability of fertility levels is remarkable. There seem to be only minimal changes from one cohort to the next. When there is any change at all, it is a slight increase in cumulative fertility up to age 30. But this may be due to a slight decrease in the mean age at childbearing.

In Latvia (*Figure 2.7*) we see a remarkable increase in cumulative cohort fertility at all ages. Since under the age pattern of Latvian fertility childbearing is essentially completed by ages 35–40, this cannot just be a timing phenomenon. A similar pattern had not appeared for period TFRs that show a decline in the mid-1960s to a low of 1.73 in 1966 and an increase thereafter reaching 2.01 in 1972 followed by another slight decline and a new maximum of 2.12 in 1984. Looking at age-specific fertility rates, we see a constant increase for ages 15–19 and 20–24 since the late 1960s. But this does not fully explain the constant increase in cohort fertility.

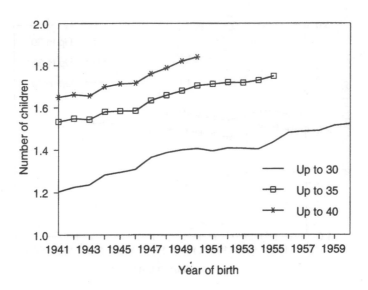

Figure 2.7. Cumulative cohort fertility in Latvia.

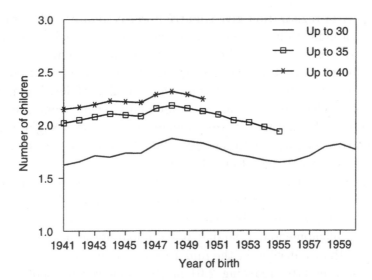

Figure 2.8. Cumulative cohort fertility in Georgia.

In Georgia (*Figure 2.8*) we see a moderate increase in cohort fertility followed by a slow decrease. This corresponds to the pattern observed for period TFR. In Kirghizia (*Figure 2.9*) we find a very steep decrease in

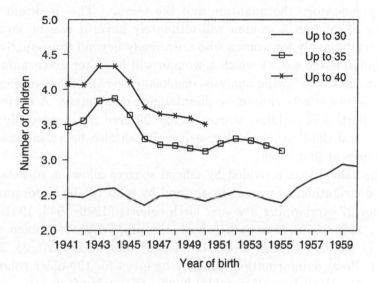

Figure 2.9. Cumulative cohort fertility in Kirghizia.

cumulative cohort fertility at ages 35 and 40, while that at age 30 remains constant and even increases from 1955. This is the typical pattern for a population in the midst of fertility transition where fertility above age 30 declines while that below age 30 may even increase.

2.7 Parity Distributions

Aside from the analysis of fertility by age, it is interesting to study fertility in respect to the number of children already borne by women. The parity-specific approach gives better insight into the process of reproduction because over the course of demographic transition it is increasingly the number of children already born, together with the number of children that women plan to have, that is the determinant of additional births. Age (within the given reproductive age span), which was the major factor in traditional societies, becomes less important than parity. The question of how evenly (or unevenly) children are distributed over all women also has far-reaching consequences on family structures and many socioeconomic aspects (see Lutz, 1989).

Looking at parity distributions of women who have not yet completed their reproductive career always combines two distinct aspects of fertility behavior: the quantum and the tempo. The quantum is the number of children a woman will ultimately have; it can be measured with certainty only for women who are already beyond reproductive age. The tempo is the age at which a woman will have her given number of children. In the following analysis, one has to be careful because in most cases the two effects cannot be disentangled completely. A decrease in the proportion of childless women at age 20 need not necessarily mean that general childlessness is decreasing. It can also be an indication of higher ages at first birth.

The tabulations provided by official sources allow us to determine parity distributions of women by age and by cohorts. The information is given for 27 overlapping five-year birth cohorts (1940–1944, 1941–1945, . . . , 1966–1970). For each cohort, the proportion of women with i births ($i = 1$ to $7+$) is given at six selected ages, a ($a = 20, 25, 30, 35, 40, 45$). Because information can only be given for the older cohorts at higher ages, the information tables have a triangular form.

For a comparative analysis of this information across cohorts and across ages, six nonoverlapping cohorts were selected (A:1941–1945, B:1946–1950, C:1951–1955, D:1956–1960, E:1961–1965, F:1966–1970). For these cohorts, parity distributions at the given ages were calculated on the following rationale: if $p1(45)$ is the proportion of women with at least one birth at age 45, $p2(45)$ is that with at least two births at age 45, and so on, then the proportion of childless women at age 45, $d0(45)$, may be derived as $1.0 - p1(45)$. The proportion of women with one child, $d1(45) = 1.0 - d0(45) - p2(45)$; the proportion with two children, $d2(45) = 1.0 - d0(45) - d1(45) - p3(45)$; and so on.

Table 2.7 gives the results of these calculations for the six cohorts selected at given ages. From this we can see that the proportion of childless women decreased at all ages from the older cohorts to the younger ones. At age 20 in the cohort born 1941–1945, 84 percent was still childless whereas in the youngest cohort, born 1966–1970, the proportion of childless was 80 percent. At higher ages the differences become more pronounced. At age 25 the proportion childless declined from 33 percent in the cohort born 1941–1945 to only 26 percent in that born 1961–1965. The data also indicate that this was not only a decrease in the age of childbearing, as described in Section 2.4, but also an increase in the parity progression ratios at parity zero as can be inferred from the

Table 2.7. Parity distributions for selected cohorts for ages 20 to 40, in the Soviet Union.

Until age	Cohorts	Parity						
		0	1	2	3	4	5	6+
20	1941–1945	0.84	0.14	0.01	0.00	0.00	0.00	0.00
	1946–1950	0.88	0.11	0.01	0.00	0.00	0.00	0.00
	1951–1955	0.85	0.14	0.01	0.00	0.00	0.00	0.00
	1956–1960	0.84	0.15	0.01	0.00	0.00	0.00	0.00
	1961–1965	0.81	0.18	0.01	0.00	0.00	0.00	0.00
	1966–1970	0.80	0.18	0.02	0.00	0.00	0.00	0.00
25	1941–1945	0.33	0.45	0.16	0.04	0.01	0.00	0.00
	1946–1950	0.29	0.51	0.16	0.03	0.01	0.00	0.00
	1951–1955	0.27	0.48	0.19	0.04	0.01	0.00	0.00
	1956–1960	0.27	0.46	0.22	0.04	0.01	0.00	0.00
	1961–1965	0.26	0.40	0.27	0.05	0.01	0.00	0.00
30	1941–1945	0.15	0.35	0.33	0.09	0.04	0.02	0.02
	1946–1950	0.11	0.40	0.35	0.08	0.03	0.02	0.01
	1951–1955	0.12	0.35	0.38	0.09	0.04	0.02	0.01
	1956–1960	0.11	0.30	0.41	0.12	0.04	0.02	0.01
35	1941–1945	0.11	0.27	0.39	0.11	0.04	0.03	0.05
	1946–1950	0.06	0.32	0.42	0.10	0.04	0.03	0.03
	1951–1955	0.07	0.26	0.43	0.13	0.04	0.03	0.03
40	1941–1945	0.09	0.26	0.40	0.12	0.04	0.03	0.06
	1946–1950	0.05	0.29	0.43	0.12	0.04	0.03	0.05

distributions at higher ages. Unfortunately for ages above 30, the number of cohorts that can be considered in the comparison gets very small. But at age 35 we still see a decline in childlessness from 11 percent to 7 percent between the cohorts of 1941–1945 and those of 1951–1955. The change in the proportions remaining childless between the two adjacent cohorts, 1941–1945 and 1946–1950, was especially remarkable.

With regard to women with one child, the picture becomes less clear. At age 20 the proportion with one child clearly increased over time, which was due to the younger childbearing age. At higher ages, however, there is some visible decrease, except for the cohort born 1946–1950, which has an unusually high proportion of women with one child. This is difficult to interpret because at parity one it is especially difficult to disentangle the effects of changes in timing and quantum.

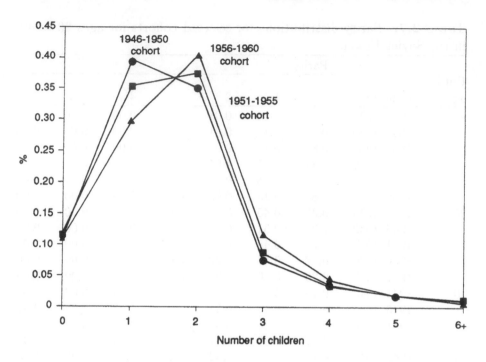

Figure 2.10. Parity distributions at age 30 for cohorts born 1946–1950, 1951–1955, and 1956–1960.

For those women with two children the picture is very clear. For all cohorts at all ages the proportion of women with two children increases. This clearly indicates a movement toward the two-child norm and a high concentration of this family size. More than 40 percent of the women born before 1950 ended up with two children, and there is every indication that for the younger cohorts a similar, if not larger, proportion of women will have two children at the end of their reproductive career.

Figure 2.10 illustrates this shift toward a higher prevalence of the two-child family. At age 30 the cohort born 1946–1950 has a distribution that peaks at parity one. For the next cohort 1951–1955, parity two is already more frequent than parity one but there is no clear peak. In the next cohort, finally, there is a clear mode at parity two.

For higher parities we have information only on the older cohorts. They show some decrease in the percentage of women with very large

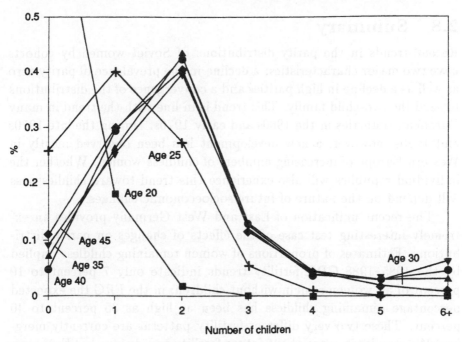

Figure 2.11. Parity distributions for most recent cohorts for which data are available for a given age.

families. So the percentage of women at age 35 with six or more children declined from 5 percent in the cohort of 1941–1945 to 3 percent in the subsequent cohorts. This also corresponds to the beginning of the fertility decline in the Central Asian republics.

Figure 2.11 shows the distributions of the cohorts for which information on the given age is available. Hence for age 20 it is the cohort born 1966–1970 whereas for age 45 it is the cohort born 1941–1945. In a way this presentation corresponds to a transversal or period approach where different cohorts at different ages are observed at the same time. The figure clearly shows that families are in the process of formation at ages 20, when the mode is at zero children, and 25, when the mode is at parity one. After age 30, the pattern does not change much and it is clearly centered at parity two. Only for very large families (with six or more children) does an older age make a difference. This is not only because women need more time to complete a large family but also because of the trend that has made large families less common.

2.8   Summary

Recent trends in the parity distributions of Soviet women by cohorts have two major characteristics: a decline in the prevalence of parity zero as well as a decline in high parities and a convergence of the distributions toward the two-child family. This trend is in line with the trend in many European countries in the 1960s and early 1970s. During the late 1970s and 1980s, however, a new development has been observed mostly in Western Europe of increasing numbers of childless women. Whether the individual republics will also experience this trend toward childlessness will depend on the nature of future socioeconomic changes.

The recent unification of East and West Germany provides an extremely interesting test case of the effects of changes on parity distributions. Estimates of proportions of women remaining childless implied by the 1985–1988 GDR fertility trends indicate only 7 percent to 10 percent of all women remain without children; in the FRG the expected percentage remaining childless has been as high as 35 percent to 40 percent. These two very different fertility patterns are currently merging. We may learn more about future fertility trends in other European populations including the populations of the USSR.

References

Coale, A.J., and T.J. Trussell, 1974. Model Fertility Schedules. *Population Index* **40**:185–258 (Erratum. *Population Index* **41**:577).

Coale, A.J., and S.C. Watkins. 1986. *The Decline of Fertility in Europe*. Princeton University Press, Princeton, NJ.

Coale, A.J., B.A. Anderson, and E. Härm, 1979. *Human Fertility in Russia Since the Nineteenth Century*. Princeton University Press, Princeton, NJ.

Henry, L., 1961. Some Data on Natural Fertility. *Eugenics Quarterly* **8**:81–91.

Lutz, W., 1989. *Distributional Aspects of Human Fertility: A Global Comparative Study*. Academic Press, New York, NY.

Lutz, W., and S. Scherbov, 1989. Modellrechnungen zum Einflußregional unterschiedlicher Fertilitätsniveaus auf die zukünftige Bevölkerungsverteilung in der Sowjetunion. *Zeitschrift für Bevölkerungswissenschaft* **15**(3/1989): 271–292.

United Nations, 1981. *Indirect Techniques for Demographic Estimation*. Population Studies No. 81; ST/ESA/SER.A/81. New York, NY.

Vishnevsky, A.G., S. Ya. Scherbov, A.B. Anichkin, V.A. Gretchoukha, and N.V. Donetz, 1988. Noveyshiye tendentsii rozhdaemosti v SSSR (Recent Fertility Trends in the USSR). *Sociologitcheskiye issledovaniya* (Sociological Studies) 3:54–67.

Chapter 3

Three Types of Fertility Behavior in the USSR

Alexandr Anichkin and Anatoli Vishnevsky

The vast differences in regional fertility levels in the republics of the former USSR and their implications for differential population growth are well known. Although groups of the population with different characteristics of reproductive behavior are distributed over the USSR by a distribution other than simply by republics, the most outstanding differentials are inter-republic variations. These can largely be explained by differences in the ethnic composition of the republics and regions which are inseparable from their culture and history.

3.1 Horizontal and Vertical Indicators

These cultural and historical peculiarities of the republics of the USSR have two dimensions. If the time axis is taken as a base they may be designated as horizontal and vertical. The horizontal dimension refers to the peculiarities of demographic behavior in societies with different cultural traditions but at approximately the same level of historical demographic development. Such horizontal differences result from various social and other factors, such as Catholic, Orthodox, or Muslim traditions. Certain factors determine the similarity of many behavioral standards across horizontal differences. One of these has always been the fact that high mortality rates necessitated the maintenance of high fertility levels, which were reflected in the cultural norms of all traditional societies. Depending on the situation this tendency is expressed differently, but, as a rule, the basic pronatalistic orientations are similar in all traditional cultures; the differences between them are of a specific, secondary nature. In this sense the horizontal differences are variations within the same historical type of fertility behavior.

In the vertical dimension one encounters a different type of variation which is determined by fundamental historical changes that affect societies of different cultural background. It is the modernization of society that leads, among other factors, to a rise of such cross-culturally valid phenomena as low mortality rates, making the traditional pronatalistic standards of behavior inappropriate. A reaction to these changes is the appearance of a new historical type of fertility behavior which gradually spreads to new sections of the population and other countries and regions. The vertical variations are those between different historical types of fertility behavior. When one such type is superseded by another this forms the basis of what is usually called fertility transition. The horizontal and vertical variations interweave; they affect each other and play a different role at different stages of the historical evolution. At the same time it is obvious that intra-type variations are larger than inter-type ones. It is thus not surprising that in the period of demographic transition the differentiation of fertility sharply increases, as can clearly be seen from the variations in fertility between the republics. Here the demographic transition brought about great changes in the reproductive behavior which had spread to the largest part of the country's population by the mid-20th century. However, the demographic transition is not yet complete in all republics. Parts of the Soviet population have for many decades differed, and still differ, in those fundamental types of fertility behavior which can readily be seen from the levels of the quantitative fertility indicators.

3.2   Three Fertility Patterns

Regardless of which fertility indicator is studied, one can see marked differences between the different parts of the USSR, with trends that sometimes diverge. There even seems to be a polarization between republics with respect to their fertility level. At one end of the scale we find republics where fertility indicators were already low in the 1950s and have remained so for the following decades (Russia, Ukraine, Byelorussia, Georgia, Lithuania, Latvia, and Estonia). Here the level and trend of fertility changes differ little from those in most developed countries. At the other end there are four Central Asian republics (Uzbekistan, Kirghizia, Tajikistan, and Turkmenistan) with an invariably high level of fertility over the entire post–World War II period which generally exceeds the level of many developing countries. Apart from these two polar groups there is an intermediate group of republics (Kazakhstan, Azerbaijan, Moldavia, and Armenia) where the level of fertility began to drop after 1960, rapidly in some cases.

The quantitative indicators of fertility in every group of republics cannot be separated from the qualitative features of demographic behavior. Together they form the basis for the distinction between the three typical fertility situations in the USSR: pre-transitional, transitional, and post-transitional.

The share of the different groups of republics in the total number of births is unequal and changes over time. In the past 40 years the contribution of the pre-transitional Central Asian republics to all births in the USSR has been steadily growing, whereas that of post-transitional republics, especially of Russia, has been decreasing (*Table 3.1*).

To get a better understanding of the specific details of each fertility regime (post-transitional, transitional, and pre-transitional) let us review in greater detail three representatives of these groups: Estonia, Azerbaijan, and Tajikistan.

Tables 3.2, 3.3, and *3.4* give some indicators of fertility levels in these republics (for comparison the figures for the USSR as a whole are given). Some obvious conclusions can be drawn from these indicators. They show how great the gap in fertility levels is between the republics which are at different stages of demographic transition. From the tables one also can see the trends in fertility over the past several decades. Even levels and trends of such rough indicators as the crude birth rates (CBRs) point to vast differences in fertility behavior in the three republics considered (*Figure 3.1*). The differences remain after the crude birth rate has been standardized by age. As is seen from the charts in *Figure 3.1*, after the effect of changes in age structure has been eliminated (as a standard the 1987 USSR population was used), the rise of

Table 3.1. Number of children born in the USSR and the proportion in post-transitional, transitional, and pre-transitional republics.

Year	Number of children born in the USSR (in 1,000s)	Post-transitional republics	Transitional republics	Pre-transitional republics
1940	6,096	86.1	7.8	6.1
1950–54	24,597	82.1	10.2	7.7
1955–59	25,738	79.0	12.0	9.0
1960–64	24,706	74.7	13.8	11.5
1965–69	20,763	70.7	14.4	14.9
1970–74	21,934	70.0	13.6	16.4
1975–79	23,595	68.6	13.6	17.8
1980–84	25,692	67.5	13.6	18.9
1985–89	21,965	65.2	13.8	21.0

Post-transitional republics: Russia, Ukraine, Byelorussia, Georgia, Lithuania, Latvia, Estonia.
Transitional republics: Kazakhstan, Azerbaijan, Moldavia, Armenia.
Pre-transitional republics: Uzbekistan, Kirghizia, Tajikistan, Turkmenistan.

the CBR in Tajikistan became even sharper and its fall in Azerbaijan deeper than in the case of the non-standardized rate.

Indirect standardization of fertility based on Coale's method (the Princeton Indexes) gives additional information on variations in fertility behavior of the population of the three republics and its trends. Coale's indexes are interpreted as a measure of deviation of the fertility level from its hypothetical maximum value. They make it possible to break down the changes in general fertility (the index of overall fertility, I_f) into changes in marital fertility (I_g) and into the proportion of married women (I_m). The data given in *Table 3.2* show that the trends of the index of overall fertility in the republics considered are very similar to those of the crude birth rate.

For Estonia it turns out that the substantial increase in the index of overall fertility in the 1980s cannot be explained with the help of the two other indexes because the main cause of this rise was a sharp increase in nonmarital fertility (in 1958 2,888 births were out of wedlock, in 1978 there were 2,698, and 5,788 in 1988).

In Azerbaijan, on the contrary, the index of overall fertility shows a significant fall: in the 1970s the drop was due to a combined decrease of I_g and I_m. In the 1960s and especially in the 1980s the situation

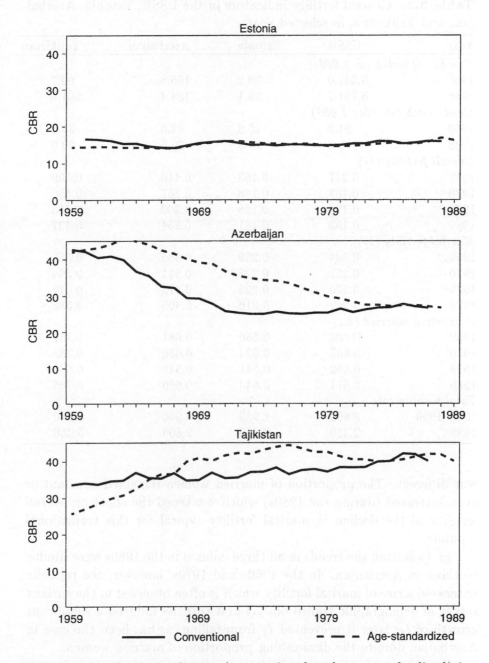

Figure 3.1. Crude birth rate (conventional and age-standardized) in Estonia, Azerbaijan, and Tajikistan, from 1959 to 1989.

Table 3.2. General fertility indicators in the USSR, Estonia, Azerbaijan, and Tajikistan, in selected years.

Year	USSR	Estonia	Azerbaijan	Tajikistan
Number of births (in 1,000s)				
1960	5,341.0	20.2	165.8	69.7
1988	5,381.1	25.1	184.4	201.9
Crude birth rate (per 1,000)				
1960	24.9	16.6	42.6	33.5
1988	18.8	15.9	26.5	40.0
Overall fertility (I_f)				
1959	0.237	0.163	0.410	0.309
1970	0.187	0.168	0.367	0.456
1979	0.194	0.165	0.269	0.455
1989	0.199	0.181	0.254	0.437
Marital fertility (I_g)				
1959	0.344	0.239	0.586	0.365
1970	0.257	0.228	0.511	0.594
1979	0.276	0.225	0.473	0.612
1989	0.267	0.216	0.405	0.580
Proportion married (I_m)				
1959	0.606	0.580	0.681	0.767
1970	0.665	0.634	0.680	0.738
1979	0.652	0.641	0.549	0.686
1989	0.671	0.643	0.609	0.705
Total fertility rate				
1958–1959	2.800	1.930	4.940	3.840
1988	2.450	2.240	2.800	5.350

was different. The proportion of married women remained constant or even increased (during the 1980s) which weakened the effect on overall fertility of the decline in marital fertility typical for this transitional republic.

In Tajikistan the trends in all three indexes in the 1980s were similar to those in Azerbaijan. In the 1960s and 1970s, however, the republic witnessed a rise of marital fertility which is often observed in the earliest stages of demographic transition. At first this rise induced a significant growth of I_f; later it prevented I_f from falling, as has been the case in Azerbaijan despite the diminishing proportion of married women.

For the total fertility rate (TFR), an indicator free from the influence of age structure, a varying picture has been observed since the late

Table 3.3. Age-specific period fertility rates (per 1,000 women) in the USSR, Estonia, Azerbaijan, and Tajikistan, in selected years.

Year	USSR	Estonia	Azerbaijan	Tajikistan
15–19 years				
1958–1959	29.2	20.1	43.0	26.1
1988	46.8	45.5	26.7	38.4
20–24 years				
1958–1959	162.2	122.3	209.6	137.8
1988	191.0	180.4	199.3	312.0
25–29 years				
1958–1959	164.8	119.1	266.5	176.5
1988	138.8	124.3	189.5	301.1
30–34 years				
1958–1959	110.1	72.9	216.1	169.7
1988	73.9	64.7	97.9	216.9
35–39 years				
1958–1959	66.6	41.9	162.7	154.4
1988	31.1	25.9	39.1	132.8
40–44 years				
1958–1959	24.1	12.0	73.5	87.4
1988	7.9	6.3	11.1	63.9
45–49 years				
1958–1959	5.0	0.9	29.6	33.2
1988	0.6	0.2	1.2	10.1

1950s. In some republics the TFR tended to rise noticeably and then to fall (Central Asian republics); in other cases (Kazakhstan, Moldavia, Azerbaijan, and Armenia) it has been strongly declining all the time; in a third group it decreased slowly with a tendency to stabilize in the second part of the period (Russia, Byelorussia, and Lithuania); and in yet another group relative stability has been registered with slight changes in either direction (Ukraine, Georgia, Latvia, and Estonia).

Against the background of this wide variety the three republics under review (Estonia, Azerbaijan, and Tajikistan) typify the trends observed over this period. The TFR in post-transitional Estonia is characterized by fluctuations from 2.0 to 2.2 births per woman. In transitional Azerbaijan it shows a constant decline from the mid-1960s; finally, in pre-transitional Tajikistan the total fertility rate increased until the mid-1970s and started a moderate decline thereafter.

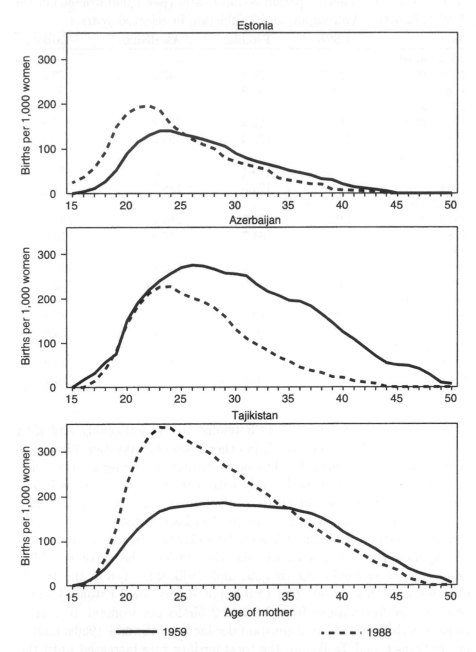

Figure 3.2. Age-specific fertility rates in Estonia, Azerbaijan, and Tajikistan, in 1959 and 1988.

Table 3.4. Selected period and cohort fertility indexes in the USSR, Estonia, Azerbaijan, and Tajikistan, in selected years.

Year	USSR	Estonia	Azerbaijan	Tajikistan
Proportion of children borne by women under age 35				
in hypothetical female cohorts (per 100)				
1958–1959	83.30	86.50	74.40	66.40
1988	91.90	92.60	90.70	80.70
Proportion of births of the first three orders				
in the total number of births (per 100)				
1959	80.20	92.10	63.50	53.30
1988	87.70	93.20	85.40	62.10
Mean age of mother in hypothetical cohorts (years)				
1958–1959	28.50	27.90	30.40	31.90
1988	26.40	25.90	27.50	29.30
Number of children borne by female cohorts under age 35				
1941	2.00	1.73	3.57	3.85
1946	1.93	1.72	3.16	4.11
1951	2.01	1.83	3.01	4.34
1955	2.06	1.90	2.63	4.31

The analysis of the changing patterns of age-specific fertility rates assists in providing a better picture of the fertility evolution over the past 30 years (*Table 3.3*). In all three republics the age curves of fertility change in the way one would expect in the course of demographic transition. Until the transition has been completed the intensity of childbearing in the older reproductive ages (45–49, 40–44, 35–39, 30–34, and even 25–29 years) decreases generally: the higher the age the greater the decrease. At younger ages (20–24 and 15–19 years) the birth intensities, in contrast, increase. The age-specific fertility patterns in the three republics at the beginning and at the end of the period are depicted in *Figure 3.2*. In 1988, as well as three decades before, distribution of births by age of mother in Tajikistan and Estonia, respectively, are good examples for pre- and post-transitional patterns. Over 30 years no fundamental changes in the curve have occurred.

In Tajikistan the 1970s were characterized by a downward trend in fertility at older age groups, especially above age 35, which is a clear indication of the onset of transitional processes. However, such changes have failed so far to alter the type of the curve. In the 20–24 and 25–29 age groups the rates stopped rising in the late 1970s, but are still very

high – higher than in the early 1960s. The age-specific fertility rate in
the 15–19 age group has practically remained unchanged over the whole
period: it is lower today in Tajikistan than in the Ukraine, Russia, or
Latvia. In Estonia the reason for the relative stability of the curve is
quite obvious: the transformation process had already been completed
in 1959, although some rejuvenation of fertility is still under way. This
may be partly due to specific circumstances in Estonia, but generally
one may say that if any increase of fertility is to be expected in Estonia,
or in other post-transitional republics, to at least replacement level, it
is not likely to originate from age groups over 30; it is much more likely
the source would be the younger age groups.

Being very distinct from both Tajikistan and Estonia, Azerbaijan
has witnessed a quick shift from one form of fertility curve to another.
Like in other republics in a fertility transition, the age-specific fertility rate kept on falling in practically all age groups, but the speed of
reduction clearly accelerated in the older age groups.

Scrutinizing the combination of changing total fertility rates and
mean ages of mothers characterized by the points in a three-dimensional
chart (see *Figure 3.3*) gives a general idea of the relationship between
the fertility level and changes in the fertility curve. We again see that
in Estonia the total fertility rate stays relatively stable (the point on
the chart hardly moves either left or right) against the background of a
somewhat reduced mean age of mother.

In Azerbaijan the picture is quite different. The main trend is represented by simultaneous movement in the three-dimensional space from
left to right and from top to bottom. It is obvious that in this case the
two movements cannot be separated because a considerable reduction in
total fertility is necessarily accompanied by a strong reduction in births
for older reproductive age groups.

And, finally, in Tajikistan the point on the graph has been moving
left (rising fertility) without noticeably moving down for a long period.
In the 1970s the movement to the left, in general, stopped (although
we cannot say that it has been followed by a clear-cut movement to the
right), but the point definitely began to move down. In other words,
there was a noticeable rejuvenation of fertility which has not yet been
accompanied by a decline in the level of fertility. When looking for an
interpretation, it must be borne in mind that the considerable growth of
fertility level in Tajikistan (as well as in other Central Asian republics)

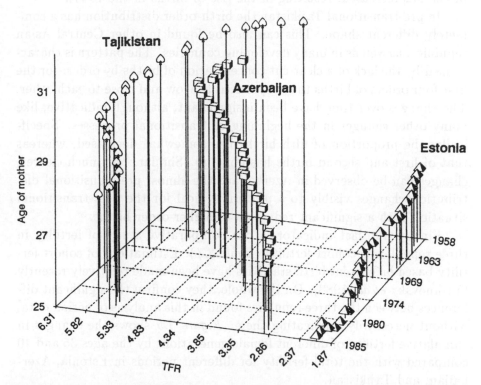

Figure 3.3. Total fertility rates and mean ages of mother in Estonia, Azerbaijan, and Tajikistan, from 1958 to 1985.

and in Azerbaijan in the 1960s, ought to be interpreted with caution as at that time births tended to be unregistered. Inadequate birth registration was gradually overcome later.

The distribution of births by order, together with the indicators of the level of fertility and the age pattern of fertility, is the third basic quantitative characteristic of the reproductive process (*Table 3.4*). In post-transitional Estonia the distribution of births by order has a form which is typical for most Western industrialized countries: a very low proportion of fourth and successive births, the overwhelming majority of births being concentrated in the first two orders (see *Figure 3.4*). No great changes in the distribution have taken place in the past three decades. At the same time it should be remembered that the proportion of first births has somewhat decreased, whereas that of second and third births has increased, resulting in the rise of births in the 1980s.

In pre-transitional Tajikistan the birth-order distribution has a completely different shape. This can also be found in other Central Asian republics, as well as in many developing countries. The pattern is characterized by the lack of a clear-cut concentration of births by order: for the first four orders of births the proportions are low and close to each other. The changes over time have been insignificant, although indicative, like many other changes in the beginning of transitional processes. Specifically, the proportion of fifth births has somewhat decreased, whereas that of first and second births has gone up. Similar, but much clearer, changes can be observed in Azerbaijan: the almost pre-transitional distribution changes visibly to a pattern typical for the post-transitional situation with a significant reduction of higher-order births.

Finally, the last indicator in our comparative study of fertility in three republics is cohort fertility (*Table 3.4*). Estimations of cohort fertility based on current birth statistics have been made relatively recently (Vishnevsky *et al.*, 1988). On the whole, they confirm the significant differences between the three republics found in this analysis of period data, without necessarily duplicating them. *Figure 3.5* shows the changes in cumulative fertility of different female generations by the ages 35 and 40 compared with the total fertility for different periods in Estonia, Azerbaijan, and Tajikistan.

In Estonia the period and cohort indexes are stable and close to each other. Here, like in many other post-transitional republics low and stable levels of fertility of present-day female generations predominate and there are no essential changes in the pace of family formation from generation to generation. In Azerbaijan both the period and cohort indexes point to a substantially decreasing level of fertility. At the same time the TFR has reduced much more than the cumulative fertility, which is obviously due to an acceleration of family formation in the cohorts born between 1940 and 1950. In Tajikistan there are more essential differences between the period and cohort indexes of fertility. Although both are high and steady, their trends are somewhat different. For example, a moderate reduction of the TFR in the early 1970s took place despite stabilized, or even slightly rising, cumulative fertility of the cohorts. Hence the fertility decline was probably a result of some recent short-term changes in family-formation patterns.

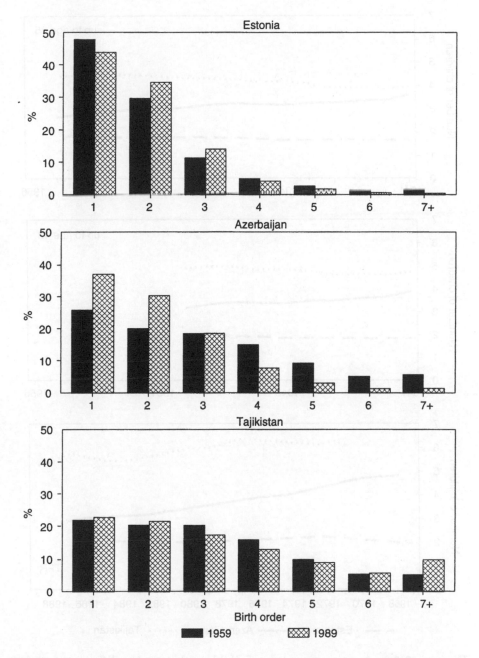

Figure 3.4. Distribution of births by order in Estonia, Azerbaijan, and Tajikistan, in 1959 and 1989.

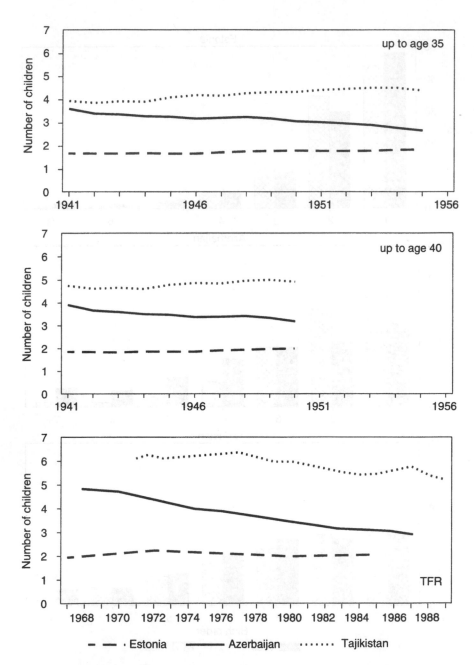

Figure 3.5. Average number of children borne by different female generations by the ages 35 and 40 and the period TFR in Estonia, Azerbaijan, and Tajikistan.

3.3   Future Fertility Patterns

What do we think about the future fertility pattern in the republics? Proceeding from the well-recognized theory of demographic transition, we expect a change toward the post-transitional type of procreative behavior in all republics. At some time in the near or more distant future Azerbaijan and even Tajikistan are going to approach Estonia both in the type of behavior of the majority of the population and in the fertility indexes. However, the theory of demographic transition generalizes the Western experience, assuming fundamental changes in the traditional pronatalistic norms in specific countries and regions. This leads to doubts about the universal nature of the theory. The critics of the transition theory generally stress the non-transient character of the horizontal cultural differences; for instance, the specifics of Islam which allegedly withstand the modernization of the demographic process.

The future will always be uncertain. We can only refer to one of the founders of the theory of demographic transition, Notestein, who said that the only way to learn about the future is to live up to that period (Notestein, 1950). Considering the past and the present, however, and in particular when looking at the Soviet experience, we may say that the empirical evidence presented in this chapter fails to support the doubts concerning the universal nature of the theory. At the beginning of the 20th century the level of fertility in Russia and the Ukraine was much higher than now in the regions of Islamic cultural traditions which could also have been explained by the cultural peculiarities of the Orthodox church. At that time the idea that Russians would in the future have the same reproductive behavior as French Catholics or Estonian Protestants would have seemed absurd. Nonetheless, it has happened. Only 30 years ago Christian Armenia differed little from the Muslim Central Asian republics in its fertility level. Today it stands closer to the European republics. Even Azerbaijan, which is strongly affected by Islamic traditions, is, as we can see, following the same route as Armenia.

References

Notestein, F., 1950. The Population of the World in the Year 2000. *Journal of the American Statistical Association* 45(251):335.

Vishnevsky, A.G., S.Ya. Scherbov, A.B. Anichkin, V.A. Gretchoukha, and N.V.
   Donetz, 1988. Noveyshiye tendentsii rozhdaemosti v SSSR (Recent Fer-
   tility Trends in the USSR). *Sociologitcheskiye issledovaniya* (Sociological
   Studies) **3**:54–67.

Chapter 4

Quantum and Timing of Births in the USSR

Leonid E. Darsky

No regular statistical data have been taken in the USSR on the interval between births. Retrospective surveys are the only source of information on birth timing. Tabulations of a 1985 survey, which polled 5 percent of the USSR population, provide a detailed picture of the situation in the late 1970s and early 1980s. This was not the first study of the timing of births using retrospective surveys. Such indicators were obtained for the first time on the basis of a 1960 survey (Sifman, 1974).

Data obtained by studying birth histories must be considered carefully. Errors could easily be introduced because of the possibility that children born many years ago who died shortly after birth may not have been counted in the data due to a loss of memory of the mother. In addition, dates of marriage and of first birth could have been shifted to conceal illegitimate births. The analysis of changes over time is only possible when the age of the mother at the time of birth is considered in addition to her current age.

4.1 Cohort Characteristics

All marriage cohorts formed before 1945 were affected by World War II and several socioeconomic and political cataclysms. Hence, one must

Table 4.1. Average intervals between births among married women, in years.

Year of marriage or previous birth	Marriage and 1st birth	1st and 2nd births	2nd and 3rd births	3rd and 4th births
1960 survey				
1920–1924	2.96	2.75	2.77	2.67
1925–1929	2.72	2.89	2.97	3.08
1930–1934	2.61	3.14	3.37	3.23
1935–1939	2.34	3.24	3.45	3.17
1940–1944	2.79	3.88	4.16	3.99
1945–1949	2.04	3.17	3.20	3.23
1985 survey				
1945–1949	2.74	4.16	3.90	3.50
1950–1954	2.30	4.14	3.88	3.36
1955–1959	1.92	4.09	3.90	3.39
1960–1964	1.78	4.59	4.13	3.48
1965–1969	1.50	4.44	3.98	3.33
1970–1974	1.39	4.30	3.81	3.14

be careful in making conclusions based on the time series presented in *Table 4.1.* However, it can be concluded that the average interval between marriage and the first birth, which changed little in prewar cohorts, began to decrease rapidly in the 1950s; at the same time the interval between the first and second births, as well as between the second and third births, in postwar cohorts was somewhat longer than in prewar cohorts, in spite of the fact that prewar cohorts' fortune was not favorable and could have caused postponement of births. In the interval between the third and fourth births, there was no significant change. The 1945–1949 cohorts in the 1960 survey are comparable with the 1970–1974 cohort of the 1985 survey. Both reflect a relatively calm period (observation period up to the moment of survey lasted 10 to 15 years, which is quite sufficient to determine an interval, yet small enough to minimize possible distortions). *Table 4.2* shows that during the period between these two cohorts, which is almost equal to the length of one generation, increased birth control had an impact on the probabilities of next birth for all orders from the second to the seventh birth. However, the intervals increased only before the second and the third births.

Table 4.2. Fertility in the 1945–1949 and 1970–1974 marital cohorts.

Previous birth order	Parity progression ratio in the cohort of		Interval between given and next births in the cohort of	
	1945–1949	1970–1974	1945–1949	1970–1974
0	0.956	0.952	2.04	1.39
1	0.828	0.741	3.17	4.49
2	0.749	0.349	3.20	3.81
3	0.711	0.476	3.23	3.14
4	0.679	0.639	3.10	3.00
5	0.724	0.630	3.10	2.87
6	0.675	0.673	3.04	2.75
7	0.698	0.667	2.82	2.68
8	0.604	0.637	2.98	2.61
9	0.531	0.599	2.75	2.60

Cohort data do not provide information on the short-term effects of conjuncture changes in fertility and on the timing of births in response to population policy. A study of this type is possible only with the help of the synthetic cohort method. Parity progression ratio increases, computed by means of demographic tables based on the probability of the next birth by interval after the previous birth, provide an explicit picture (*Table 4.3*). Of interest is the way fertility levels reacted to population-policy measures which were not adopted simultaneously in the country.

4.2 Population-Policy Measures

In the early 1980s, several population-policy measures for mothers and children were adopted (*Okhrana materinstva i detstva v SSSR*, 1986). These benefits included an increase in the length of maternity leave and an increase in family allowances. These benefits were adopted in different parts of the country in three stages beginning in regions with low fertility: November 1981, November 1982, and November 1983. Studying the fertility trends during these stages we see that fertility increase was, at least partly, connected with population-policy measures (*Table 4.4*). But the data also show that fertility had started to increase before the measures could have had any effect.

Table 4.3. Indicators of fertility tables of married women in 1980–1984.

Number of years	Number of N-parity women at t years after birth of				
	1st child	2nd child	3rd child	4th child	5th child
0	10,000	10,000	10,000	10,000	10,000
1	8,952	9,594	9,419	9,210	9,315
2	7,088	8,636	7,724	6,746	6,874
3	5,810	8,032	6,687	5,205	5,357
4	4,793	7,628	6,157	4,499	4,750
5	3,974	7,308	5,810	4,102	4,420
6	3,353	7,046	5,575	3,837	4,229
7	2,860	6,838	5,418	3,681	4,110
8	2,517	6,677	5,309	3,585	4,030
9	2,279	6,555	5,237	3,520	3,980
10	2,096	6,458	5,187	3,477	3,949
15	1,752	6,276	5,104	3,393	3,867
20	1,724	6,250	5,078	3,364	3,842
Parity progression ratio	0.828	0.375	0.492	0.664	0.616
Mean interval between births	3.81	3.91	2.96	2.70	2.60

However, there are valid reasons for considering some part of this increase to be the effect of the population-policy measures. The increase of the indicators follows the adoption of population-policy measures that had been introduced at different times in the territories (*Table 4.4*).

As expected the fertility increase was temporary. Today, the previous trends toward decline and stabilization of fertility at a low level have been re-established. But questions remain: Was this increase real? Though temporary, was it an increase in the number of births in families, or only a decrease in the interval between births?

The average birth intervals in *Table 4.4* do not show any systematic changes synchronous with the probabilities of family growth. Birth-interval distribution, obtained from fertility tables, practically did not change; the increase in parity progression ratios was independent of the interval after the previous birth. This fact challenges the theory that every short-term effect of the policy measures to increase fertility is connected with the change of birth intervals. It may well be that the interrelations are the effect of the synthetic cohort method, but theoretical consideration cannot be the reason for such a conclusion. A

Table 4.4. Indicators of fertility tables by parity in 1980–1984.

Year	Parity progression ratio				Mean interval (years) between given and next births			
	1	2	3	4	1	2	3	4
1st group of territories (privileges adopted in 1981)								
1979	0.792	0.279	0.362	0.480	4.24	4.63	4.70	3.11
1980	0.805	0.301	0.326	0.486	4.20	4.67	3.28	2.85
1981	0.822	0.294	0.329	0.558	4.16	4.59	3.72	3.01
1982	0.861	0.354	0.383	0.561	3.99	4.76	3.61	2.89
1983	0.868	0.381	0.393	0.547	3.90	5.09	3.66	2.81
1984	0.870	0.375	0.387	0.596	3.95	5.16	3.97	2.80
2nd group of territories (privileges adopted in 1982)								
1979	0.729	0.235	0.295	0.492	4.41	4.57	3.70	3.11
1980	0.741	0.229	0.286	0.484	4.33	4.51	3.56	3.09
1981	0.744	0.228	0.287	0.492	4.31	4.52	3.49	3.04
1982	0.769	0.249	0.319	0.513	4.29	4.65	3.56	2.92
1983	0.813	0.278	0.327	0.522	4.17	4.61	3.59	2.94
1984	0.815	0.289	0.286	0.468	4.16	4.87	3.66	2.98
3rd group of territories (privileges adopted in 1983)								
1979	0.921	0.700	0.716	0.765	2.48	2.79	2.53	2.47
1980	0.932	0.704	0.720	0.767	2.47	2.82	2.50	2.47
1981	0.930	0.702	0.698	0.737	2.46	2.82	2.53	2.48
1982	0.939	0.703	0.705	0.741	2.42	2.88	2.61	2.49
1983	0.931	0.699	0.690	0.738	2.47	2.94	2.74	2.60
1984	0.946	0.752	0.718	0.768	2.19	3.01	2.70	2.64

final conclusion can be made only at the end of the century, when it will be possible to measure intervals in real cohorts. Now we can only analyze the differentiation of birth timing among the women of the main nationalities of the Soviet population. Nationality is important first of all because its role is significant in differential fertility.

4.3 Cohort Fertility by Nationality

Deliberate childlessness is not widely practiced in the USSR. The interval between marriage and the first birth is inadequately registered in the survey. Nevertheless, a decrease in the average interval among all selected populations can clearly be seen. The effects of growth in the number of premarital conceptions, increased premarital sexual activity,

Table 4.5. Average interval between marriage and first child, in years.

Nationality	Year of marriage					
	1945–1949	1950–1954	1955–1959	1960–1964	1965–1969	1970–1974
Total population	2.74	2.30	1.92	1.78	1.50	1.39
1. Russians	2.35	2.07	1.83	1.76	1.51	1.38
2. Ukrainians	2.55	2.20	1.86	1.75	1.46	1.37
3. Byelorussians	2.33	2.03	1.68	1.63	1.39	1.34
4. Uzbeks	5.35	4.07	2.70	2.10	1.54	1.50
5. Kazakhs	4.65	3.70	2.61	2.05	1.45	1.33
6. Georgians	2.80	2.46	1.96	1.89	1.59	1.54
7. Azerbaijanis	4.12	3.41	2.48	2.04	1.64	1.58
8. Lithuanians	3.34	2.86	2.39	2.10	1.69	1.53
9. Moldavians	3.40	2.92	2.27	2.00	1.66	1.45
10. Latvians	2.73	2.66	2.37	2.18	1.81	1.49
11. Kirghiz	6.37	4.99	3.38	2.67	1.58	1.45
12. Tajiks	4.30	3.49	2.46	2.00	1.46	1.58
13. Armenians	2.92	2.47	1.97	1.77	1.51	1.45
14. Turkmen	6.94	5.09	3.34	2.25	1.55	1.60
15. Estonians	2.77	2.44	2.02	1.88	1.53	1.44
16. Tatars	2.77	2.20	1.77	1.63	1.35	1.32
17. Jews	2.76	2.28	2.16	2.04	1.79	1.67

improved health conditions of those getting married, and possible distortions in the information caused by disregarding the number of some children born many years ago who died shortly after birth are apparent. Mean marriage duration at first birth differs mostly by nationalities, and this differentiation decreases in time (*Table 4.5*).

Figure 4.1 displays the interval between the second and third births at different years for the Russian women surveyed. When comparing such data, ethnic heterogeneity must be considered; the ethnic composition of cohorts is different, and this may have an impact. From 1945 to 1969 the practice of postponing the third birth was rather widely spread among married Russian women. Among postwar cohorts (those who gave birth to a second child between 1945 and 1949) half bore a child three years later. Among those who gave birth to a second child between 1970 and 1974 only 32 percent bore a third child three years later. Among those who do not postpone births (Tajiks, Uzbeks, and Turkmens) 95 percent of women who bore a second child bore a third

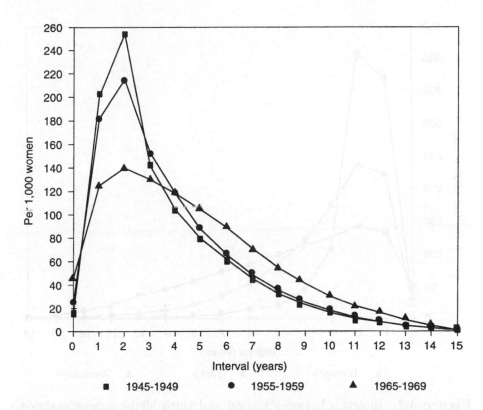

Figure 4.1. Interval between second and third births among Russian women.

child. In addition, 75 percent to 77 percent of women had a third child within three years. On the other hand, more than half of the Russian women who bore a third child postponed the third birth.

The postponement of births is unequally practiced among population groups: more widespread in urban than in rural areas; more often in large cities than in small ones; more among those with a high level of education than among those with a low level. However, the largest differences in reproductive behavior are still connected with differences in nationality. In *Figure 4.2* the interval between second and third births is presented for three nationalities. The comparison with *Figure 4.1* shows that the distributions correspond to different stages of transition. Thus in the process of demographic transition, interval distributions

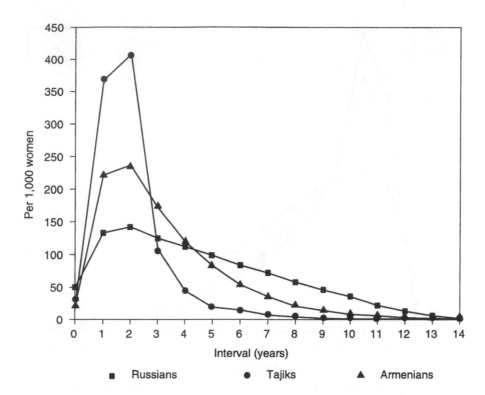

Figure 4.2. Intervals between second and third births among married women of three nationalities, second child born between 1970 and 1974.

undergo a definite evolution. The same regularities are seen for the interval between the first and the second births and between the third and the fourth births. While the probability of the next birth decreases, the share of those who postpone a birth – increasing birth interval – rises. This relation is more pronounced when parity progression ratios and mean intervals for women of different nationalities are compared. The data are presented in *Table 4.6* and are ordered by the increase of interval between the first and the second births.

Figure 4.3 shows that, as far as the second birth is concerned, all of the nationalities are divided into two groups. The first group comprises nationalities that do not postpone the second birth; the probability of the second birth ranges between 0.85 for Georgians (6) and 0.98 for Tajiks (12), and the interval between the first and the second births is

Table 4.6. Parity progression ratios and mean birth intervals for married women of major nationalities married in 1970–1974.

Nationality[a]	Parity progression ratios			Duration of marriage at first birth	Mean interval (years) between		
	1st to 2nd child	2nd to 3rd child	3rd to 4th child		1st and 2nd births	2nd and 3rd births	3rd and 4th births
12. Tajiks	0.978	0.957	0.924	1.58	2.36	2.45	2.53
4. Uzbeks	0.974	0.945	0.890	1.50	2.36	2.58	2.67
7. Azerbaijanis	0.943	0.860	0.732	1.58	2.36	2.82	2.88
14. Turkmen	0.967	0.947	0.875	1.60	2.43	2.53	2.62
11. Kirghiz	0.968	0.931	0.890	1.45	2.50	2.67	2.74
5. Kazakhs	0.950	0.869	0.831	1.33	2.58	2.93	2.91
13. Armenians	0.898	0.587	0.346	1.45	2.67	3.65	3.50
6. Georgians	0.848	0.410	0.253	1.54	2.70	3.78	3.53
15. Estonians	0.752	0.276	0.266	1.44	3.55	4.44	3.47
10. Latvians	0.683	0.261	0.315	1.49	3.94	4.31	4.02
9. Moldavians	0.848	0.529	0.461	1.45	4.01	4.09	3.68
8. Lithuanians	0.746	0.285	0.306	1.53	4.10	4.58	3.82
16. Tatars	0.816	0.390	0.366	1.32	4.14	4.56	3.96
3. Byelorussians	0.785	0.252	0.253	1.34	4.37	4.35	3.76
2. Ukrainians	0.744	0.240	0.250	1.37	4.64	4.50	3.83
1. Russians	0.691	0.193	0.219	1.38	4.93	4.89	3.94
17. Jews	0.569	0.117	0.257	1.67	5.56	4.79	3.60

[a]Nationalities are numbered as in *Table 4.5*.

from 2.4 years for Tajiks (12) and Uzbeks (4) to 2.7 years for Georgians (6) and Armenians (13).

The second group includes eight nationalities also, but the probability of the second birth ranges between 0.68 for Latvians (10) and 0.8 for Tatars (16); the average interval is from 4.1 years for Tatars (16) to 4.9 years for Russians (1). Only Jews (17) are beyond these groups ranges. They are characterized by particularly low fertility with the probability of the second birth being 0.57 and by an interval of 5.6 years.

Figure 4.3 shows that the situation for the third birth is similar to the second birth but includes an intermediate group in addition to the two extreme groups. This group consists of Georgians (6), Moldavians (9), and Armenians (13). Indicators of the fourth birth also give two compact groups, and only Azerbaijanis (7) are somewhat different from the first group.

Generally speaking, the lower the order of birth, the greater the differentiation by interval. Among those nationalities where the postponement of births is not widely practiced, the average interval, irrespective of the parity in the 1970–1974 cohort, almost does not differ and is equal to approximately 2.5 years. This interval corresponds to the biological norm. Among the nationalities that do not deliberately postpone births, the interval decreases from cohort to cohort (*Table 4.7*). The decrease is small though noticeable, especially for the interval between the first and the second orders. This may partly be explained by the decrease of lactation period as a result of more intensive women's employment outside the home, changes in nutrition patterns, and increasing fecundity due to improvements in mothers' health.

In the nations where birth control is common, the average interval rose slightly from cohort to cohort (*Table 4.7*); however, the proportion of children born in the three years after the previous birth decreased by 1.5 times. Simultaneously, with the decrease of probabilities of the second and third births, a concentration in second births around the desired interval – about five years – took place. The share of births with short and long intervals decreased, and the third and the fourth births were often postponed.

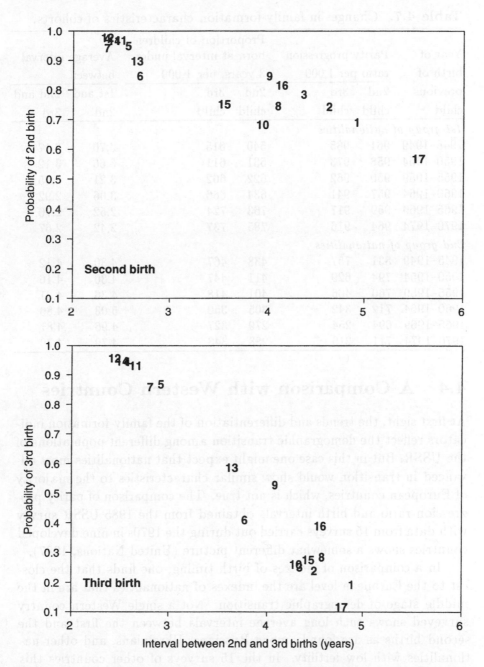

Figure 4.3. The connection of next birth probability and inter-birth interval. The numbers 1 to 17 correspond to the major nationalities in *Table 4.5*.

Table 4.7. Changes in family-formation characteristics of cohorts.

Year of birth of previous child	Parity progression ratio per 1,000		Proportion of children born at interval under 3 years, per 1,000		Average interval between	
	2nd child	3rd child	2nd child	3rd child	1st and 2nd	2nd and 3rd
1st group of nationalities						
1945–1949	961	985	540	615	3.70	3.17
1950–1954	958	973	531	611	3.66	3.12
1955–1959	960	962	622	662	3.21	2.95
1960–1964	957	941	634	666	3.06	2.92
1965–1969	959	917	763	724	2.52	2.70
1970–1974	964	916	785	737	2.42	2.67
2nd group of nationalities						
1945–1949	831	757	438	467	4.30	4.12
1950–1954	794	629	411	447	4.30	4.16
1955–1959	760	466	401	418	4.36	4.31
1960–1964	712	342	305	350	5.03	4.86
1965–1969	694	254	279	327	4.96	4.87
1970–1974	711	216	288	343	4.79	4.70

4.4 A Comparison with Western Countries

At first sight, the trends and differentiation of the family-formation indicators reflect the demographic transition among different populations of the USSR. But in this case one might expect that nationalities most advanced in transition would show similar characteristics to the majority of European countries, which is not true. The comparison of parity progression ratio and birth intervals obtained from the 1985 USSR survey with data from 15 surveys carried out during the 1970s in nine developed countries shows a somewhat different picture (United Nations, 1987).

In a comparison of indexes of birth timing, one finds that the closest to the European level are the indexes of nationalities that are in the middle stage of demographic transition. Not a single Western country surveyed shows such long average intervals between the first and the second births as are found among Russians, Ukrainians, and other nationalities with low fertility. In the 15 surveys of other countries this interval varied from 2.5 years in the USA in 1976 to 3.3 years in Hungary in 1977. Among the nations studied in the Soviet survey four had an interval shorter than the interval in the USA and nine had an interval longer than that in Hungary among nine nationalities. Only four nationalities – Kirghiz, Kazakhs, Armenians, and Georgians – have intervals within the range of Western variation.

The Baltic nationalities (Latvians, Estonians, and Lithuanians, traditionally considered as closest to the European culture and life-style) have indexes that are in the middle of the 17 nationalities and rather similar to Western Europe. This is more evident in the pattern of parity progression ratios than in the pattern of birth intervals. Apparently, there are some peculiarities in the timing of birth in the USSR, which cause the relationship between mean interval and parity progression ratios to be different from the relationship in other industrialized countries.

Measured by the total fertility rate in a synthetic cohort, as well as by the mean number of children borne by birth or marriage cohorts, fertility trends and differences within the USSR fit well into the framework of demographic transition. It is suggested that the European demographic pattern will sooner or later occur in all regions.

References

Okhrana materinstva i detstva v SSSR (Maternity and Childhood Protection in the USSR), 1986. Yuridicheskaya Literatura, Moscow.

Sifman, R.I., 1974. *Dinamika rozhdaemosti v SSSR* (Fertility Dynamics in the USSR). Statistika, Moscow.

United Nations, 1987. Fertility and Family Planning in the Developed Countries. In *Fertility Behavior in Context of Development*. New York, NY.

Chapter 5

Ethnic-Territorial Differences in Marital Fertility: A 1985 Survey

Galina Bondarskaya

Territorial and ethnic differences in demographic processes, fertility in particular, are important characteristics of the demographic pattern in the USSR. Urlanis (1963) was one of the first to notice the interrelation between territorial differences and nationalities in fertility levels. Since then fertility studies based on surveys conducted in 1960, 1967–1968, 1972, 1975, 1981, and 1985 show that territorial differences in fertility level in the USSR are highly significant and are the result of fertility differences among nationalities (ethnic groups).

Ethnic differences in fertility in the USSR result from different stages and irregularities in the demographic transition of some populations due to peculiarities of their economic and sociocultural development. Territorial and ethnic differences in fertility in the USSR, their interrelation and dynamics, and the role of ethnic composition of the population in determining the fertility level in certain territories were studied by Soviet demographers in the 1970s and 1980s (Bondarskaya, 1970, 1977; Belova, 1976; Belova *et al.*, 1977, 1983, 1988; Vishnevsky and Volkov, 1983; Bondarskaya and Darsky, 1988).

71

The level of total fertility in the USSR was and still is determined almost entirely by marital fertility. In the late 1980s almost 90 percent of all babies were born to legally married parents. Hence, it is the reproductive behavior of married couples that will determine the total fertility level at least in the near future. For this reason it is important to study marital fertility. Most attention should be given to the analysis of differences in reproductive behavior and reproductive intentions of young married couples because they will determine the fertility level in the beginning of the next century.

This chapter describes fertility rates for marriage cohorts, obtained from the 1985 sociodemographic survey. Each woman age 18 or above was asked about the number of children she has already borne, and each married woman between 18 and 44 years of age was asked about the number of children she is going to have.

The marital fertility indicator chosen combines the number of children born with the number of births expected. It is denoted as completed marital fertility, that is, the mean number of children borne by each married woman in a marriage cohort at the end of a reproductive period (Darsky, 1972, 1985). Only first marriages are considered. For real cohorts past childbearing age (cohorts formed in the period 1945–1959), completed marital fertility is calculated as the mean number of births; for cohorts still within reproductive age (cohorts formed in the period 1960–1984), the number of total births (already born and expected) is used.

5.1   Trends and Ethnic–Territorial Differences in Marital Fertility

In the USSR, as a whole, cohort marital fertility has been decreasing over the postwar period. This aggregate trend, however, results from different changes of marital fertility in each republic and nationality (*Table 5.1*).

The majority of the Soviet population (79.2 percent in the 1989 census) lives in the republics with a low or relatively low fertility level, where the fertility transition is essentially completed. A smooth, though slowed in the last marriage cohorts, fertility decrease can be observed only in the republics with the highest fertility level in the first postwar cohort (Moldavia, Georgia, Byelorussia, and Lithuania). In the two largest republics (Russia and the Ukraine) marital fertility is stable at the level of 1.9 to 2.0 children for each married woman. In Estonia and Latvia the level of fertility is the lowest in the marriage cohort of 1945–1949, but a small increase in fertility has been recorded beginning with the marriage cohorts of 1960s. Women married in the period 1980–1984 in Estonia and Latvia intended, on average, to have more children than women in the marriage cohort of 1945–1949.

Marital fertility trends also vary in the republics with high fertility levels. This group includes four Central Asian republics (Uzbekistan, Kirghizia, Tajikistan, and Turkmenistan) and Kazakhstan, Azerbaijan, and Armenia. In 1989 20.8 percent of country's population lived in these republics and almost 35 percent of the total number of births occurred in these republics.

A decrease of marital fertility in Kazakhstan, Azerbaijan, and Armenia has occurred during the entire postwar period; this decrease is the largest in this group of republics. The number of children in the families of the 1980–1984 marriage cohort is expected to be smaller by 1.5 times in Kazakhstan and smaller by 1.4 times in Azerbaijan and Armenia than the number of children in the families of the 1945–1949 cohort.

In the Central Asian republics the tendency toward a decrease in fertility is first apparent in the marriage cohorts formed in the 1960s. The decrease of completed marital fertility in the 1980–1984 marriage cohort, in relation to the 1965–1969 cohort (which had the highest fertility in all the Central Asian republics), was significant: the decline ranged between 14.6 percent in Turkmenistan and 20.6 percent in Tajikistan. However, during the first 20 years of the postwar period, completed marital fertility was increasing. Such an early increase in fertility rates can be explained by positive changes in the country – its moving to peaceful life after the war. Fertility growth could have been influenced by improvements in life conditions, women's health, prenatal and natal care, and so on, as well as the tradition of a large family.

The analysis of variation in completed marital fertility in successive marriage cohorts shows that, because of differences in fertility trends in the republics, territorial differences in marital fertility were increasing until the late 1960s. In the late 1960s the 1965–1969 marriage cohort had the maximal level of fertility differentiation. The diminishing territorial difference in fertility in later cohorts was due to a fertility decrease in

Table 5.1. Completed fertility in first marriage cohorts by year of marriage in the USSR, republics, and titular nationalities of the republics.[a]

Republic[a]	Expected number of children borne by women, by the year of first marriage							
	1945–1949	1950–1954	1955–1959	1960–1964	1965–1969	1970–1974	1975–1979	1980–1984
USSR	2.83	2.64	2.57	2.38	2.42	2.40	2.39	2.35
Republics with low fertility level								
Latvia								
Total	1.87	1.78	1.77	1.80	1.84	1.85	1.94	1.91
Titular nation.	2.00	1.86	1.84	1.89	1.92	1.93	2.04	2.02
Estonia								
Total	2.02	1.84	1.82	1.91	1.95	1.97	2.05	2.05
Titular nation.	2.10	1.96	1.92	2.03	2.05	2.07	2.14	2.16
Ukraine								
Total	2.15	2.07	2.05	1.98	1.99	1.99	1.96	1.88
Titular nation.	2.23	2.14	2.13	2.04	2.04	2.03	2.00	1.91
Lithuania								
Total	2.43	2.23	2.19	2.13	2.08	2.06	2.03	1.95
Titular nation.	2.53	2.28	2.19	2.15	2.09	2.07	2.04	1.97
Russia								
Total	2.58	2.35	2.20	2.02	1.98	2.00	2.00	1.93
Titular nation.	2.33	2.13	2.01	1.87	1.88	1.92	1.93	1.86
Byelorussia								
Total	2.76	2.58	2.47	2.26	2.13	2.05	2.01	1.93
Titular nation.	2.81	2.59	2.44	2.22	2.12	2.05	2.02	1.94
Georgia								
Total	2.76	2.61	2.62	2.55	2.57	2.60	2.64	2.71
Titular nation.	2.59	2.49	2.48	2.45	2.50	2.55	2.59	2.66
Moldavia								
Total	3.28	2.98	2.94	2.64	2.49	2.51	2.46	2.28
Titular nation.	3.48	3.23	3.17	2.88	2.68	2.67	2.56	2.34

[a]The republics are arranged in the order of the increase in the index for total population in 1945–1949 cohort (line Total).

the republics with a high fertility level and stabilization and even small growth in the fertility rate in republics with a low fertility level.

Actual and projected fertility tendencies are to a great extent determined by the reproductive behavior of each republic's titular nationalities. The trends of territorial differences in marital fertility are partly the reflection of ethnic differentiation. Territorial differences are lightly smoothed by the heterogeneous ethnic composition of the population in these republics, especially among the urban population of Central Asia, where the share of nationalities with a low fertility level is relatively high.

Table 5.1. Continued.

Republic[a]	Expected number of children borne by women, by the year of first marriage							
	1945–1949	1950–1954	1955–1959	1960–1964	1965–1969	1970–1974	1975–1979	1980–1984
USSR	2.83	2.64	2.57	2.38	2.42	2.40	2.39	2.35
Republics with high fertility level								
Armenia								
Total	4.08	3.77	3.53	3.17	3.01	2.93	2.92	2.87
Titular nation.	3.67	3.34	3.18	2.92	2.83	2.78	2.77	2.71
Kazakhstan								
Total	4.08	3.67	3.78	3.42	3.32	3.11	2.87	2.66
Titular nation.	5.77	5.80	6.10	5.64	5.10	4.53	3.88	3.47
Kirghizia								
Total	4.64	4.56	4.68	4.50	4.80	4.41	4.06	3.92
Titular nation.	6.09	6.28	6.56	6.51	6.59	5.90	5.23	4.80
Azerbaijan								
Total	4.66	4.56	4.66	4.36	4.10	3.71	3.45	3.22
Titular nation.	5.21	5.12	5.07	4.72	4.41	3.95	3.60	3.31
Turkmenistan								
Total	4.94	5.39	5.64	5.97	6.18	5.77	5.38	5.28
Titular nation.	5.44	6.18	6.51	6.92	7.04	6.50	5.96	5.82
Uzbekistan								
Total	5.10	5.31	5.52	5.35	5.55	5.19	4.81	4.70
Titular nation.	5.69	6.03	6.28	6.25	6.34	5.78	5.26	5.04
Tajikistan								
Total	5.62	5.87	6.13	6.00	6.35	5.66	5.21	5.04
Titular nation.	6.28	6.58	6.87	6.85	7.04	6.27	5.57	5.21
Range of variation								
Total	3.75	4.09	4.36	4.20	4.51	3.92	3.44	3.40
Titular nation.	4.28	4.72	5.03	5.05	5.16	4.58	4.03	3.96

5.2   Fertility of the 1980–1984 Marriage Cohort

The fertility level of the youngest married couples surveyed is of special interest. It is their reproductive behavior that must be considered in the hypotheses of fertility trends in the near future. The 1980–1984 marriage cohort is one of the marriage cohorts that will determine the fertility level up to the year 2000. By that time its reproductive activity will generally be completed. The expected total number of children in this cohort may serve as a reference point in fertility forecasts but on condition that the married couples do not change their intentions under the impact of changes in the political, economic, and social spheres of society.

Under present conditions, the expectations of the population groups with stable low fertility level and groups where the transition to a small family is near completion will probably be fulfilled. But, in population groups where the practice of birth control is in its initial stage yesterday's, or even today's, reproductive intentions of families may not be realized. Their intentions may be exceeded because of the slow spread of birth-control practices. Nevertheless, the analysis of fertility differences in the 1980–1984 marriage cohort by sociodemographic groups gives an idea of probable future reproduction patterns in different groups.

5.2.1   Fertility trends

One can estimate the changes in completed marital fertility in time by comparing the indicators of two cohorts separated by the period of one generation. They are the 1955–1959 and 1980–1984 cohorts. They may be called *maternal* and *filial*, correspondingly. Changes in the total expected family sizes between these two cohorts are compared in *Table 5.2*.

On the average, in the USSR, among the women married for the first time in 1980–1984, the expected number of children is only 8.6 percent less than among the women married for the first time 25 years earlier. In urban areas there are no differences. Among rural populations, fertility is declining; women who married in the early 1980s are going to have slightly fewer children than women who married in the late 1950s.

Regardless of this small decrease in marital fertility in the filial cohort, a great diversity in the variations of reproductive behavior of married couples in certain republics and of certain nationalities is expected. An increase in expected fertility is observed in only three republics – Latvia, Estonia, and Georgia. This increase will be determined by the reproductive behavior of young urban women: married women of the filial marriage cohort expect to have more children than women of the maternal cohort.

In the other republics a fertility decline is expected. In the European republics (Russia, the Ukraine, Lithuania, and Moldavia) the decline is expected only in the rural population, but in Armenia, Azerbaijan, Kazakhstan, and the Central Asian republics a significant fertility decline is expected in both urban and rural populations. At the same time a large discrepancy can be observed between rates of increase in completed marital fertility in the urban population of Kazakhstan

Table 5.2. Relative increase of mean expected number of children in the 1980–1984 marriage cohort (filial) compared with the 1955–1959 marriage cohort (maternal) in the USSR and the republics, in percent.

Republic[a]	Urban & rural population		Urban population		Rural population	
	Total	Titular nat.	Total	Titular nat.	Total	Titular nat.
USSR	−8.6		−0.5		−9.4	
Latvia	+7.9	+9.8	+18.1	+22.2	−4.5	−1.8
Estonia	+12.6	+12.5	+17.2	+19.4	+2.3	+3.2
Ukraine	−8.3	−10.8	+2.3	+1.1	−16.3	−17.3
Lithuania	−11.0	−10.0	+1.1	+1.6	−19.1	−17.8
Russia	−12.3	−7.5	+0.0	+1.1	−23.8	−19.5
Byelorussia	−21.9	−20.5	−5.6	−4.1	−27.0	−26.7
Georgia	+3.4	+7.3	+13.3	+12.2	−0.3	+6.4
Moldavia	−22.4	−26.2	+1.5	−2.8	−26.4	−27.1
Armenia	−18.7	−14.8	−11.1	−8.7	−25.4	−20.6
Kazakhstan	−29.6	−43.1	−18.2	−42.3	−35.0	−41.7
Azerbaijan	−30.9	−34.7	−22.9	−29.7	−32.6	−34.1
Kirghizia	−16.2	−26.8	−3.9	−34.3	−17.8	−22.0
Uzbekistan	−14.9	−19.7	−13.3	−25.8	−18.0	−17.8
Turkmenistan	−6.4	−10.6	−10.9	−22.4	−5.2	−6.8
Tajikistan	−17.8	−24.2	−19.1	−32.3	−20.0	−21.2

[a]Republics are arranged in the order of the increase in the mean expected number of children of the 1955–1959 cohort.

and the Central Asian republics and among their titular nationalities. The sharp decline in marital fertility rates among the Kazakh (42 percent), Kirghiz (34 percent), Tajik (32 percent), and Turkmenian (22 percent) populations is leveled by the presence of several ethnic groups with much lower fertility levels. Particularly noticeable is the impact of the national structure of the urban population on the level and trends in Kirghizia and Kazakhstan, where the percentages of the Kirghiz and Kazakh populations in 1989 were only 29.9 and 39.7, correspondingly.

5.2.2 Differences in territories

Assuming that replacement fertility must be at the level of about 2.28 children per married woman (with no illegitimate births), potential fertility in marriages in the USSR contracted in the early 1980s may be considered sufficient for ensuring the full replacement of a generation. This average level is reached because of the high fertility levels in rural populations; in urban populations, completed fertility in young marriage cohorts is below replacement level.

However, average rates of the country do not reflect the fertility rate that is expected in certain territories and among certain population groups (*Figure 5.1*). This study is limited to the situation in republics of the USSR, though fertility differences exist within republics as well (especially in Russia, the Ukraine, and Kazakhstan).

The republics can be divided into fertility groups. On the one hand, there are republics where completed fertility of the 1980–1984 marriage cohort cannot ensure generation replacement. This group comprises Russia, the Ukraine, Byelorussia, Latvia, Lithuania, and Estonia; Moldavia may also be considered part of this group. On the other hand, the Central Asian republics represent a group with a steady high level of fertility. In Kazakhstan, Azerbaijan, and Armenia the process of transition to low fertility is relatively fast. They are in the intermediate position; Georgia may also be considered part of this group by its fertility level, but not by its trend where fertility is stable at the level above replacement.

The highest completed marital fertility is expected in the Central Asian republics; the lowest, in the republics of the European part of the USSR (*Figure 5.1*). The maximum value of the mean expected number of children in the 1980–1984 marriage cohort (5.28 in Turkmenistan) is almost triple the minimum value (1.88 in the Ukraine). In urban populations these differences are less than in rural populations: maximal and minimal values of these estimates differ, correspondingly, by two and three times and occur in the same republics – Turkmenistan and the Ukraine.

The expected number of children of the 1980–1984 marriage cohort in certain territories depends on the reproductive behavior of married couples in different social groups of the population (particularly in different ethnic groups), in different types of residence (urban or rural communities, small or big cities), and with different educational levels.

5.2.3 Differentiation by nationalities

Reproductive intentions among the women of different nationalities vary significantly (*Table 5.1*). The largest number of children in the family

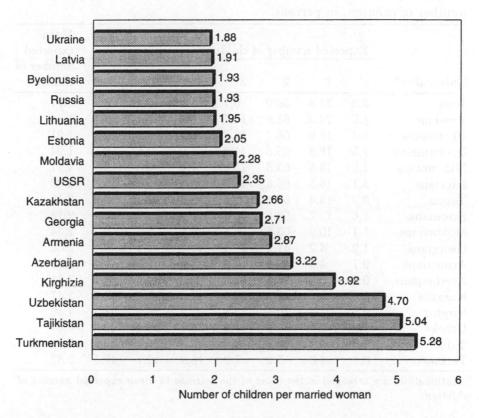

Figure 5.1. Expected number of children of the 1980–1984 marriage cohort in the USSR and the republics.

is expected by women of indigenous nationalities in the Central Asian republics ; the smallest, by women in the European republics. The range of variation in the number of children is rather wide: from 1.86 among Russian women to 5.82 among Turkmenian women. Among urban populations the maximal value (4.72 among the Turkmenian population) exceeds the minimal value (1.81 among the Russians) by 2.6 times; and among rural populations the maximal value is exceeded by 3.0 times (6.20 among the Turkmenian population and 2.06 among the Russians). In comparison with the maternal cohort, there is some decrease in the range of rate variation among the nationalities of the filial cohort in both urban and rural areas. The decrease of these differences is, mainly, the

Table 5.3. Distribution of women married in 1980–1984 by expected number of children, in percent.

Nationality[a]	Expected number of children							Mean expected number of children
	0	1	2	3	4	5	6+	
Jews	3.8	31.8	56.0	6.7	1.2	0.4	0.1	1.71
Russians	1.5	24.9	61.6	10.7	0.8	0.4	0.1	1.86
Ukrainians	1.4	19.9	66.7	10.8	0.8	0.3	0.1	1.91
Byelorussians	1.5	18.8	66.6	11.4	1.0	0.6	0.1	1.94
Lithuanians	1.9	18.6	63.6	13.6	1.4	0.8	0.1	1.97
Latvians	3.1	16.5	60.5	17.3	1.5	0.7	0.4	2.02
Tatars	0.9	16.4	57.6	20.4	3.0	1.3	0.4	2.15
Estonians	1.6	13.7	58.6	22.1	2.4	1.2	0.4	2.16
Moldavians	1.1	10.9	53.5	26.1	5.5	2.5	0.4	2.34
Georgians	1.2	4.2	41.0	39.1	11.0	2.8	0.7	2.66
Armenians	0.7	4.5	40.7	37.3	13.0	2.7	1.1	2.71
Azerbaijanis	0.3	2.9	26.8	31.9	22.6	10.4	5.1	3.31
Kazakhs	0.9	5.6	28.0	25.1	15.2	15.5	9.7	3.47
Kirghiz	0.4	1.9	10.5	12.8	22.2	24.9	27.3	4.80
Uzbeks	0.3	1.0	7.2	9.9	26.4	23.4	31.8	5.04
Tajiks	0.2	1.1	6.5	8.9	23.7	22.8	36.8	5.21
Turkmen	0.1	1.0	5.2	7.2	16.9	20.1	49.5	5.82

[a]Nationalities are arranged in the order of the increase in mean expected number of children.

result of changes in the reproductive intentions among young women of Central Asian nationalities accompanied by a relatively small decrease in the expected number of children in comparison with older cohorts of women of European nationalities.

Among urban women ethnic fertility differentiation is somewhat lower than among rural women. The rural life-style and the homogeneous structure of rural population are more favorable for maintaining family and cultural traditions (including those restraining the transition to a small family) and, hence, for continuing differences in reproductive attitudes among the women of certain nationalities.

Table 5.3 presents data on the distribution of married women of some nationalities in the 1980–1984 marriage cohort by expected number of children. Included in the table are women of titular nationalities of the republics, as well as Tatar and Jewish women. These 17 nationalities practically cover the spectrum of fertility levels in the USSR.

The expected marital fertility of eight nationalities presented in *Table 5.3* (Jewish, Russian, Ukrainian, Byelorussian, Lithuanian, Latvian, Tatar, and Estonian women) in the 1980–1984 cohort is not sufficient for their natural (excluding ethnic assimilation) replacement. (For the 1955–1959 marriage cohort, Byelorussian and Tatar women were not among the nationalities that could not ensure replacement level.) The marital fertility level of urban Moldavian women is below replacement level. Armenian, Azerbaijani, and Kazakh women are moving in the same direction rather fast. It is expected that among Azerbaijani and Kazakh women this process will not go as far as among Russian, Ukrainian, and Jewish women, and women of the Baltic states, because of the restraining influence of Muslim traditions.

To a certain extent the principal differences in family preferences that determine fertility pattern in marriage cohorts of different nationalities can be judged by the distribution of women by the number of children they are going to have. Data in *Table 5.3* show that a childless family is not popular among the women married in 1980–1984 in the groups with both low and high fertility levels. The proportion of women intending to remain childless varies rather significantly in both urban and rural areas. It is so small (never more than 4 percent) that it cannot have any noticeable impact on the final result of reproductive activity in the cohort under consideration.

In groups of nationalities with a low fertility level (Russian, Ukrainian, Byelorussian, Jewish, Tatar, Latvian, Lithuanian, and Estonian women), both urban and rural residents prefer to have a two-child family.

Among nationalities with a higher fertility level, where the rate and the intensity of the process of changing reproductive behavior varies, there are more variations in the preferred family size. Many urban Armenian, Azerbaijani, and Kazakh women intend to limit their family to two or three children; among the Kirghiz, a family with four children is popular; Uzbek and Tajik women prefer a family with four children; and the Turkmens prefer six or more children.

In rural areas the transition to a lower fertility level is lagging. Moldavian women in both rural and urban areas often prefer a family with two children; Georgian and Armenian women prefer three children. Obvious stratification by reproductive behavior is seen among rural Kazakh women: a little less than half of them intend to limit their family to two or three children, and approximately the same number are going to have four or more children.

The percentage of couples not intending to limit their family size (families that expect to have six or more children) is the highest among rural women of Central Asian nationalities: from 33.6 percent among Kirghiz women to 57.4 percent among Turkmenian women.

Reproductive behavior of women of certain nationalities and different social groups is determined by their life-style. Such factors as religion, national traditions, and customs also have their impact on the process of consciousness formation.

5.2.4 Differentiation by place of residence

In the USSR, today, the conditions of life depend on the type of settlements: urban or rural areas, cities or towns. In addition, the expected fertility level of the 1980–1984 marriage cohort depends on the place of residence of a young family – in urban or rural areas. Under rural conditions women's reproductive intentions are higher than under urban conditions, being rather significant among some nationalities.

Urban settlements, as a factor in determining reproductive behavior of the population, are far from equal. Thus, the conditions and mode of life of the citizens in major cities of the republics differ from the district centers and the urban-type settlements. Having no data on population structure in different types of cities, we could only compare fertility rates in cities according to their size.

As the analysis shows, marital fertility is higher in smaller cities than in larger cities. These differences are very insignificant for groups of nationalities with a low fertility level and are rather essential for groups of nationalities with a high fertility level (see *Figure 5.2*).

Table 5.4 presents the data on the expected number of children among married women of 17 nationalities who were married in 1980–1984. By the extent of differences in the number of children, to which women of different nationalities and living in cities of different sizes intend to limit their families, the activity of the process of changing reproductive behavior and the extent of completion of transition to low fertility can be determined. The smallest differences in reproductive intentions among women in cities of different sizes is both among

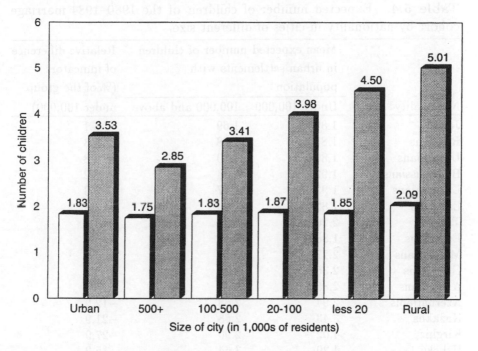

Figure 5.2. Expected number of children per married woman of the 1980–1984 marriage cohort, in groups with low and high levels of fertility in settlements of different type and in cities of different sizes, in thousands of residents: light, Russians, Ukrainians, Byelorussians, Latvians, Lithuanians, Estonians, Moldavians; shaded, Uzbeks, Tajiks, Turkmen, Kirghiz, Kazakhs, Azerbaijanis.

nationalities with practically completed transition and among nationalities in the very beginning of transition. Most intensive is the process of transition to low fertility in urban areas among Kirghiz, Kazakh, and Tajik women. Urban Turkmenian women are the least affected by this process.

The intensity of the process of transition to a lower fertility level among urban women of Central Asian nationalities depends on their proportion in the population of these cities, as well as on their sociocultural characteristics, among which educational level is of first priority.

Table 5.4. Expected number of children of the 1980–1984 marriage cohort by nationality in cities of different size.

Nationality[a]	Mean expected number of children in urban settlements with population		Relative difference of indicators (% of the group under 100,000)
	Under 100,000	100,000 and above	
Jews	1.80	1.69	− 6.1
Russians	1.86	1.78	− 4.3
Ukrainians	1.89	1.80	− 4.8
Byelorussians	1.92	1.84	− 4.2
Lithuanians	1.93	1.77	− 8.3
Latvians	1.82	1.89	+ 3.8
Tatars	2.20	1.98	−10.0
Estonians	1.88	2.17	+15.4
Moldavians	2.17	2.04	− 6.0
Georgians	2.35	2.38	+ 1.3
Armenians	2.85	2.41	−15.4
Azerbaijanis	2.94	2.58	−12.2
Kazakhs	3.43	2.68	−21.9
Kirghiz	4.02	2.90	−27.9
Uzbeks	4.29	3.64	−15.2
Tajiks	4.39	3.50	−20.3
Turkmen	4.63	4.26	− 8.0
Range of variation	2.83	2.57	

[a]Nationalities are arranged in the order indicated in *Table 5.3*.

5.2.5   Differentiation by educational level

Educational level is an important characteristic that has an influence on family reproductive behavior. To a great extent it defines a woman's social status, professional orientation and skill, family welfare, cultural level, and the overall life-style.

Data in *Table 5.5* show the influence of education on fertility. There is no clear association between education and fertility in low fertility areas. The pronounced inverse relation between educational and fertility levels found in the studies of the 1960s and 1970s has almost disappeared. Women with higher education are not different in their expected number of children from women with a lower educational level, among Latvian and Tatar women, for instance, or even intend to have more children than women with secondary education (Estonian women).

Table 5.5. Marital fertility of the 1980–1984 marriage cohort by nationality and educational level.

Nationality[a]	Higher, incomplete higher, secondary vocational	Secondary	Incomplete secondary
Jews	1.68	1.84	[b]
Russians	1.84	1.89	1.92
Ukrainians	1.89	1.96	1.93
Byelorussians	1.92	1.99	1.96
Lithuanians	1.95	2.01	2.04
Latvians	2.02	2.02	2.07
Tatars	2.14	2.15	2.24
Estonians	2.19	2.13	2.30
Moldavians	2.25	2.41	2.38
Georgians	2.56	2.75	2.78
Armenians	2.57	2.81	2.67
Azerbaijanis	2.94	3.42	3.52
Kazakhs	3.26	3.59	3.71
Kirghiz	4.28	5.06	4.71
Uzbeks	4.44	5.20	5.09
Tajiks	4.23	5.32	5.51
Turkmen	5.22	5.96	5.74
Range of variation	3.54	4.12	3.82

[a]Nationalities are arranged in the order indicated in *Table 5.3*.
[b]Insignificant because of a small number of women.

Among nationalities with a higher fertility level, the differences in the expected number of children associated with educational level are still rather significant. Thus, among the titular nationalities of Central Asia only women with higher and secondary vocational education tend to have small families. Among the women of these nationalities, the differences between the women with secondary and incomplete secondary education are small.

Listing the nationalities in the same order as in *Tables 5.3* and *5.5* does not change the regularity of increase in the value of indicators. This confirms the conclusion that educational level, at least in the 1980–1984 cohort, is not important to marital fertility level differentiation.

A more detailed analysis of the influence of these factors upon the reproductive behavior of women of different nationalities and upon the

trends of ethnic fertility differentiation in the USSR can be found in
Bondarskaya (1977) and Belova *et al.* (1977, 1988).

5.3   Conclusion

The impact of ethnic affiliation on reproductive behavior was and still is
one of the most significant social characteristics, more important than
educational level, place of residence, women's employment, and so on.

Marital fertility tendencies show that the differences in fertility levels
attributed to nationality have peaked. In the future a decline in ethnic
differences in fertility should be expected.

In the near future ethnic differences in marital fertility in the USSR
will remain. Its level and dynamics will depend on the perspectives of
change in the reproductive behavior of peoples with a high fertility level,
which is determined by socioeconomic development.

Territorial fertility levels are influenced by territorial population ethnic structure. This factor is particularly important with regard to heterogeneous populations. In such cases territorial fertility dynamics is
under the impact of the fertility level within certain nationalities and
the population's ethnic structure. For instance, the migration flow of
nonnative populations with a low fertility level from the Central Asian
republics will slow down the fertility decrease in these republics though
fertility decline for titular nationalities of these republics has already
started.

Territorial differences in marital fertility defined as uneven demographic development of different population groups exists now, and will
remain. The level of this differentiation makes doubtful the expediency
and correctness of using averages of indicators to analyze the entire
USSR.

References

Belova, V.A., 1976. *Chislo detei v sem'e* (The Number of Children in the
    Family). Statistika, Moscow.

Belova, V.A., G.A. Bondarskaya, A.G. Vishnevsky, L.E. Darsky, and R.I. Sif-
    man, 1977. *Skolko detei budet v sovetskoy sem'e: rezultaty obsledovaniya*
    (How Many Children Will Be in a Soviet Family: Survey Results). Statis-
    tika, Moscow.

Belova, V.A., G.A. Bondarskaya, and L.E. Darsky, 1983. Dinamika i differ-
    entsiatsiya rozhdaemosti v SSSR: Po materialam obsledovaniya (Fertility

Dynamics and Differentiation in the USSR: By Survey Data). *Vestnik statistiki* (Herald of Statistics) **12**.

Belova, V.A., G.A. Bondarskaya, and L.E. Darsky, 1988. Perspectivy rozhdaemosti v SSSR (Fertility Perspectives in the USSR). In A.G. Volkov, ed. *Metodologiya demograficheskogo prognoza* (Backgrounds of Demographic Forecasting). Nauka, Moscow.

Bondarskaya, G.A., 1970. Rol' etnicheskogo faktora v formirovanii territorialnykh razlichiy rozhdaemosti (The Role of Ethnic Factor in the Formation of Territorial Fertility Differences). In A.G. Volkov, L.E. Darsky, and A.Ja. Kvasha, eds. *Voprosy demografii* (Problems of Demography). Statistika, Moscow.

Bondarskaya, G.A., 1977. *Rozhdaemost'v SSSR: Etno-demograficheskiy aspect* (Fertility in the USSR: Ethnic-Demographic Aspect). Statistika, Moscow.

Bondarskaya, G.A., and L.E. Darsky, 1988. Etnicheskaya differentsiatsiya rozhdaemosti v SSSR v gody, primykayuschie k perepisyam naseleniya (Ethnic Differentiation of Fertility in the USSR in the Years Adjacent to Censuses). *Vestnik statistiki* (Herald of Statistics) **12**:16–21.

Darsky, L.E., 1972. *Formirovanie sem'i* (Family Formation). Statistika, Moscow.

Darsky, L.E., 1985. Produktivnost'braka (The Completed Marital Fertility). In *Demograficheskiy entsiklopeditcheskiy slovar'* (Demographic Encyclopedia). Sovietskaya Entsiclopedija, Moscow.

Urlanis, B.Ts., 1963. *Rozhdaemost'i prodolzhitel'nost'zhizni v SSSR* (Fertility and the Length of Life in the USSR). Gosstatizdat, Moscow.

Vishnevsky, A.G., and A.G. Volkov, eds. 1983. *Vosproizvodstvo naseleniya SSSR* (Reproduction of the Population of the USSR). Finansy i statistika, Moscow.

Chapter 6

Fertility Transition in Estonia, Latvia, and Lithuania

Kalev Katus

This article focuses on the fertility transition in the Baltic states – Estonia, Latvia, and Lithuania. The most complete data for the region are available on Estonia. For the study of the regional differences, the analysis is mainly based on the Princeton European Fertility Project indexes at the county level. For the comparison of Estonia, Latvia, and Lithuania, some other indexes have also been used.

Historically, the Baltic states politically served the Russian Empire as its outposts in the West: Estonia carried out this role for two centuries; Lithuania, for one century. However, demographically the Baltic states could have been defined as the West's outposts in the East. This study of fertility transition supports this latter view.

6.1   Pre-transitional Fertility

Demographic research has not been systematically carried out for the pre-transitional period in the Baltic states. The most detailed study is by Palli who investigated the demographic trends in some Estonian parishes in the 18th century using the method of family reconstruction

Table 6.1. Age-specific marital fertility rates in three Estonian parishes.

Age	Rouge female cohort born in 1661–1675	Otepää female cohort born in 1716–1749	Karuse female cohort born in 1712–1760
15–19	–	392	–
20–24	385	458	429
25–29	412	420	390
30–34	358	364	386
35–39	–	375	342
40–44	–	208	–
45–49	–	43	–
TMFR	(10.90)	11.30	(10.95)

Sources: Palli, 1973, 1980, 1988.

(Palli, 1973, 1980, 1988). *Table 6.1* gives the age-specific marital fertility rates and total marital fertility rate (TMFR) in three parishes in the 17th and 18th centuries. The level of marital fertility in these Estonian parishes implies somewhat more than 10 births for married women by the end of their reproductive age. No indication for parity-specific fertility control could be found by Palli or others in 18th-century Estonia. Comparisons of the findings of other European countries show similarities in fertility pattern between Estonia and other European nations (Palli, 1974).

The first symptoms of a change in traditional reproductive behavior emerged in the late 18th century. Most significant is the spread of a new marriage pattern, characterized by a later age at marriage and a high proportion of women remaining unmarried. It is called the Western European marriage pattern (Hajnal, 1965). In the Estonian parish of Oteppä, for example, the average age at first marriage for women was 22.8 years in 1750–1774 and 24.2 years in 1775–1799; the proportion of married women was 55.8 percent in 1765 (Palli, 1988).

Since Hajnal's paper (1965) systematic attention has been paid to the Western European marriage pattern. Usually the transition to a Western European marriage pattern is not regarded as part of fertility transition. The common argument is that the phenomenon is not related to parity-specific fertility control. Nevertheless, general fertility had been considerably lowered by the change of the marriage pattern up to the beginning of marital fertility transition. Coale points out the importance of a balance between fertility and mortality in the long run (Coale, 1986). He, as well as others, also points out that once the parity-specific fertility control has begun, it cannot be stopped until a low level of fertility is reached (Coale, 1986; Knodel, 1977). If both statements are correct, the transition to the Western European marriage pattern in Estonia was a first reaction to the mortality decrease. In this case the transition to the Western European marriage pattern should be interpreted in a closer connection with the marital fertility transition than is usually done. In most Western European countries this nuptiality transition occurred in pre-statistical times, but there is evidence that in Finland the transition occurred not before the late 18th century and was an initial reaction to population growth (Lutz, 1985).

If our speculation on the role of the Western European marriage pattern in Estonia is wrong and the phenomenon is linked to some other social developments, it is still worth underlining that the Baltic states formed the eastern boundary of the spread of this pattern in Europe. In Estonia and northern and western Latvia, the Western European type of marriage was clearly taking shape during the 18th century when the territory was already part of the Russian Empire where the phenomenon was not experienced. In eastern Latvia (Latgale) and Lithuania the pattern was evident later.

The demographic consequences of the transition to the Western European marriage pattern resulted in a relatively low level of general fertility before parity-specific fertility control started. Undoubtedly, the course of this marital fertility transition was influenced by the preceding change in nuptiality. This clearly distinguished the Baltic states, especially Estonia and Latvia, from Russia and many other Eastern European countries.

6.2 Fertility Transition

So far few papers have been published on the demographic transition in the Baltic states (Katus, 1982, 1987; Stankuniene, 1989; Zvidriņš, 1978, 1983). The data presented have been rather selective and unsystematic. No family reconstruction method has been used for the 19th century except in a study of the Estonian Swedes (Hyrenius, 1942). The official demographic statistics on the Russian Empire have been relatively poor, and the first modern census of Estonia and Latvia was conducted in 1881.

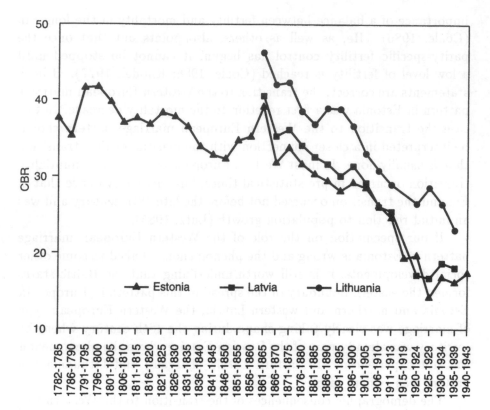

Figure 6.1. Crude birth rates in the Baltic states, 1782–1943.

The only fertility indicator available for a longer period for every Baltic state is the crude birth rate (CBR) which is discussed in Section 6.2.1.

6.2.1 Crude birth rate

The crude birth rate started to decline during the first half of the 19th century (*Table 6.2* and *Figure 6.1*) in the Baltic states. This becomes especially clear when the time series are compared with the Russian and Eastern European levels. Pre-transitional CBRs fluctuated around 35 per 1,000 women. It can be assumed that the main factor causing the early decrease in the 19th century was the transition to the Western European marriage type.

The decrease of CBRs in Estlandia, Livlandia, and Kurlandia, covering the territory of Estonia and most of Latvia, had for some decades been ahead of the trends in Vitebsk and Vilno gubernias. The trend in Kauno was only slightly ahead of the last two regions. The differences remained noticeable after World War I, when the CBR in Lithuania was clearly higher than in Estonia and Latvia. According to Zvidriņš (1986) 20 percent to 25 percent of Latvians lived in Latgale, Vitebsk gubernia.

6.2.2   Princeton indexes

Fertility indexes for the census years 1881, 1897, 1922, and 1934 may be calculated for Estonian counties (maakond) as defined and used by the Princeton European Fertility Project (see *Table 6.3*). The calculation of indexes is based on some data adjustments. Technical details as well as some changes in county boundaries are discussed in the Appendix. For Lithuania the data necessary for the calculation of the fertility indexes by counties are available from the census of 1923 (*Table 6.4*). Latvia is represented by indexes from 1925, 1930, and 1935 but only for the country as a whole (*Table 6.5*). Some details on data adjustments for the calculations for these regions are also described in the Appendix. Although the Lithuanian and Latvian data are much less detailed than the Estonian data, they are sufficient to make some general observations on fertility trends in the Baltic states.

Coale *et al.* (1979) present the same indexes on the Russian gubernias covering the territory of the Baltic states during the years 1870 and 1897 and on three Baltic states in 1926 (*Table 6.6*).

Figure 6.2 compares the overall fertility indexes (I_f) and the marital fertility indexes (I_g) of different European countries with the Estonian level in 1881. The European figures were taken from Coale and Treadway (1986) for the closest available year to the Estonian data. *Figure 6.3* gives analogous data for the mid-1930s; this period marks the cessation of the fertility transition in the demographically advanced countries. These figures also include data on Latvia and Lithuania. Despite the dependency of the fertility indexes on the population age structure, they serve as a good tool for comparisons when analyzing great changes in fertility levels in different countries.

In 1881 the I_f level was lower in France and Ireland than in Estonia. The Estonian level was close to that of Sweden, Switzerland, and Norway. As Estonia had a relatively high proportion of married women (I_m) compared with those countries, except for France, the I_g level in Estonia was even lower.

Table 6.2. Crude birth rates in the Baltic states, 1730–1943.

Year	Estonia[a]	Latvia[b]	Lithuania[c]
1730–1734	40.7[d]		
1772	42.5[e]		
1782–1785	37.6		44.5[f]
1786–1790	35.3		43.0[g]
1791–1795	41.6		
1796–1800	41.7		
1801–1805	39.7		
1806–1810	37.0		
1811–1815	37.6		
1816–1820	36.8		
1821–1825	38.3		
1826–1830	37.8		
1831–1835	36.5		
1836–1840	33.9		
1841–1845	32.8		
1846–1850	32.3		
1851–1855	34.7		
1856–1860	35.2		

Year	Estlandia	Livlandia	Kurlandia	Vitebsk	Vilno	Kauno
1861–1865	39.1	40.6	36.2	48.0	50.2	42.3
1866–1870	31.8	33.2	31.0	47.3	45.8	38.3
1871–1875	33.7	34.7	31.0	48.2	45.8	38.8
1876–1880	31.6	33.8	29.0	43.5	41.0	36.0
1881–1885	30.3	31.5	28.8	41.5	39.5	34.1
1886–1890	29.4	29.6	27.8	42.8	43.0	34.8
1891–1895	28.1	27.7	26.5	41.7	44.0	33.6
1896–1900	29.2	29.4	28.6	40.6	38.0	35.2
1901–1905	28.5	26.8	27.5	37.6	33.5	32.5
1906–1910	26.2	23.9	23.4	35.6	32.3	29.3
1911–1913	24.6	22.6	24.6	33.3	30.6	27.3

[a]Estonia covers the territory of Estlandia and northern Livlandia of the Russian Empire. Data for Estonia, 1982–1860, Rakvere county.
[b]Latvia covers Kurlandia, southern Livlandia, and northwest Vitebsk.
[c]Lithuania covers Kauno and northwest Vilno.
[d]28 Estonian parishes.
[e]10 parishes in Tartu county.
[f]Lithuania, the average of years 1786 and 1790.
[g]Vilnius region, the average of years 1783 and 1784.

Table 6.2. Continued.

Year	Estonia	Latvia	Lithuania
1915–1919	19.3	–	17.0
1920	18.4	17.0	22.7
1921	20.3	19.7	24.6
1922	20.2	21.8	27.3
1923	20.1	21.9	28.4
1924	19.2	22.3	29.4
1925	18.3	22.2	28.9
1926	17.9	22.0	28.5
1927	17.7	22.1	29.4
1928	18.0	20.6	28.8
1929	17.1	18.8	27.2
1930	17.4	19.8	27.4
1931	17.4	19.3	26.8
1932	17.6	19.3	27.3
1933	16.2	17.8	25.7
1934	15.4	17.2	24.8
1935	15.9	17.6	23.4
1936	16.1	18.1	24.2
1937	16.1	17.7	22.3
1938	16.3	18.4	22.7
1939	16.3	18.5	22.4
1940	16.5	23.0	
1941	18.4		
1942	18.7		
1943	15.9		

Sources: Ekonomika, 1976; Katus, 1990a; Latvijas, 1940; Lietuvos, 1940; Ozelis, 1934; Palli, 1980; Rachin, 1956; Rahvastikuprobleeme, 1937; Stankuniene, 1989; Vahtre, 1973.

In the early 1880s only France and Hungary had a lower I_g than Estonia. In comparison with its neighbors, Estonia had considerably lower fertility than Russia (including the neighboring region of Pskov) and Finland. Among the neighboring countries Sweden bears the closest resemblance to the Estonian case, although Swedish marital fertility tends to remain a little higher.

The decline of I_f and I_g, until reaching a subreplacement level in the 1930s, appears to be close to linear in Estonia, taking into account four specific years during the period. The relative changes in all

Table 6.3. Princeton indexes in Estonia, 1881, 1897, 1922, and 1934.

County	I_f	I_g	I_h	I_m
_1881_				
Virumaa	0.3573	0.6682	0.0318	0.5119
Jarvamaa	0.3471	0.6159	0.0389	0.5345
Harjumaa	0.3112	0.5984	0.0317	0.4933
Laanemaa	0.3307	0.6235	0.0349	0.5025
Saaremaa	0.3269	0.6120	0.0397	0.5019
Parnumaa	0.3363	0.6181	0.0349	0.5169
Viljandimaa	0.3110	0.6124	0.0372	0.4764
Tartumaa	0.3099	0.6239	0.0319	0.4699
Vorumaa	0.3553	0.6809	0.0387	0.4932
Estonia	0.3279	0.6278	0.0346	0.4947
_1897_				
Virumaa	0.3253	0.5966	0.0386	0.5140
Jarvamaa	0.3218	0.5707	0.0478	0.5245
Harjumaa	0.2845	0.5520	0.0381	0.4799
Laanemaa	0.2918	0.5530	0.0405	0.4906
Saaremaa	0.2875	0.5891	0.0427	0.4483
Parnumaa	0.2858	0.5556	0.0370	0.4800
Viljandimaa	0.2634	0.5283	0.0409	0.4571
Tartumaa	0.2841	0.5712	0.0387	0.4612
Vorumaa	0.3114	0.6146	0.0401	0.4727
Estonia	0.2924	0.5691	0.0397	0.4778

I_f = Fertility index. I_h = Illegitimate fertility index.
I_g = Marital fertility index. I_m = Proportion of married women.

three indexes and the index for illegitimate fertility (I_h) are shown in _Figure 6.4_.

During the 1920s and 1930s the lowest fertility level in the Baltic region measured by I_f as well as by I_g was found in Estonia (_Figures 6.3_ and _6.5_). In Latvia the fertility level was a little higher, probably because of the impact of the high level in Latgale. The decomposition of I_f shows a similar pattern in both countries. Lithuanian fertility as measured by I_f was remarkably high in the 1920s. It was closer to the Estonian level in 1897 than in 1922. There is no doubt that the fertility transition began later in Lithuania, probably by two to three decades, than in the two neighboring northern countries. At the same time the proportion married (I_m) was the lowest in Lithuania, which implies a higher level of marital fertility (I_g) in this state than in Estonia and

Table 6.3. Continued.

County	I_f	I_g	I_h	I_m
1922				
Virumaa	0.2027	0.4211	0.0231	0.4513
Jarvamaa	0.1764	0.3683	0.0242	0.4423
Harjumaa	0.1575	0.3219	0.0222	0.4513
Laanemaa	0.2160	0.4611	0.0310	0.4301
Saaremaa	0.2358	0.5549	0.0409	0.3791
Parnumaa	0.1987	0.4149	0.0245	0.4463
Viljandimaa	0.1728	0.3765	0.0269	0.4173
Tartumaa	0.1830	0.3952	0.0256	0.4259
Valgamaa	0.2028	0.4487	0.0229	0.4227
Vorumaa	0.2041	0.4411	0.0281	0.4263
Petserimaa	0.3424	0.5956	0.0203	0.5599
Estonia	0.1943	0.4079	0.0253	0.4417
1934				
Virumaa	0.1847	0.3395	0.0295	0.5007
Jarvamaa	0.1893	0.3435	0.0424	0.4879
Harjumaa	0.1091	0.2246	0.0185	0.4397
Laanemaa	0.1933	0.3595	0.0378	0.4832
Saaremaa	0.2111	0.4370	0.0434	0.4258
Parnumaa	0.1713	0.3205	0.0352	0.4773
Viljandimaa	0.1651	0.3128	0.0389	0.4608
Tartumaa	0.1540	0.3173	0.0288	0.4339
Valgamaa	0.1714	0.3369	0.0346	0.4525
Vorumaa	0.1873	0.3723	0.0331	0.4546
Petserimaa	0.2542	0.4126	0.0335	0.5824
Estonia	0.1642	0.3189	0.0300	0.4646

Calculated on data sources: Rahvastik, 1930; Rahvastiku, 1935; Rahvastikuprobleeme, 1937; Sündivus, 1925; Tuhande, 1924, Vol. 3–11.

Latvia. Concerning heterogeneity within the states, it is remarkable that differences between fertility indexes in counties were greater in Lithuania than in Estonia. Moreover, the differences within Lithuania could to a lesser extent be explained by the degree of urbanization or ethnic and cultural background variables than they could be in the case of Estonia.

In the 1930s fertility dropped below replacement level in Estonia and Latvia. At that time both regions had one of the lowest fertility levels in the world (*Figure 6.3*). The populations of Eastern Europe and Southern Europe became even more different from the populations in Estonia and Latvia than they had been half a century earlier. In

Table 6.4. Princeton indexes in Lithuania, 1923.

County	I_f	I_g	I_h	I_m
Alytaus	0.3329	0.6971	0.0210	0.4614
Birzu	0.2238	0.5639	0.0156	0.3798
Kauno	0.2658	0.5938	0.0241	0.4243
Kedainiu	0.2586	0.6010	0.0253	0.4053
Kretingos	0.3255	0.8013	0.0393	0.3757
Mariampoles	0.2728	0.6408	0.0269	0.4006
Mazeikiu	0.2874	0.6531	0.0389	0.4048
Panevezio	0.2248	0.6101	0.0171	0.3505
Raseiniu	0.2842	0.6470	0.0320	0.4102
Rokiskio	0.2467	0.6138	0.0174	0.3847
Seinu	0.2986	0.6588	0.0176	0.4383
Sakiu	0.2308	0.6245	0.0284	0.3396
Siauliu	0.2403	0.5884	0.0250	0.3823
Taurages	0.3061	0.7249	0.0399	0.3889
Telsiu	0.3123	0.7687	0.0482	0.3668
Traku	0.2860	0.5947	0.0161	0.4665
Utenos	0.2781	0.6032	0.0177	0.4449
Vilkaviskio	0.2351	0.5719	0.0163	0.3938
Ukmerges	0.2620	0.5939	0.0223	0.4195
Ezerenu	0.3023	0.6517	0.0262	0.4417
Lithuania	0.2692	0.6326	0.0256	0.4015

I_f = Fertility index. I_h = Illegitimate fertility index.
I_g = Marital fertility index. I_m = Proportion of married women.
Calculated on data sources: Lietuvos, 1926, 1927.

Table 6.5. Princeton indexes in Latvia, 1925, 1930, and 1935.

Year	I_f	I_g	I_h	I_m
1925	0.2118	0.4526	0.0253	0.4365
1930	0.1843	0.3773	0.0283	0.4470
1935	0.1745	0.3229	0.0318	0.4903

I_f = Fertility index. I_h = Illegitimate fertility.
I_g = Marital fertility index. I_m = Proportion of married women.
Calculated on data sources: Ceturta, 1937; Latvijas, 1940; Otra, 1926; Treta, 1931.

particular fertility in the 1930s was significantly lower in Estonia and Latvia than in neighboring Russia and in Finland.

Table 6.6. Princeton indexes in the Baltic states for selected years.

Region	Year	I_f	I_g	I_h	I_m
Estlandia	1870	0.348	0.669	0.053	0.479
	1897	0.310	0.596	0.032	0.493
Livlandia	1870	0.335	0.670	0.043	0.466
	1897	0.290	0.580	0.036	0.467
Kurlandia	1897	0.295	0.548	0.026	0.515
Vitebsk	1870	0.502	0.883	0.024	0.568
	1897	0.444	0.751	0.040	0.568
Kauno	1870	0.432	0.825	0.036	0.502
	1897	0.394	0.752	0.033	0.502
Vilno	1870	0.516	0.870	0.047	0.570
	1897	0.428	0.722	0.038	0.570
Estonia	1926	0.202	0.411	0.025	0.458
Latvia	1926	0.167	0.359	0.026	0.423
Lithuania	1926	0.292	0.561	0.028	0.498

I_f = Fertility index. I_h = Illegitimate fertility index.
I_g = Marital fertility index. I_m = Proportion of married women.
Source: Coale *et al.*, 1979, pp. 137–139.

6.2.3 Completed cohort fertility of married women

Information on completed cohort fertility is available for Estonian counties based on the 1922 census data on the number of children born. The available cross-tabulations give statistics on women distributed by parity and 10-year age groups. The last parity includes women with five or more children.

Table 6.7 presents the average number of children born to four cohorts of married women: (1) before 1853, (2) 1853–1862, (3) 1863–1872, and (4) 1873–1882. The first three cohorts were at age 70 or older, 60 to 69, and 50 to 59, respectively, in 1922 (i.e., they had completed their fertility). The fourth cohort was at age 40 to 49 at the time of the census, which may result in a slight undercount of cohort fertility. While computing the average number of children born, it was assumed that women with five or more children had given birth to seven children on average. This assumption is rather problematic. Undoubtedly, during the fertility transition important changes occur in the composition of married women of higher parities. At the initial stages of transition these changes may even be of greater importance than changes in distribution at lower parities. This assumption tends to understate the real

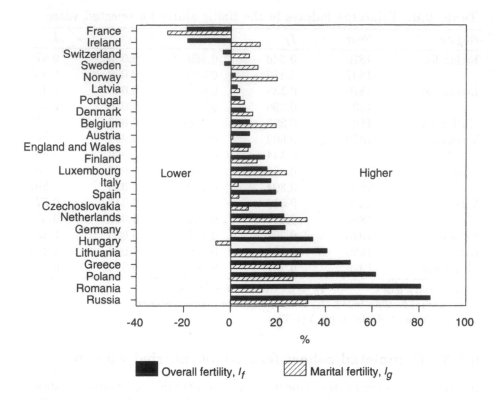

Figure 6.2. Overall fertility indexes, I_f, and marital fertility indexes, I_g, of European countries compared with Estonian level, 1881.

fertility decrease of cohorts. The figures in *Table 6.7* therefore give a low estimate of actual fertility decline.

The average number of children born to married women decreases from one cohort to the next (*Figure 6.6*). The completed fertility given for the oldest marriage cohort does not seem to be high enough to be classified as pre-transitional. It is quite possible that this cohort includes some fertility controllers. A clearly decreasing trend can be observed in all Estonian counties ranging between the lowest fertility county of Viljandimaa and Petserimaa which had the highest number of children born.

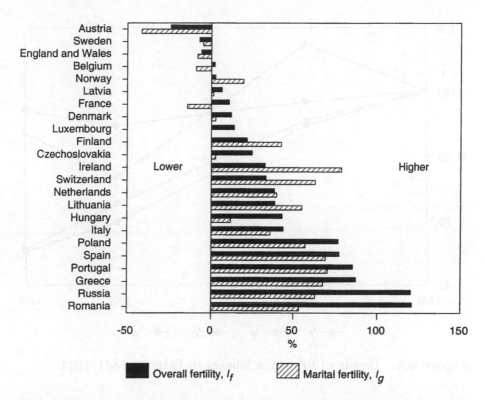

Figure 6.3. Overall fertility indexes, I_f, and marital fertility indexes, I_g, of European countries compared with Estonian level, 1934.

6.2.4 Age-specific fertility rates

Data on the age of women at birth are not available for the initial stages of the fertility transition in the Baltic states. Age-specific fertility patterns are given in *Figure 6.7* for the 1930s when the fertility transition was in full progress in Lithuania and was ending in Estonia and Latvia.

It is remarkable that despite the fact that fertility was very low in the 1930s, even below replacement level in Latvia and Estonia, corresponding fertility curves are not concave as is usually expected for controlled fertility populations. The decline in all three curves after the peak at age 30 to 34 is very close to linear. From this it may be inferred that the reproductive behavior of the population was very heterogeneous: a large number of women remained childless while many still had several

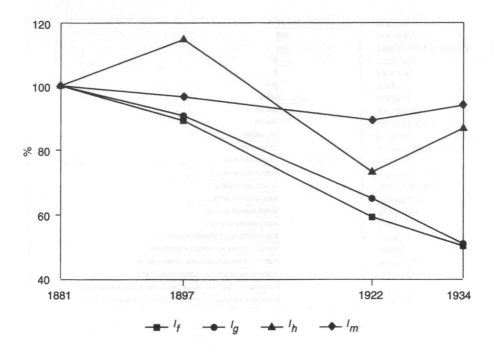

Figure 6.4. Trends of Princeton indexes in Estonia, 1881–1934.

children. In the case of Estonia, there is more direct proof for such heterogeneity. In the 1930s high fertility differentials by social groups can
be found despite the low average level of fertility at that time. These
fertility differentials are discussed in Katus (1987).

The age-specific fertility curves of Latvia and Estonia are generally
very similar, but the fertility curve of Latvia is slightly higher. The
Lithuanian curve of age-specific fertility not only is higher but also peaks
in the 30 to 34 age group instead of the 25 to 29 age group as in Estonia
and Latvia.

As mentioned before, the shape of the age-specific fertility curve in
all three republics is far from the typical modern type despite the low
level of fertility. A decrease in the mean age of childbearing and the
formation of a clearly concave curve after the maximum age occurred
later in history associated with even an increase in the level of fertility.

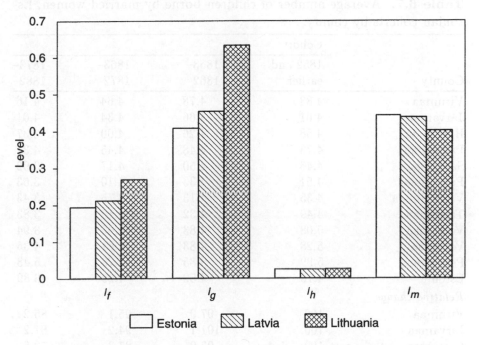

Figure 6.5. Princeton indexes in Estonia, 1922; Latvia, 1925; and Lithuania, 1923.

6.3   Discussion

When exactly the fertility transition in the Baltic states began is uncertain. In 1881 the process was already in progress in Estonia and presumably in most parts of Latvia. In Viljandimaa, an Estonian region well advanced in the transition, the trend of CBRs showed some increase during the 1820s and 1830s (Nóges, 1925). The decreasing trend began in the 1840s. Following a slow decline during the first two decades, the decrease accelerated after the 1860s. The growth in birth rate preceding the decline is similar to the increase that occurred in other Estonian counties at a somewhat later period and also in Latvia (Livlandia and Kurlandia) in the early 1870s (Zvidrinš, 1983). Knodel (1986) finds a similar growth of marital fertility in German villages. A slight fertility increase before the beginning of the fertility transition seems to be a common feature in many countries. It may be assumed that the start of

Table 6.7. Average number of children borne by married women, Estonian cohorts by county.

| County | Cohort | | | |
	1852 and earlier	1853– 1862	1863– 1872	1873– 1882
Virumaa	4.88	4.78	4.64	4.16
Jarvamaa	4.61	4.66	4.34	4.01
Harjumaa	4.58	4.26	4.00	3.37
Laanemaa	4.73	4.48	4.45	4.31
Saaremaa	4.46	4.50	4.17	3.92
Parnumaa	4.51	4.33	4.10	3.63
Viljandimaa	4.35	4.13	3.75	3.43
Tartumaa	4.49	4.32	4.15	3.83
Valgamaa	5.08	4.83	4.45	3.94
Vorumaa	5.28	4.83	4.61	4.35
Petserimaa	5.99	5.85	5.73	5.48
Estonia	4.70	4.52	4.29	3.89
Relative change				
Virumaa	100	97.9	95.1	85.2
Jarvamaa	100	101.1	94.2	87.2
Harjumaa	100	93.0	87.3	73.5
Laanemaa	100	94.6	94.0	91.0
Saaremaa	100	100.9	93.6	88.0
Parnumaa	100	96.0	90.9	80.5
Viljandimaa	100	95.0	86.2	78.7
Tartumaa	100	96.1	92.3	85.3
Valgamaa	100	95.0	87.6	77.4
Vorumaa	100	91.4	87.4	82.4
Petserimaa	100	97.7	95.6	91.6
Estonia	100	96.1	91.2	82.7

Calculated on source: Tuhande, 1924.

family limitation in Estonia and presumably in Latvia (excluding Latgale) dates back to the 1850s or at least to the 1860s. In Lithuania the transition began later.

The fertility decline seems to have been a rather slow process, not a revolutionary reduction in one or two generations. During the first phase, the fertility decrease was almost parallel to the decline in mortality. In an international comparison, Zvidrinš (1983) points out that the demographic transition in Latvia is similar to the pattern in France.

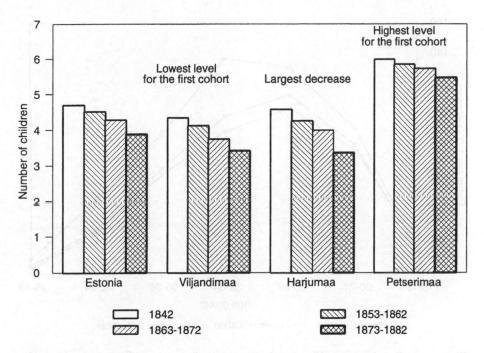

Figure 6.6. Average number of children, cohorts of married women of Estonia, 1842–1882.

Katus (1982) makes the same observation for Estonia. The relatively low population growth in Estonia and Latvia is the consequence of this almost simultaneous decline in mortality and fertility.

An interesting topic for future research is the study of county differences in the fertility transition. The process has been remarkably homogeneous among the regions of Estonia, except for Petserimaa where the level of marital fertility, especially overall fertility, in 1922 is comparable to that of other counties in 1881. In Petserimaa the majority of the population was Russian, belonging to the Orthodox church. Since Estonians as well as Latvians were ruled by Germans and Russians, they did not have their own upper class in the mid-19th century. It is also important to consider that in the beginning of fertility transition the urban populations consisted mostly of non-Estonians and non-Latvians. In Latvia the Catholic Latgale county clearly differed from other regions of the country in its later transition. Lithuania is also characterized by high regional heterogeneity in the process of fertility transition.

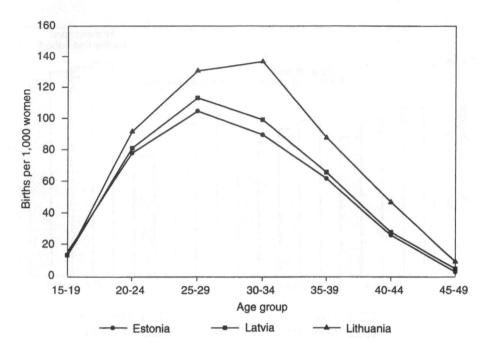

Figure 6.7. Age-specific fertility rate in Estonia, 1935; Latvia, 1935; and Lithuania, 1937.

The indexes I_f, I_m, I_g, and I_h given here do not correspond exactly to the indexes calculated by the Princeton European Fertility Project (Coale *et al.*, 1979). Because of different years considered (1871 and 1881) and different data sources the two calculation are not compatible. Compared with the calculations presented here for nine Estonian counties for 1897, the Princeton European Fertility Project calculations for the same time show slightly higher I_g and lower I_h and I_m in Estlandia and Livlandia. Nevertheless, the differences are minor for 1871–1881, and the indexes of the overall fertility (I_f) correspond exactly.

For the 1920s the differences are considerably greater. In the case of Estonia the indexes, except for I_h, determined by the Princeton European Fertility Project for 1926 are higher than those presented here for 1922. Taking into account the short-lived baby boom of the early 1920s, which ended by the middle of the decade, the differences between the two sets of indexes become even greater. In the case of Latvia, on the other hand, I_f and I_g are remarkably lower according to the Princeton

European Fertility Project. There are some differences for Lithuania as well, but both data sets reflect that the Lithuanian fertility level was clearly higher than that of its northern neighbors.

However, the differences between the two sets of indexes lead to contradicting conclusions. According to the Princeton European Fertility Project the fertility level in Latvia was the lowest among the Baltic states. My calculations show the lowest level in Estonia, at least in the 1920s and 1930s.

No doubt the timing of the fertility transition in Estonia and Latvia was much closer to Western Europe than to Russia. The same is true for the fertility level between the two world wars. After World War II fertility trends in Estonia and to a lesser extent in Latvia diverged from the general European trend (Katus, 1990a, 1990b; Puur and Vikat, 1990). In a sharp contrast to the European trend, Estonia had no increase in fertility during the postwar period. No baby boom was observed, and fertility remained steadily below replacement level. In the 1950s Estonian fertility was probably the lowest in the world.

This trend changed in the late 1960s when surprisingly fertility increased. Measured by period, the TFR rose by approximately 18 percent in a few years: from 1.94 in 1965 to 2.26 in 1970. This short-term jump turned out to be a long-term change. A comparison of the period from 1968 to 1988 with the period from 1928 to 1968 reveals that fertility was approximately 15 percent higher during the 1968 to 1988 period. This is an unusual situation in Europe: since having risen above replacement level in the late 1960s, period fertility has not fallen below this level. This is in sharp contrast to most countries, especially those with an early fertility transition reaching a very low level of fertility before World War II (Calot and Blayo, 1982; Festy, 1984; van de Kaa, 1987; Klinger, 1988).

6.4 Conclusion

Estonia and Latvia (except for Latgale) have experienced a rather early fertility transition. The pre-transitional fertility situation, as well as the timing of the fertility decline in these Baltic states, has differed from the pattern in Eastern Europe and, especially, in Russia. The case of Lithuania is closer to Eastern Europe than to neighboring Latvia and Estonia. This specific history of the fertility transition in the Baltic states was one factor contributing to the great heterogeneity of the populations in the Soviet Union.

During the 1970s and 1980s Estonia had a rather stable replacement fertility which is exceptional for demographically advanced countries. This makes Estonia an interesting subject of future demographic research, the results of which might be useful in formulating population policies in many European countries. Recent political independence should be a further reason for demographic studies on the Baltic countries as they approach stronger integration with the rest of Europe.

Acknowledgment

I would like to thank Vlada Stankuniene, Inta Kruminš, and Juris Kruminš for providing the Lithuanian and Latvian data.

Appendix: Data Adjustment and County Boundaries for the Estimation of Princeton Indexes

County Boundaries

County boundaries in Estonia did not change in the periods of 1881–1897 and 1922–1934 but did in the period of 1897–1922. Three main changes occurred. Two new counties were added by 1922; before that the territory of Petserimaa was not included in the nine Estonian counties. Valgamaa was formed from parts of Tartumaa, Viljandimaa, Virumaa, and Valgamaa (mostly Latvian regions at the time). Virumaa was enlarged toward the east to include the city of Narva in 1922.

The calculations of Lithuanian fertility indexes exclude three counties and the city of Klaipeda because the census had not been carried out in 1923 in these areas. The county boundaries correspond to those at the time of the census. In the case of Lithuania, as well as Estonia, the county figures include the urban population.

Census Data on Women's Marriage and Age Structure

In the case of Estonia the data on female five-year age groups, except for married women, are available for all four censuses at the county level. For 1881 and 1897 the marriage structure is available by 10-year age groups added to the first group at ages 15 to 19. For 1934 the marriage structure is mixed: three five-year groups are supplemented by 10-year groups of ages 30 to 39 and 40 to 49. In the case of Latvia and Lithuania five-year age groups are used.

Table 6A.1. Comparison of total number of births by calculation method.

Method used	Number of births	
	1881	1897
CBR	27,761	27,328
SFR	27,961	27,893
Reverse-survival	28,040	27,111

The variables I_g, I_h, and I_m were calculated using only 10-year age groups of married women; the Hutterites' fertility is adapted to the same age intervals.

Data on Births

The 1881 and 1897 data on the number of births in Estonian counties have been calculated on the reverse-survival method (*Indirect*, 1983) for five years preceding the census, correspondingly, 1876–1880 and 1892–1896. Single-year age groups taken from censuses and the corresponding life table up to the age of five have been used. The latter data for 1881 were calculated using the ICM program, Mortpak package, and based on mortality parameters for Estonia during 1880–1884 (Rahvastiku, 1935). The single-year probabilities of mortality for 1897 are directly derived from life tables for Estonians determined by Ptoukha (1960).

The total number of births in nine Estonian counties can be calculated using the crude and special (per 1,000 women in fertile age, SFR) fertility rates available for Estonia (Rahvastiku, 1935). The result is compared to the total number of births by counties calculated by reverse-survival method. The data sets match surprisingly well (*Table 6A.1*).

In other cases the data on births are available from the annual registration system of the three countries. With the exception of Estonia, five annual data sets of the number of births during the corresponding census were averaged for the calculations. For the calculations of Estonia, only three years were used: 1921, 1922, and 1923.

References

Calot, G., and Ch. Blayo, 1982. Recent Course of Fertility in Western Europe. *Population Studies* 36(3):349–72.

Ceturta tautas skaitisana, 1937. Latvija 1935 gada, III, Riga.

Coale, A.J., 1986. The Decline of Fertility in Europe Since the 18th Century as a Chapter in Demographic History. In A.J. Coale and S.C. Watkins, eds.

The Decline of Fertility in Europe. Princeton University Press, Princeton, NJ.

Coale, A.J., and R. Treadway, 1986. Summary of the Changing Distribution of Overall Fertility, Marital Fertility, and the Proportion Married in the Provinces of Europe. In A.J. Coale and S.C. Watkins, eds. *The Decline of Fertility in Europe*. Princeton University Press, Princeton, NJ.

Coale, A.J., B. Anderson, and E. Härm, 1979. *Human Fertility in Russia Since the Nineteenth Century*. Princeton University Press, Princeton, NJ.

Ekonomika i kultura Litovskoi SSR v 1975 g. (Economics and Culture in Lithuania in 1975), 1976. *Statistitcheskiy sbornik* (Statistical Collection). Vilnius.

Festy, L., 1984. *Fertility in Western Countries from 1870 to 1970*. UN, Bangkok.

Hajnal, J., 1965. European Marriage Patterns in Perspective. In D. Glass and D. Eversley, eds. *Population in History*. Aldine, Chicago, IL.

Hyrenius, H., 1942. *Estlands Svenskarna*. Lund, London.

Indirect Techniques for Demographic Estimation, 1983. United Nations, New York, NY.

Kaa, van de D.J., 1987. Europe's Second Demographic Transition. *Population Bulletin* **42**(1).

Katus, K., 1982. *Dolgosrotchnye tendentsii razvitii i upravlenii demografitcheskoy sistemy, Dissertacia* (Long-term Development of Demographic System, unpublished dissertation). Moscow University, Moscow.

Katus, K., 1987. Evolucia rozhdaemosti v Estonii za trista let (Fertility Evolution in Estonia in Three Hundred Years). *Sociologitcheskije issledovaniya* (Sociological Studies) **1**:54–61.

Katus, K., 1990a. Fertility Trend in Estonia. Paper presented at the Estonian-Finnish Demographic Seminar, Tallinn.

Katus, K., 1990b. Demographic Trends in Estonia throughout the Centuries. *Yearbook of Population Research in Finland* **28**:50–66.

Katus, K., forthcoming. Sündimuse langus Eestis mõõdetuna sündimusindeksite kaudu. Rahvastiku-uuringud, Series B. Tallinn.

Klinger, A., 1988. The Future of Reproduction. In *Future Changes in Population Age Structures*. IIASA Conference Papers, Laxenburg.

Knodel, J., 1977. Family Limitation and the Fertility Transition: Evidence from the Age Patterns of Fertility in Europe and Asia. *Population Studies* **31**(2):219–249.

Knodel, J., 1986. Demographic Transition in German Villages. In A.J. Coale and S.C. Watkins, eds. *The Decline of Fertility in Western Europe*. Princeton University Press, Princeton, NJ.

Latvijas PSR statistikas tabulas, 1940. Riga.

Lietuvos gyventojai, 1926. Kaunas.

Lietuvos statistikos metrastis 1924–1925 m., 1927. Kaunas.

Lietuvos statistikos metrastis 1931 m., 1932. Kaunas.

Lietuvos statistikos metrastis 1939 m., 1940. Vilnius.

Lutz, W., 1985. *Finnish Fertility Since 1722*. Väestötatkimus laitos, Helsinki.

Nóges, M., 1925. Rahvamuutused Viljandi maakonnas 1801–1923. Demograafiline töö. Viljandi.

Otra tautas skaitisana, 1926. Latvija 1925 gada 10. Februari, III. Riga.

Ozelis, K., 1934. Naturalnis ir mechanims gyventojis Keitimasis Lietuvoje 1915–1933 m. Kaunas.

Palli, H., 1973. Ajaloolise demograafia probleeme Eestis. Tallinn.

Palli, H., 1974. O razvitii narodonaseleniya Estonii v sravnitelno-istoritcheskom plane (Estonian Population Development in Comparative Historical Perspective). *Izvestija AN ESSR, Obchestvennye nauki* (Proceedings of Academy of Science of ESSR, Social Sciences) **23**(4).

Palli, H., 1980. *Estestvennoje dvizhenie selskogo naseleniya Estonii 1650–1799* (Demographic Reproduction of Estonian Rural Population 1650–1799). Vols. 1–3. Eesti Raamat, Tallinn.

Palli, H., 1988. Otepää rahvastik aastail 1716–1799. Tallinn.

Ptoukha, M.V., 1960. *Otcherki po statistike naseleniya* (Essays on Population Statistics). Gosstatizdat, Moscow.

Puur, A., and A. Vikat, 1990. Elutsüklilised erisused Tallinna perekondade demograafilises koosseisus ja ainelises kindlustatuses. Rahvastiku- uuringud. Series A, No. 24. Tallinn.

Rachin, A., 1956. Naselenije Rossii za sto let (1811–1913) [Population of Russia over a Hundred Years (1811–1913)]. Gosstatizdat, Moscow.

Rahvastik ja tervishoid Eestis, 1930. Eesti demograafia, Vihk IV. Tallinn.

Rahvastiku koostis ja korteriolud, 1935.

Rahvastikuprobleeme Eestis, 1937. 1.III 1934 rahvaloenduse andmed, Vihk IV Tallinn.

Stankuniene, V., 1989. Vosproizvodstvo naseleniya Litvy (Population Reproduction in Lithuania). In *Demografitcheskoje razvitie Litvy* (Demographic Development of Lithuania). Institute of Economics of the Lithuanian SSR, Vilnius.

Sündivus, surevus, abielluvus ja rahvaliikumine 1921–1923, 1925. Eesti demograafia, Vihk II. Tallinn.

Treta tautas skaitisina, 1931. Latvija 1930 gada, IV. Riga.

Tuhande üheksasaja kahekümnene teise aasta üldrahvalugemise andmed, 1924. Vihk I–XI. Tallinn.

Vahtre, S., 1973. Eestimaa talurahvas hingeloenduste andmeil. Tallinn.

Zvidrinš, P., 1978. O demografitcheskom perehode v Latvii (On Demographic Transition in Latvia). *Izvestija AN Latv.SSR* (Proceedings of Academy of Science of Latvian SSR). Riga.

Zvidrinš, P., 1983. Demographic Transition in Latvia. In P. Khalatbary, ed. *Demographic Transition*. Akademie-Verlag, Berlin.

Zvidrinš, P., ed. 1986. *Naseleniye Sovetskoi Latvii* (Population of Soviet Latvia). Zinatne, Riga.

Chapter 7

Changes in Spatial Variation of Demographic Indicators in Russia

Sergei V. Zakharov

Time and space are two integral parts of the measurement of any process. Until recently, much greater attention has been paid to the time dimension in the study of demographic transition (such as a staged process of the changes in population reproduction registered at the level of fundamental shifts of the mean values of demographic characteristics). Beginning with the works by Landry and Notestein in the 1930s, this staged concept of demographic dynamics has become a stable research tradition based upon solid and continuously expanding empirical knowledge.

A different situation exists in the study of the spatial (regional) regularities of the demographic transition. Territorial divergences of demographic indicators have been studied since demography became a field of research. Comparative studies had been conducted long before the knowledge of the evolution of the demographic system as a whole. But the concept of the regional demographic disparities (regional heterogeneity of the demographic system) has not yet evolved in demography.

Moreover, publications dealing with the evolution of the regional homogeneity – heterogeneity of demographic variables as an independent phenomenon in a historical context – are few. Most studies on regional demography use a descriptive language similar to the language used by geographers to characterize large areas with complex landscapes, where more or less pronounced differences of the environmental conditions of life, human behavior, and modes of economic activities may be found. There might be the impression of an artificial autonomy of theory in the field of regional demography concerning demographic evolution. But the empirical material accumulated in many countries should have facilitated the development of empirically based theoretical generalizations (Blum, 1989; Brass and Kabir, 1977; Coale and Watkins, 1986; Kannisto, 1986; Kintner, 1988; Mosk, 1980; Rele, 1988).

In this connection, Watkins (1990), using material from the Princeton European Fertility Project, is one of the first to present the demographic transition in spatial *and* time projections. This chapter dealing with the theoretical and empirical aspects of demographic transition follows the same course (Zakharov, 1987a, 1987b, 1989, 1990, 1991).

The characteristics of the demographic transition in the USSR as a whole and in the individual republics have been discussed in Soviet and non-Soviet literature (Coale *et al.*, 1979; Vishnevsky, 1977; Vishnevsky and Volkov, 1983); but no special research has been done on the evolution of regional differentiation of the demographic reproduction regime at the Russian provincial level and its correspondence to the stages of demographic transition.[1] This chapter focuses on the regional variation in fertility and mortality in the provinces of the Russian Federation.

7.1   Specific Features of Methods of Analysis

In this study the territorial definition of *Russia* is limited to the total territory of the Russian Empire, which later formed the RSFSR. To retain a comparative analysis at various stages of the demographic transition during multiple administrative and territorial changes, especially intensive in the 1920s and 1930s, and to fill in gaps where a lack of information exists, several types of communities were studied.

- The set of relatively comparable provinces of European Russia (without the North Caucasus), comprising 31 gubernia that subsequently became oblast and autonomous republics of Russia, whose borders may be considered constant throughout the entire transition.

- The set of Russian administrative units at the provincial level for which we have two types of data available: pre–World War II data for 42 territories, with the exception of the North Caucasus, Siberia, and the Far East; post–World War II data for all 73 territorial units.

In the mid-1920s, the USSR Central Statistical Board compared the 1925–1926 borders and areas of the USSR territories with the 1912–1913 borders and areas (Tsentralnoye statisticheskoye upravlenije SSSR, 1926). According to that study the comparability of the areas with the 31 regions of prerevolutionary Russia was very good for 13 provinces (90 to 100 percent of the area of the old region was included in the province according to the 1925–1926 administrative division), good for 11 (60 to 80 percent), and poor for 5 (less than 50 percent). The correspondence trends, not of the absolute but of the relative values of demographic indicators, were studied. This measure of comparability of territories may be regarded as satisfactory. After 1940 data for fully comparable, contemporary borders of the RSFSR oblasts were used.

In addition to the crude birth rate (CBR) and crude death rate (CDR) for the years 1897, 1926, 1940, 1959, and 1979, the following indicators were calculated for each territory: total fertility rate (TFR), probability of a mother surviving to mean age of childbearing (PS), and the net reproduction rate (NRR). As these indicators cannot be obtained in a direct way for the earlier stages of demographic transition, estimations were made. Coale's index of overall fertility (I_f) was used for the retrospective assessment of TFR (for the years 1897, 1926, and 1940), calculated on the basis of censuses and vital statistics by regions. The indicator PS, as the ratio between the gross and net reproduction rates for the same years, was assessed independently based on reported infant mortality rates (IMRs). Thus two equations of linear regression were used in the calculations:

$$\text{TFR} = 13.83592 \times I_f - 0.28174 \qquad (7.1)$$

$$\text{NRR/GRR} = -1.85921(\text{IMR} \times 0.001) + 0.99218 . \qquad (7.2)$$

In equation (7.1), R^2 was 0.98; in equation (7.2) it was 0.94. The logical justification of equation (7.1) is sufficiently clear, as the TFR and I_f are, essentially, age-adjusted characteristics of fertility, changing in time practically in a parallel way. The rationale of equation (7.2) is the steady high (about 50 percent) life-table proportion of female deaths under the age of 1 in the total number of women dying by age 30. Thus, this proportion for European Russia was 52.6 percent in 1896–1897 and 51.7 percent in 1926–1927, and for Russia as a whole 50.9 percent in 1958–1959 and 49.1 percent in 1988.

The equation parameters were estimated based on available data of Western and Eastern European countries, as well as the USSR and union republics over a long period (for fertility, 24 countries, 160 points; for mortality, 31 countries, 235 points). The calculations were done for the 31 relatively comparable provinces and for other regions of Russia. The comparison of the values obtained by the indirect technique with the values determined by the established method shows that the error for 1979 did not exceed 5 percent for all territories of Russia and was less than 1 percent for the relatively comparable provinces, which is evidence of the quality of the retrospective model, when applied to the regional level. This model may be used only for a crude assessment of indicators and has a number of well-known drawbacks. In particular, it poorly fits the jumping-up variants and reacts weakly to the conjuncture conditioned changes of indicators, which is especially important in the post-transitional period. But the model does not lead to considerable errors when global shifts are investigated; this was verified with the data from various countries.

Because the analysis focuses on the discovery of statistical boundaries of regional variations in the characteristics of demographic reproduction regime, a peculiar graph-analytical language was employed, which is used in analytical geometry and multidimensional statistics. It had as its key notions the *zone of point concentration* and the *ellipse of point dispersion* generally called upon to characterize the geometric place of the points in a system of coordinates. They are the fields of variation of the elements of regions set in the multidimensional space of parameters. In this case it means the presentation on a plane and the respective statistical geometric interpretation of the evolution of the regional differentiation in demographic variables. This evolution is linked to the change of reciprocal location of the point regions (their form of distributions) in the space constructed by the axes of the coordinate system. The values of demographic indicators forming the reproduction regime – fertility and mortality – are put on the axes.

Each area of variation or zone of point concentration has its most probable boundaries. For the description of the boundaries for each period the following algorithm was chosen: the mean value of the variable of interest was determined, for instance, the TFR; for that mean TFR value, those regions were selected that conform to the minimum and maximum IMR values in interval of the most probable variation of that indicator, covering 95 percent of all points; the mean value for the second variable – the IMR – was determined; for that mean value, the regions were selected that correspond to the minimum and maximum TFR values of the interval of the most probable variation.

Thus, statistical boundaries of the regional variation were presented in the form of boundary values, that is, four poles of the zone of points concentration. Regions that are closest to the center (closest to the mean values of both variables simultaneously) were called the *most typical regions*. We also identified the *vanguard* and *rearguard* groups situated at the ends of the imaginary axis corresponding to the vector of transition from the highest levels of mortality and fertility to the lowest levels.

7.2 The Demographic Transition in Russia

The following features are regarded as the main characteristics of the demographic transition in Russia:

- The late start of the transition as compared with the European countries. In Russia it began in the late 19th century and early 20th century; a later transition was observed in Europe only in Bulgaria, Romania, Portugal, and Albania.
- A rapid transition, completed by the mid-1960s. The transition was more rapid in non-European countries such as Japan.
- A practically simultaneous beginning of the irreversible decline of mortality and fertility that was not an exception to Europe, but, as a rule, was not observed in Third World countries.
- The absence of a pronounced population explosion due to the factors mentioned above and a series of social cataclysms that coincided with the possible realization of the demographic growth potential.
- Clearly unique transitional trajectories of fertility and mortality in Russia in terms of TFR, IMR, and life expectancy as compared with the European countries, which is connected with the specific nature of the initial levels and the speed of their transformation at different stages.[2]

If the indicators for the European countries and Russia are presented in a comparable transitional time scale, despite some reduction in divergences due to the elimination of the impact of the starting date of transition, the peculiarities of the trajectory of the change in demographic indicators in Russia can clearly be seen (Zakharov, 1989, 1991). During most of the transition period up to the 1950s, Russia moved along the periphery of transition of the reproduction regime in Europe. *Table 7.1* lists the field of the transition of demographic regimes in Europe. The estimated net reproduction rate is presented in the course of transition in 22 European countries. As can be seen in *Table 7.2* Russia does not follow this typical pattern of change.

The sequence of wars, revolutions, and other discontinuities that shook Russia during the early 20th century played a significant role in the demographic transition. These, on the one hand, accelerated the process of demographic modernization that ultimately brought about the closing of the demographic gap between the European countries and Russia and, on the other hand, made the demographic peculiarities of Russia more pronounced. The high speed of the transition and its respectively short duration after World War II, helped to edge Russia into the ranks of the European countries. But since the 1950s the peculiarity of Russia's demographic trajectory remains pronounced, being a consequence of mortality trends that are not typical for Europe. It is accepted that the demographic transition has been completed when TFR equals 2.2 and IMR equals 25 (there is a 0.95 probability that a newborn girl will replace her mother). Therefore, the duration of the transition in Russia was less than 70 years (from the late 19th century to the mid-1960s).

According to our estimate the value of the transition multiplicator (Chesnais, 1979), which shows the extent of possible population growth due to the cumulative effect of annual rates of natural increase, was between 2.1 and 2.2, which is a little lower than the average European level. But the average speed of the growth due to the natural increase during the time of transition is high – about 1.1 percent annually (the geometric mean of rates) – which shows a higher potential demographic explosion in Russia than in Europe. At the same time, it should be pointed out that this mean value was formed by strongly oscillating rates of natural increase, including the negative impact of crisis years. This impact, which was reflected in the undulating nature of the cumulative

Table 7.1. Estimated values of NRR as a combination of TFR and IMR during the transition in Europe.

IMR	TFR									
	6.0	5.5	5.0	4.5	4.0	3.5	3.0	2.5	2.0	1.5
240	1.60	1.47	1.33							
230	1.65	1.52	1.38	1.24						
220	1.71	1.57	1.42	1.28						
210	1.76	1.61	1.47	1.32						
200		1.66	1.51	1.36	1.21					
190		1.71	1.56	1.40	1.25					
180		1.76	1.60	1.44	1.28	1.12				
170			1.65	1.48	1.32	1.15	0.99			
160			1.70	1.53	1.36	1.19	1.02			
150			1.74	1.57	1.39	1.22	1.04	0.87		
140			1.79	1.61	1.43	1.25	1.07	0.89		
130			1.83	1.65	1.46	1.28	1.10	0.92		
120				1.69	1.50	1.31	1.13	0.94	0.75	
110				1.73	1.54	1.35	1.15	0.96	0.77	
100					1.57	1.38	1.18	0.98	0.79	
90					1.61	1.41	1.21	1.01	0.81	
80					1.65	1.44	1.23	1.03	0.82	
70					1.68	1.47	1.26	1.05	0.84	
60					1.72	1.50	1.29	1.07	0.86	
50					1.76	1.54	1.32	1.10	0.88	
40						1.57	1.34	1.12	0.90	
30						1.60	1.37	1.14	0.91	
20							1.40	1.17	0.93	0.70
10								1.19	0.95	0.71

Table 7.2. Trends of selected demographic indicators in Russia.

Year	IMR	TFR	NRR
1897	326.0[a]	7.490[a]	1.40[a]
1926	191.0[a]	6.110[a]	1.93[a]
1940	205.0[a]	4.260[a]	1.27[a]
1958–1959	41.0	2.615	1.186
1969–1970	23.8	1.992	0.934
1978–1979	23.0	1.885	0.882
1988	18.9	2.130	1.005

[a]Estimated mean value for the 31 provinces in European Russia.

function of the multiplicator, resulted in only a partial achievement of the tendency toward the multiplication effect and surfaced in the first decades of the transition. The actual size of the Russian population increased 1.9 times from 1897 to 1966 (in absolute figures, the difference between the value of the multiplicator and the last figure amounts to about 22 million people).

The demographic transition in Russia was most similar to the transition in Bulgaria, Romania, Portugal, and Greece. The transitions in Sweden, Denmark, Belgium, Italy, and France differed the most.

The demographic transition in Russian was a relatively late but rapid transition. It had its own trajectory of change in mean values of demographic indicators and a partially realized tendency toward a substantial multiplicative effect. Detailed analysis of the peculiar features of Russia's demographic trajectory can be found in Zakharov (1989, 1990, 1991).

7.3 Stages of Regional Differentiation

Let us consider the changes of regional variation in CBRs and CDRs. *Figure 7.1* shows the transition in Russia as a sequence of zones of concentration of the point regions (31 territories). The diagonal line corresponds to zero natural increase. The process comprises several stages.

The pre-transitional stage is characterized by disordered distribution of regions with random variation of demographic indicators, which is difficult to interpret. The concentration zone for 1861–1880 is a circle: the CBR ranges between 40 and 60 births per 1,000 residents; and

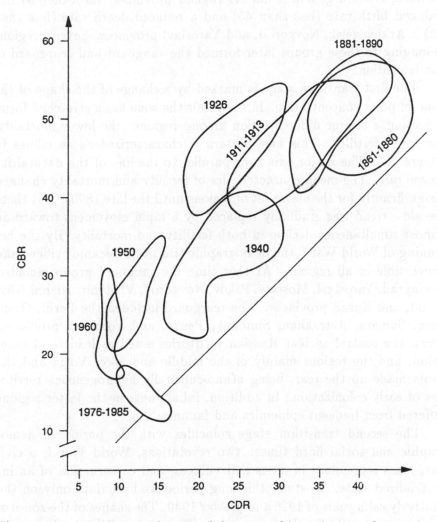

Figure 7.1. Evolution of regional demographic variation: the sequence of zones of point concentration of CBRs and CDRs for 31 Russian provinces.

the CDR ranges between 30 and 50 deaths per 1,000 residents, which generally constitutes a positive natural increase at the level of about 10 per 1,000 residents. Two groups of territories (within the borders of present-day provinces) can be distinguished: one with a high birth rate (more than 55) and a high death rate (more than 40) – Perm, Orel,

Samara, Orenburg, and Tula, Astrakhan provinces; the other with a reduced birth rate (less than 45) and a reduced death rate (less than 40) – Arkhangelsk, Novgorod, and Yaroslavl provinces. Several regions belonging to these groups later formed the vanguard and rearguard of the transition.

The first transition stage is marked by a change of the shape of the zone of point concentration. In the 1890s the zone has a stretched form, reflecting a clearer differentiation among regions: the lower mortality, the lower fertility. This new pattern is characterized as an ellipse in *Figure 7.1*. The major axis runs parallel to the line of the natural increase rate. The mean characteristics of fertility and mortality changed insignificantly for the entire set of regions until the late 1890s. But then, the slow trend was gradually replaced by a rapid movement toward an almost simultaneous decline in both fertility and mortality. By the beginning of World War I, the demographic transition became evident and irreversible in all regions. At that time the vanguard group included Leningrad, Yaroslavl, Moscow, Pskov, Novgorod, Vladimir, Nizhni Novgorod, and Kazan provinces. The rearguard included the Perm, Orenburg, Samara, Astrakhan, Simbirsk, Penza, and Voronezh provinces. Thus, the central ancient Russian territories acted as leaders of transition, and the regions mainly of the middle and lower Volga and the Urals made up the rear, being ethno-culturally heterogeneous territories of early colonization. In addition, inhabitants in the latter regions suffered from frequent epidemics and famines.

The second transition stage coincides with the period of demographic and social hard times: two revolutions, World War I, a civil war, mass repressions in cities and villages, and construction of an industrialized state. To study this long period we have data only on the relatively calm years of 1926 and prewar 1940. The shapes of the zones of point concentration for these two years differ substantially. In the second stage, the composition of the leading and lagging groups changed considerably. Thus, in the 1920s and 1930s, northern regions in the rearguard were the Karelian ASSR, Arkhangelsk, Vologda, and Kirov provinces; as a result, the sequence of regions on the way to demographic transition that was established in the preceding stage was disturbed. The accelerated mortality decline led to a considerable increase in natural growth rate in all regions in 1926. Subsequently, fertility began declining more intensively; by 1940 a sharp decrease of the natural increase became apparent. There was a tendency toward a temporary slowdown of the transition and consolidation of regions. It is apparent in the flexed elliptic shape of the point concentration zone for 1940. World War II intensified the process of the reduction of heterogeneity.

The third transition stage comprises two periods. The first period was the period of final formation of the vanguard and rearguard groups (the 1950s). The central provinces – Leningrad, Novgorod, Pskov, Tver, Moscow, Ryazan, Smolensk, Tula, and Yaroslavl – strengthened their positions in the leading group, squeezing out the Blackearth Voronezh, Kursk, and Orel provinces. Three groups of territories were clearly in the lagging group: northern Karelian ASSR, Arkhangelsk, Vologda, and Kirov provinces; the Urals group – Bashkir ASSR, Perm, and Orenburg provinces; and the Volga group – the Tatar ASSR, Ulyanovsk, and Astrakhan provinces. The second period (the 1960s) was a period during which the CBR decline became much stronger than the CDR decline. By the early 1960s, the regional differentiation of the natural increase was determined mainly by the variation in the birth rate (the irregular shape stretched along the CBR line). There was a simultaneous slowdown of the CDR decline and even an increase caused by the evolution of the age structure.

The fourth transition stage marks the completion of the evolution of the regional diversity: the tempo slowed substantially and the demographic homogeneity of regions sharply increased. In the 1970s, mortality increased and life expectancy declined almost everywhere in the RSFSR, but these processes were manifested in different ways in different regions. In the 1970s and 1980s a tendency of steady depopulation in the vanguard territories was evident – Pskov, Tver, Tula, Novgorod, Ivanovo, Ryazan, Kursk, and Tambov provinces. The rearguard group was represented by Bashkir ASSR, Karelian ASSR, Perm, Arkhangelsk, Orenburg, Vologda, Kirov, and several Volga provinces. At this transition stage the rearguard lost its initial importance because the demographic transition in these territories may be considered to be complete.

Until now this description of the evolution of the regional differentiation of demographic indicators in Russia was based on the analyses of rates that are strongly dependent on the age composition of the population. Will the elimination of the influence of the age structure and the use of more accurate indicators result in other conclusions regarding the evolution of the regional differentiation?

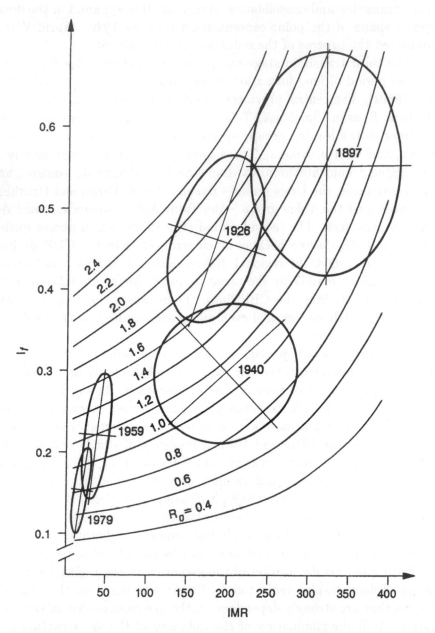

Figure 7.2. Distribution of the European Russian provinces according to Coale's index (I_f) and infant mortality rate (IMR).

Table 7.3. Evolution of regional variation of selected demographic indicators for 31 provinces in Russia.

Indicator	1897	1926	1940	1959	1979
IMR					
Mean	323.0	187.0	211.0	40.9	22.0
Standard deviation	47.0	26.0	42.0	7.3	3.3
Coefficient of variation	14.6	14.1	20.1	17.9	14.8
TFR					
Mean	7.4	6.1	3.8	2.7	1.9
Standard deviation	1.0	0.8	0.6	0.6	0.2
Coefficient of variation	13.0	12.3	16.5	21.1	9.9
NRR					
Mean	1.4	1.9	1.1	1.2	0.9
Standard deviation	0.4	0.2	0.2	0.2	0.1
Coefficient of variation	21.2	12.0	16.2	20.1	9.8

Figure 7.2 shows ellipses of the distribution in 31 Russian territories using Coale's index (I_f) and IMR. For each zone of point concentration the length of the major and minor axes of the ellipse corresponds to the doubled standard deviation of the values of the indicated variables on the axes of the graph. The tilt of slope of the major axes of the ellipses was determined on the basis of the angular coefficients of the regression equations between the variables.

It is not difficult to notice that basically *Figure 7.2* is in accordance with *Figure 7.1*. A substantial increase of regional homogeneity in the absolute deviation from the mean values may clearly be traced. The evolution in terms of relative deviation (coefficient of variation) is more complex than the absolute deviation (see *Table 7.3*). The increase of the variation coefficient at the middle stages of the transition is possible and even natural.

The actual distribution of the Russian provinces by the reproduction-regime component is shown in *Figures 7.3* and *7.4*. They display all provinces for which data are available. Each pair of figures is made in comparable scale and proves the increase in the regional homogeneity of the reproduction regime. *Table 7.4* contains information on the most probable regional variation of demographic indicators and the names of typical Russian provinces at the main stages of the evolution.

Table 7.4. The range of the most probable regional variation of demographic indicators and regions of typical demographic regimes in Russia in 1897, 1926, 1958–1959, and 1978–1979.

Indicator	At mean TFR	At mean IMR	Regions
1897 (31 provinces)			
TFR	7.7	5.9–8.7	Simbirsk,
IMR	250–410	310	Tambov,
NRR	1.0–1.9	1.2–1.7	Tver
1926 (41 provinces)			Vladimir,
TFR	5.9	5.5–6.9	Voronezh,
IMR	125–235	190	Nizhni Novgorod,
NRR	1.6–2.1	1.7–2.2	Ryazan
1958–1959 (73 provinces)			A^a: Belgorod,
TFR	2.9	2.1–3.5	Krasnodar,
IMR	32–52	41	Nizhni Novgorod,
NRR	1.28–1.34	0.94–1.70	B^a: Tatar ASSR,
			Udmurt ASSR,
1978–1979 (73 provinces)			Komi ASSR,
TFR	2.05	1.6–2.1	Vologda,
IMR	16–30	23	Volgograd,
NRR	0.931–0.98	0.77–1.07	Novosibirsk

[a]The point indicating mean values for RSFSR as a whole in 1958–1959 is between the two large point clusters with their own centers of gravity. A TFR = 2.5–2.6, IMR = 37–38, NRR = 1.1–2.2; B TFR = 3.5–3.6, IMR = 46–49, NRR = 1.5–1.6.

7.4   Conclusion

Russia's geographic position on the border of Europe has affected its demographic history. About 100 years ago, relatively late according to European standards, Russia found itself involved in demographic transition. Russia's demographic transition, reflected in the change in mortality, fertility, and nuptiality, differed from that of the European countries. Nevertheless, the demographic transition as a global transformation of the reproduction regime may be considered to be complete at the level of mean characteristics by the late 1960s. At that time a noticeable similarity of Russia's demographic parameters to the parameters of other European countries was observed. From the analysis of the data on Russian provinces, a number of regular features of the evolution of regional variation may be determined:

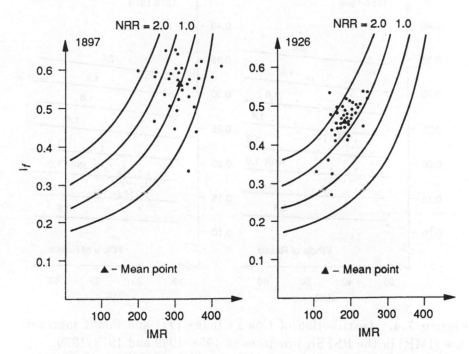

Figure 7.3. Distribution of Coale's index (I_f) and infant mortality rate (IMR) in the European Russian provinces in 1897 and 1926.

- The pre-transitional period is characterized by a great regional variation in fertility and mortality rates. In the beginning of transition some order is introduced into the distribution of regions that points at the direction of change: the lower mortality, the lower fertility.
- The degree of the variation and the nature of the change are determined by the time of the initiation of transition (the emergence of vanguard and rearguard groups) and the pace of transition in different regions (the change of the composition of regions in these groups).
- The tendency toward a constancy in geographic location of the leading and lagging regions is mainly retained throughout the transition, though each region's level of demographic parameters may change substantially. The analyzed set of relatively comparable territories of European Russia is characterized by the progress of the central provinces. The most consistent members of the vanguard group are

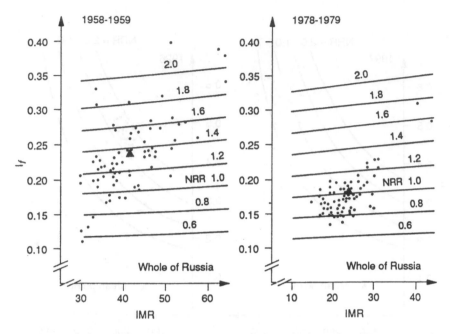

Figure 7.4. Distribution of Coale's index (I_f) and infant mortality rate (IMR) in the RSFSR provinces in 1958–1959 and 1978–1979.

the Novgorod, Yaroslavl, Leningrad, and Moscow provinces; of the rearguard, the Perm, Orenburg, and Astrakhan provinces.

Thus, the expansion of the demographic transition in Russia moved from the center to the periphery and, for Russia, from the west to the south and east. Today, the national–cultural autonomies of the Northern Caucasus, Kalmyk ASSR, and Siberia (the Tuva, Yakut, and Buryat ASSRs, where the IMR reaches 30 or more and the TFR is 3 or more) occupy a stable place in the rearguard. The process to increase the statistical homogeneity of provinces under the conditions of low fertility and mortality is sufficiently intensive in Russia, leading to the inevitable simplification of the post-transitional regional topological structure.

It is not difficult to verify the historical indispensability and irreversibility of this process by analyzing the experience of other countries (see, for instance, Watkins, 1990; Zakharov, 1989, 1990). Achieving this regional homogeneity of demographic indicators, as well as increasing demographic homogeneity in other aspects, is a distinctive feature and, to a certain extent, may serve as a criterion of the completion of transition of the demographic regime at the regional and country levels.

Notes

[1] Henceforth regional differentiation, regional heterogeneity, and regional variation are used to denote the statistical diversity of the indicators.

[2] It is well known that the pre-transitional late 19th-century characteristics of fertility, mortality, and nuptiality in Russia greatly differed not only from those that existed at that time in Western and Northern Europe, but also from those that existed nearby – in Eastern Europe and in the Baltic states and, to a certain extent, in the Ukrainian and Byelorussian regions (see Coale *et al.*, 1979; Ptoukha, 1960; Vishnevsky, 1977; Vishnevsky and Volkov, 1983). Moreover, the demographic characteristics in Russia were fundamentally different from those observed in the pre-transitional period in each European country.

References

Blum, A., 1989. Les disparités régionales de fécondité en France. In G. Calot, A.G. Vishnevsky, and L.L. Rybakovskiy, eds. *Natilité et Famille*. Editions du Progrès, Moscow.

Brass, W., and M. Kabir, 1977. Regional Variations in Fertility and Child Mortality during the Demographic Transition in England and Wales. In J. Hobcraft and Ph. Rees, eds. *Regional Demographic Development*. Croom Helm, London.

Chesnais, J.C., 1979. L'effet multiplicatif de la transition demographique. *Population* 6:1138–1144.

Coale, A.J., and S.C. Watkins, eds. 1986. *The Decline of Fertility in Europe*. Princeton University Press, Princeton, NJ.

Coale, A., B. Anderson, and E. Härm, 1979. *Human Fertility in Russia Since the Nineteenth Century*. Princeton University Press, Princeton, NJ.

Kannisto, V., 1986. *Geographic Differentials in Infant Mortality in Finland in 1871–1983*. Central Statistical Office of Finland, Helsinki.

Kintner, H.J., 1988. Determinants of Temporal and Areal Variation in Infant Mortality in Germany, 1871–1933. *Demography* 25(4):597–609.

Knox, P.L., 1981. Convergence and Divergence in Regional Patterns of Infant Mortality in the United Kingdom from 1949–1951 to 1970–1971. *Social Science and Medicine* 15d(3):323–328.

Mosk, C., 1980. Rural–Urban Fertility Differences and Fertility Transition. *Population Studies* 34(1):77–90.

Ptoukha, M.V., 1960. *Otcherki po statistike naseleniya* (Essays on Population Statistics). Gosstatizdat, Moscow.

Rele, J.R., 1988. *70 Years of Fertility Change in Korea: New Estimates from 1916 to 1985*. Reprints of the East–West Population Institute, No. 231. East–West Center, Honolulu, HI.

Stokowski, F., 1983. The Pattern of Demographic Transition from the Aspect of Regional Differentiation. In P. Khalatbari, ed. *Demographic Transition*. Akademie-Verlag, Berlin.

Tsentralnoye statisticheskoye upravlenije SSSR (Central Statistical Board of the USSR), 1926. *Administrativno-territorialny sostav SSSR v sopostavlenii s dovoennim deleniem Rossii* (Administrative-Territorial Composition of the USSR in Comparison with Prewar Structure of Russia). Moscow.

Vishnevsky, A.G., ed. 1977. *Brachnost' rozhdaemost' i smertnost' v Rossii i v SSSR* (Nuptiality, Fertility, and Mortality in Russia and in the USSR). Statistika, Moscow.

Vishnevsky, A.G., and A.G. Volkov, eds., 1983. *Vosproizvodstvo naseleniya SSSR* (Reproduction of the Population of the USSR). Finansy i statistika, Moscow.

Watkins, S.C., 1990. From Local to National Communities: The Transformation of Demographic Regimes in Western Europe, 1870–1960. *Population and Development Review* 16(2):241–272.

Zakharov, S.V., 1987a. Istoricheskoye ponimaniye regionalnogo faktora evolutsii vosproizvodstva naseleniya (Historical Conception of Regional Factor in Evolution of Human Reproduction). In L.L. Rybakovskiy, ed. *Demograficheskiye protsessy: voprosy izucheniya* (Demographic Processes: Problems under Study). Institute of Sociology, Moscow.

Zakharov, S.V., 1987b. O sootnoshenii evolutsii vosproizvodstva naseleniya vo vremeni i v prostranstve (Evolution of Population Reproduction Regime through Time and Space: The Question of Interconnection). In *Demograficheskije aspekty uskoreniya socialno-economicheskogo razvitiya: Vsesoiuznaya konferentsiya* (Demographic Aspects of Accelerating of Socioeconomic Development: All-Union Conference), Vol. 1. Kiev.

Zakharov, S.V., 1989. Demographic Transition in Russia: Estimation of the Regional Variation. Paper presented at "Reconstitution et Dynamiques des Populations du Passe." Paris.

Zakharov, S.V., 1990. *Evolutsiya regionalnich osobennostej vosproizvodstva naseleniya Rossii: Avtoreferat dissertatsii* (Evolution of Regional Demographic Differentials in Russia: Essay on Dissertation). Institute for Socioeconomic Studies of Population, Moscow.

Zakharov, S.V., 1991. Demograficheskiy perekhod v Rossii i evolutsiya regionalnich demograficheskich razlichiy (Demographic Transition in Russia and the Evolution of Regional Demographic Differentials). In A.G. Vishnevsky, ed. *Demografija i sociologija: Sem'ia i semeynaya politika* (Demography and Sociology: Family and Family Policy). Institute for Socioeconomic Studies of Population, Moscow.

Chapter 8

Contraception and Abortions: Trends and Prospects for the 1990s

Alexandr Avdeev

Over the past decade the problems of family planning began to cause interest in the USSR. Three mutually related questions have been occupying the attention of the mass media, demographers, and sociologists:

- Why does the USSR have the highest level of induced abortions in the world?
- What are the trends and factors associated with abortion and contraception?
- Is it possible to modify the situation and in what way?

It is difficult to find answers to these questions because official abortion statistics were not published between 1930 and 1987, and even these official statistics do not have enough detail for an in-depth analysis.

During the first liberalization of the abortion law (1920–1936) in the USSR there were very good abortion statistics, providing information on age, social characteristics, marital status, number of children, deliveries, and previous abortions of women who obtained an abortion in a clinic. In 1936, when the abortion law was restricted, the system of collecting abortion statistics was almost completely abolished, and it was not restored after 1955 when the Soviet government liberalized abortion again.

Between 1955 and 1965 the number of abortions was rapidly increasing and reached its maximum of 8.3 million in 1965. It then slightly decreased and during the next two decades it stabilized at the level of 7 million per annum, in spite of the Ministry of Health efforts to spread modern contraceptive methods.

This chapter aims to estimate the basic indicators of abortion and contraception from the incomplete data available, to analyze the causes of abortion-level stabilization during the 1980s in the USSR, and to anticipate the future problems and effects of family-planning activity concerning abortion and contraception.

8.1   Why Has Induced Abortion Become the Main Method of Family Planning in the USSR?

The beginning of fertility decline in the course of demographic transition occurs at the same time as the start of family planning, which includes such important elements as the attitudes of society, social institutions, families, and individuals toward deliberate birth control, abortion, and contraception.

Throughout the world, there are three basic types of family planning. The origins of the first were typical for Western Europe: the decline of birth rate had begun at a time when contraception was underdeveloped and abortion was prohibited and considered a crime. Under these conditions any information on contraception was well received by the population and the freedom to obtain such information was supported by a majority of litigators who were eager to have birth control. Thus, the contraceptive way of family planning was created. Typical of this way are a negative attitude to abortion and a high motivation for contraceptive practice, even after the liberalization of abortion law.

The second type of family planning may be found in countries in which the decrease in birth rate had begun during the period of the contraceptive revolution in the 1960s. Modern contraception was supplied by the governments of many countries, especially within the framework of family-planning programs, often significantly exceeding the demand. This was an inversion of the European type but resulted in the same consequence: the spread of contraception as the main method of birth control.

In the USSR the birth-rate decrease had begun under conditions of liberal abortion policies without wide supply of contraception. This forms the third type of family planning.

The birth-rate decline started in Russia in the late 19th century, but was significantly distorted during the Russian–Japanese war, the revolution of 1905, the First World War, the revolution of 1917, and the following civil war. Legalization of abortion in 1920 under conditions of hunger had rapidly made it a widespread method of birth control. So after the increase in the birth rate in the mid-1920s, abortion became more accessible and soon became the main method of family planning.

The main aim of abortion policy in the 1920s was to decrease the number of abortions outside hospitals. This policy was successful, but it resulted in the further spread of abortion as the principal method of family planning because contraception was in limited supply.

The use of abortion for family-planning purposes was based on the ideological doctrine, which existed for many years, that a decrease in birth rate (and thus abortion) was a temporary phenomenon not typical of socialism. It was believed that the improvement in economic conditions of life would automatically result in increased birth rates and decreased abortion levels. This ideological doctrine prevented interest in modern contraception and hindered its dissemination.

In the early 1930s abortion policy in the USSR experienced a crisis. Abortion data over different time periods demonstrate that, though the quality of life improved, the number of abortions continued to rise. Simultaneously, a fertility decrease and a reduced population growth caused by forced collectivization and hunger was observed. Abortions were forbidden in 1936 as a result of this.

It may have been thought that the restriction of abortion law could have reoriented family-planning practice in the USSR to the "Western" way. However, this process was slowed down by the Second World War distorting the demographic development of the USSR.

The next liberalization of abortion law in 1955 was the revival of the abortion policy which had existed between 1920 and 1936. As before, abortion and fertility decline were considered as temporary phenomena. Only in the 1960s was it accepted that a general change in reproductive patterns, resulting in low fertility, had occurred in the USSR. It was understood that low fertility is a necessary reaction to low mortality. As a consequence of this understanding the attitude to abortion changed too. It became clear that it was impossible to reduce abortions significantly without widespread use of modern contraception and without changing family-planning activities in general.

In the 1960s and 1970s health planners believed that a growth in contraceptive practice, even in less effective methods, should decrease the abortion level. But they did not take into account the fact that families did not worry about the global problem of reducing the number of abortions. The main problem within a family was how to avoid unwanted childbearing, and if contraception was not 100 percent effective the motivation to use it was very low (Verbenko *et al.*, 1968).

So, the USSR Ministry of Health paid increased attention to the problem of contraception, giving a priority to intrauterine devices (IUD) and oral contraceptives (OC), with the intention of achieving a change in the abortion level in the 1980s. However, this did not happen. Section 8.2 explains the reasons for this.

8.2   Framework and Data for Analysis

At present, Soviet official statistics only include data on the annual number of registered abortions occurring in hospitals, grouped by territorial units. However, this indicator depends significantly on factors such as the number of women of reproductive age, the proportion who are married, the age distribution of births, and the contraceptive prevalence. It is difficult to analyze changes in abortion rates on the basis of official statistical information. Thus the main analytical task is to identify the influence of different demographic factors on abortion level.

First, the abortion level depends on the proportion of women of reproductive age (WRA) in the population. One may exclude the influence of this factor by calculating the general abortion rate (GAR, or abortion rate) equal to the number of abortions per 1,000 WRA. The impact of age structure on the WRA can be removed by standardizing GAR or by calculating the total abortion rate (TAR) as a sum of age-specific abortion rates (ASAR), by analogy with the total fertility rate (TFR).

As the age grouping of abortions is not given in the Soviet statistics, we can determine the TAR value by dividing GAR by 1,000 and multiplying by 35 – the duration of the reproductive period in years. This procedure is sometimes used for the comparison of indicators in different countries or for TAR use in demographic models. However, for the analysis of abortion dynamics such simplification is not viable because the differences in age-specific distribution of fertility are ignored.

The other method of estimating TAR, which is more accurate and more appropriate for countries with high abortion levels, is based on the assumption that the hypothetical age-specific distribution of abortion number is proportional to the age-specific number of averted births. This procedure provides an estimate of ASAR so that a researcher may use the wide spectrum of modern analytical methods. The calculation is carried out by comparing any standard natural fertility model and observed marital fertility in the ratio of age-specific fertility rate (ASFR) to proportion married. Marital fertility in the age group 15–19 was taken to be equal to 75 percent of marital fertility in the group 20–24 (Bongaarts and Kirmeyer, 1980).

The method for the estimation of ASAR is given in the following equation:

$$a(x) = \frac{m(x)[n(x) - mf(x)]A}{\sum_{x=15}^{49} \{W(x)m(x)[n(x) - mf(x)]\}}, \tag{8.1}$$

where $a(x)$ is the age-specific abortion rate, $n(x)$ is the age-specific natural fertility rate, $mf(x)$ is the age-specific marital fertility rate (ASMFR), $m(x)$ is the proportion of currently married women of age x, $W(x)$ is the number of women of age x, and A is the annual number of induced abortions.

The calculations showed that TAR changed slightly from 1979 to 1986, it significantly decreased in 1987, and in 1988 it returned almost to the initial level.

The estimation of ASAR for five-year groups was carried out using this procedure, and then the standardized abortion rate was calculated for the period from 1979 to 1989. The estimated ASAR for 1979 was used as a standard, and the numbers of women in 1979, 1987, and 1989 were taken from state statistics. The numbers of women in other years were interpolated.

Abortion and contraception are two types of family-planning methods. It is very important to estimate the impact of each one on the formation of fertility level. The model of intermediate or proximate determinants of fertility (PDF), suggested by Bongaarts (1978), is a good instrument for this.

Equation (8.2) generalizes the essential relationships of this model:

$$\text{TFR} = \text{TF}C_m C_c C_a C_i \ , \tag{8.2}$$

where TFR is the total fertility rate; C_m is the index of the proportion of women who are married; C_c is the index of contraception; C_a is the index of abortion; C_i is the index of postpartum infecundability; and TF is the total fecundity, i.e., an average number of children who could be borne by a woman in the absence of such inhibitors as marriage, contraception, abortion, and lactation, which has an inhibitory effect on ovulation. The values of indexes which estimate the impact of proximate determinants on fertility vary from one (the factor is absent) to zero (the factor blocks fertility completely).

In assessing the values of proximate determinants of fertility the main problem is determining the value of the total abortion rate (TAR) and the contraceptive prevalence (u), equal to the proportion of currently married women of reproductive age (MWRA) using contraception.

There are no data on contraceptive prevalence in Soviet state statistics. The only information available was on the absolute number of registered abortions and on the manufacturing and distribution of some kinds of contraceptives through the medical institutions and commercial network, as well as the published data on fertility, nuptiality, and population age structure. The variable u was found from the regression equation describing the relationship between u and crude birth rate (CBR):

$$\text{CBR} = 46.9 - 42.0u \quad R^2 = 0.91$$

or

$$u = \frac{(46.9 - \text{CBR})}{42.0} \ . \tag{8.3}$$

The parameters of this equation were estimated by Nortman for the populations of 32 developing countries where the prevalence of induced abortion was minimal (Nortman and Hofstatter, 1980). In this case the term u should be estimated under a very high level of induced abortions. Thus, to use equation (8.3) for analysis it is necessary to calculate the corrected CBR resulting from the hypothesis that all the births that were estimated to have been averted that year by induced abortions had occurred. It is known that an induced abortion always averts less than one birth, because the maximum duration of pregnancy in this case is equal to three months, in general, and ovulation is resumed after one or two months – much sooner than after giving birth. In the absence of contraception, an induced abortion averts about 0.4 births, while about 0.8 births are averted when moderately effective contraception is used. The births averted per induced abortion, b, may be estimated with the equation:

$$b = 0.4(1 + u) \ . \tag{8.4}$$

To be exact, u should equal the proportion of women protected by contraception who have had an induced abortion. Since this information is usually unavailable, it would not be a gross error to use the proportion of MWRA currently using contraception as u (Bongaarts, 1978). The corrected CBR can be calculated using the formula

$$\text{CBR}^* = \text{CBR} + 0.4(1 + u)\text{CAR} \ , \tag{8.5}$$

where CBR^* is the crude birth rate corrected for induced abortions and CAR is the crude abortion rate equal to the number of induced abortions per 1,000 people per year. Substituting equations (8.4) and (8.5) in equation (8.3) yields the following formula for estimating the value of u associated with a given level of abortions:

$$u = \frac{46.9 - (\text{CBR} + 0.4\text{CAR})}{(42.0 + 0.4\text{CAR})} \ . \tag{8.6}$$

In 1979 the observed CBR was 18.2. This corresponds with the theoretical contraceptive prevalence of 68 percent. The adjusted CBR was equal to about 33–35, which agrees with the real contraceptive prevalence of about 34 percent.

The decrease of postpartum infecundability must lead to fertility increase or to contraceptive prevalence growth or to abortion-level rise. For this reason the estimate of the postpartum infecundability index (C_i) for the USSR and the republics is a very important but very difficult task, because the data on breast-feeding duration and postpartum sterility are absent in the Soviet statistics.

According to the data of the sample survey carried out in 1985 and 1986 by the gynecologist Yun for the random sample of WRA in Moscow and Uzbekistan (Katkova *et al.*, 1989), the average duration of lactation after the first delivery was 6.8 months in Moscow and 13 months in rural Uzbekistan. After the second delivery this decreased to 6.3 months in Moscow and it increased almost to 15 months in Uzbekistan. The mean duration of postpartum infecundability, i, in Moscow was 4.2 months after the first delivery and 3.1 months after the second one. In Uzbekistan the figures were 7.7 months after the first delivery, 9.4 after the second one, 6.5 after the third delivery, and 10.4 after the fourth.

It was thought that these data corresponded with two limiting points of the regression line describing the linear relationship between CBR and i. The crude estimation of this relationship is

$$i = 0.202\text{CBR} + 0.372 \ . \tag{8.7}$$

We can now obtain the set of values of i for all the republics and then calculate C_i using the PDF model:

$$C_i = \frac{20}{(18.5 + i)} \ . \tag{8.8}$$

The calculation of the proportion of married women is not difficult, as there are complete marital statistics in the USSR. The only difficulty is determining the proportion of those living in unions (married or marriage-like) in age group 15–19. The marital fertility rate for teenagers will be unrealistically high if it is calculated by dividing the ASFR by the proportion of people married obtained from official marital statistics. There are many incidents of premarital and extramarital conceptions in this age group, so if age when married and age of first birth are compared, both contained in official statistics, unnaturally short intervals between marriage and first birth are observed. In reality there are many more married teenagers or teenagers in sexual relationships than are registered in the statistics. Assuming that the marital fertility rate in this group is 75 percent of that of the 20–24-year-old age group, the proportion married before age 20 may be corrected as follows:

$$m^*(15\text{–}19) = \frac{m(20\text{–}24)f(15\text{–}19)}{0.75f(20\text{–}24)} \ , \tag{8.9}$$

Table 8.1. Proximate determinants of fertility in the USSR and the
republics in ascending order of total fertility rate, 1985 to 1986.

	TFR	C_i	C_m	C_c	C_a	TF	TAR
USSR	2.46	0.87	0.49	0.65	0.55	16.14	3.65
Ukrainia	2.07	0.91	0.55	0.49	0.51	16.53	3.30
Byelorussia	2.08	0.91	0.49	0.49	0.59	16.40	2.42
Latvia	2.09	0.91	0.54	0.48	0.54	16.50	2.93
Estonia	2.11	0.91	0.56	0.48	0.52	16.52	3.24
Russia	2.11	0.89	0.49	0.62	0.47	16.39	4.62
Lithuania	2.13	0.90	0.53	0.40	0.69	16.44	1.49
Georgia	2.36	0.88	0.50	0.50	0.66	16.26	2.06
Armenia	2.54	0.84	0.42	0.62	0.74	15.72	1.64
Moldavia	2.75	0.86	0.49	0.69	0.49	15.92	3.63
Azerbaijan	2.91	0.82	0.43	0.67	0.81	15.29	1.33
Kazakhstan	3.08	0.83	0.49	0.74	0.74	15.55	3.07
Kirghizia	4.17	0.79	0.55	0.89	0.74	14.60	3.26
Uzbekistan	4.68	0.75	0.56	0.94	0.83	13.77	2.18
Turkmenistan	4.72	0.76	0.59	0.88	0.87	13.92	1.64
Tajikistan	5.60	0.73	0.67	1.00	0.88	13.02	1.79

where $m^*(15\text{--}19)$ is the adjusted proportion of married people in the age group 15–19, $m(20\text{--}24)$ is the proportion of married people in the age group 20–24, and $f(15\text{--}19)$ and $f(20\text{--}24)$ are age-specific fertility rates.

Table 8.1 shows the estimated values of PDF for the USSR with the republics arranged in ascending order of TFR. An increase in the index of the proportion of those who are married (C_m) is observed. This signifies the decrease of the impact of this factor on fertility. The abortion index (C_a) is positively associated with TFR, as is the change in contraception index (C_c). Contraception appears to prevail over abortion at some stage of demographic transition.

The logarithmic transformation of equation (8.2) gives a better result for the proximate determinant:

$$\frac{\ln(C_m) + \ln(C_c) + \ln(C_a) + \ln(C_i)}{\ln(C_m C_c C_a C_i)} = 1 , \tag{8.10}$$

where 1 or 100 percent is the sum of the relative impacts of all proximate determinants on TFR. It is simple to determine the values of relative weights for every proximate determinant.

As *Table 8.2* shows, the relative occurrence of abortion is also significant in republics which joined the USSR in 1940. The liberalization

Table 8.2. Relative impact of proximate determinants on fertility in the USSR and the republics, 1985–1986.

	TFR	$C_i(\%)$	$C_m(\%)$	$C_c(\%)$	$C_a(\%)$	TF	TAR
USSR	2.46	7.40	37.92	22.90	31.78	16.14	3.65
Ukrainia	2.07	4.54	28.76	34.31	32.39	16.53	3.30
Byelorussia	2.08	4.60	34.82	34.82	25.76	16.40	2.42
Latvia	2.09	4.58	29.90	35.62	29.90	16.50	2.93
Estonia	2.11	5.65	34.58	23.17	36.60	16.52	3.24
Russia	2.11	4.57	28.12	35.59	31.71	16.39	4.62
Lithuania	2.13	5.20	31.31	45.19	18.30	16.44	1.49
Georgia	2.36	6.62	35.92	35.92	21.53	16.26	2.06
Armenia	2.54	9.57	47.64	26.25	16.54	15.72	1.64
Moldavia	2.75	7.74	36.61	19.04	36.61	15.92	3.63
Azerbaijan	2.91	12.00	51.04	24.22	12.74	15.29	1.33
Kazakhstan	3.08	12.41	47.50	20.05	20.05	15.55	3.07
Kirghizia	4.17	18.84	47.78	9.31	24.07	14.60	3.26
Uzbekistan	4.68	25.78	51.97	5.55	16.70	13.77	2.18
Turkmenistan	4.72	25.67	49.35	11.96	13.03	13.92	1.64
Tajikistan	5.60	37.15	47.28	0.47	15.09	13.02	1.79

of the abortion law in 1955 was the first for these republics in which the fertility level was already quite low. The same situation occurred in some regions of Ukraine and Byelorussia. It proves that abortion culture can not only begin and develop spontaneously, but penetrate quickly into other countries under similar conditions. If the influences of natural fertility (index C_i) and exposure (index C_m) on fertility are excluded, the relative impact of abortion and contraception on birth control can be determined.

On ranging the republics in descending order of abortion impact on fertility (*Table 8.3*), the Asian republics occupy the initial places. Thus, the belief that Islamic cultures do not accept abortion as a family-planning method is erroneous. Probably, the expansion of abortion culture to this region took place as a consequence of fertility decline.

It is difficult to explain the very low relative impact of abortion on fertility in the Transcaucasian republics. It seems that many abortions in this region are not reported. In the European republics the relative impact of contraception on fertility is greater than that of abortion. It is supposed that an abortion culture is some intermediate stage in family-planning development which would then be replaced by a contraceptive

Table 8.3. Relative impact of abortion and contraceptives in the USSR
and the republics, 1985–1986 (total impact equals 100 percent).

	TFR	C_c(%)	C_a(%)	TF	TAR
USSR	2.46	41.88	58.12	16.14	3.65
Tajikistan	5.60	3.04	96.96	13.02	1.79
Uzbekistan	4.68	24.93	75.07	13.77	2.18
Kirghizia	4.17	27.90	72.10	14.60	3.26
Moldavia	2.75	34.22	65.78	15.92	3.63
Russia	2.11	38.77	61.23	16.39	4.62
Turkmenistan	4.72	47.86	52.14	13.92	1.64
Kazakhstan	3.08	50.00	50.00	15.55	3.07
Ukrainia	2.07	51.44	48.56	16.53	3.30
Estonia	2.11	52.88	47.12	16.52	3.24
Latvia	2.09	54.36	45.64	16.50	2.93
Byelorussia	2.08	57.48	42.52	16.40	2.42
Armenia	2.54	61.35	38.65	15.72	1.64
Georgia	2.36	62.52	37.48	16.26	2.06
Azerbaijan	2.91	65.52	34.48	15.29	1.33
Lithuania	2.13	71.18	28.82	16.44	1.49

culture. Special programs and laws regarding birth control will only
enhance or inhibit natural processes. As is shown in *Table 8.3*, in the
mid-1980s abortions had a stronger influence on fertility in the republics
with high fertility than contraception. However, if fertility declines,
abortion is partially replaced by contraception.

8.3 Factors Affecting Abortion and Contraceptive Prevalence Dynamics in the 1980s

From the early 1980s the Ministry of Health enforced family-planning
measures to reduce abortions. In 1980 about 2 million IUDs were man-
ufactured in the USSR, and in 1987 the figure had reached 4.8 million.
In addition to this IUDs were imported from Yugoslavia from 1983, and
in 1987 1 million IUDs were imported from Finland. In 1988 the manufacture of Soviet IUDs commenced. The import of hormonal pills from
Eastern European countries was slowly growing and reached 6.5 million
packets in 1989. So the proportion of married women of reproductive

age (MWRA) who had access to this kind of contraception was about 2 percent. In 1990 the number imported increased abruptly and the proportion of MWRA who had access to pills reached about 10 percent. From 1986 to 1990 condom manufacture increased from 195 to 350 million. In addition to this, 657 million condoms were imported in 1990.

It may be expected that such expansion of contraceptive availability would change the abortion situation. In reality, during the first half of the decade the number of abortions decreased slightly, but in 1985 and in 1986 it exceeded 7 million again. An abrupt decrease occurred in 1988, when the published annual number of abortions was equal to 6.088 million. This could be taken as a radical change in the trend, because the annual number of abortions was reduced by about 1 million. However, it was discovered to be a statistical trick: 1.46 million vacuum-aspirations (terminations of pregnancy up to five weeks) had not been included in abortion statistics. Thus, in 1988 the number of abortions reached a record value for the decade: 7.528 million.

Why did such active efforts from the Ministry of Health not have the expected results and why, in the early 1980s, was a decrease in abortion rate observed? The analysis of the contraceptive-use pattern and fertility dynamics answers these questions.

The government's program of family support which began in the USSR in November 1981 helped to increase fertility. The rise of CBR had occurred mainly due to numerous cohorts born between 1960 and 1965 entering childbearing age. In addition, an increase in TFR was induced by a synchronization of reproductive activity of different women's generations. This influenced the dynamics of abortion indicators.

If the abortion level decline was a consequence of fertility-level increase, it should begin approximately nine months earlier. In 1981, the year before the fertility leap, observed abortion rate deviated from the standardized one (see *Table 8.4*). The maximal increase in TFR occurred in 1987, but in 1986 the observed abortion rate began to rise and in 1988 it came near the standardized one.

This situation was enhanced by women aged 25–34, because this age group was characterized by the greatest number of abortions. Women who were 25–29 years old in 1982–1988 caused a significant growth in age-specific fertility rates. However, when they were 20–24 and 15–19 years old, the situation was the opposite: they had fertility rates lower

Table 8.4. Abortion rate in the USSR, 1979–1988.

Year	Number of Abortions	Abortion rate Standardized	Observed	GFR	TAR	Contraceptive prevalence
1979	7,009,000	102.5	102.5	70.3	3.65	34.8
1980	7,003,000	103.2	102.3	70.9	3.70	34.7
1981	6,834,000	103.8	99.6	72.3	3.56	33.7
1982	6,921,000	104.6	100.6	74.1	3.56	32.1
1983	6,765,000	105.3	97.8	78.0	3.43	31.3
1984	6,780,000	106.1	97.5	77.4	3.38	31.1
1985	7,034,000	107.2	101.3	77.4	3.65	30.2
1986	7,116,000	108.4	102.5	80.8	3.65	29.0
1987	6,818,000	108.6	97.2	79.6	3.11	29.1
1988	7,528,000[a]	110.2	107.4	76.8	3.58	31.4

[a]Including 1,460,000 abortions performed by vacuum-aspiration.
Abortion rate – number of abortions per 1,000 WRA; GFR (general fertility rate) – births per 1,000 WRA; TAR (total abortion rate) – number of abortions per woman, calculated as sum of age-specific abortion rates (analogous to TFR); Contraceptive prevalence – proportion of currently married women ages 15–49 currently using contraception.
Sources: State Committee of the USSR on Statistics (Goskomstat), 1988, 1989; *Vestnik statistiki*, 1990.

than the average. So, in 1982–1988 they had caught up. The same is true, but to a lesser extent, for women who were 30–34 years old in 1982–1988.

In 1988–1989 the same cohort, characterized by its high fertility in the previous decade, became the most likely group to receive abortions. Thus we can expect the regeneration of the abortion growth trend.

The calculations conducted on the basis of the age-specific version of the PDF model (Bongaarts and Kirmeyer, 1980) showed interesting results concerning contraceptive prevalence. In 1987 it decreased to 29 percent, compared with 35 percent in 1979. Thus, the growth in fertility had induced not only a decrease in the number of abortions, but also a decrease in contraceptive prevalence.

What happened to the millions of IUDs? According to Ministry of Health data, the annual number of IUDs inserted was 1.11 million in 1982, 1.436 million in 1984, 2.929 million in 1987, and 3.363 million in 1988. We had tried to restore the contraceptive mix in the USSR

Table 8.5. Model estimation of contraceptive prevalence in the USSR (share of MWRA currently using contraception, in percent).

	1979	1980	1981	1982	1983	1984	1985	1986	1987	1988
IUD	1.48	3.09	4.58	5.86	7.16	8.57	9.73	10.69	12.09	13.14
Pill	0.50	0.57	0.62	0.65	0.69	0.75	0.79	0.81	0.87	1.40
Condom	2.00	2.29	2.51	2.66	2.85	3.10	3.27	3.38	3.64	4.07
Vaginal spermicide	0.79	0.80	0.79	0.76	0.75	0.75	0.74	0.72	0.73	0.80
Other	30.00	27.94	25.25	22.22	19.83	17.95	15.70	13.41	11.80	12.01
All	34.77	34.69	33.75	32.15	31.28	31.12	30.23	29.01	29.13	31.42

in 1979–1988, using the data on the number of inserted IUDs, manufacturing and import of the other contraceptives, and the estimates of contraceptive prevalence.

The data in *Table 8.5* show that from 1979 to 1988 modern contraceptive methods competed with traditional methods, not with abortions. The proportion of IUDs in the contraceptive mix increased by a factor of about 10, but this led to a decrease in the share of the traditional methods (coitus interruptus, rhythm, etc.), not to increased contraceptive prevalence. In 1979 traditional methods made up over 80 percent of all methods. In 1987 their share decreased to 40 percent, and the proportion of IUD, in contrast, increased from 4.3 percent in 1979 to 41.5 percent in 1987. Change in the contraceptive mix structure resulted in a growth of average effectiveness but did not affect the abortion level.

8.4   Prospects for Family Planning in the 1990s

The threat of AIDS will drastically affect the method of family planning in the 1990s. The situation where over 10 percent of women of reproductive age are forced into having a surgical operation is associated with a high risk of AIDS infection. Therefore, the prevention of unwanted pregnancy by using modern prophylactic contraception should be considered as an important element for the prevention of AIDS. This will require a significant increase in expenditure.

In the current structure of the contraceptive mix, established in the early 1990s, the traditional methods still occupy a significant position. Therefore, their replacement by modern methods will continue.

This process will undoubtedly exert an inhibiting effect on contraceptive prevalence growth and on abortion-level decrease.

In the 1980s the use of IUDs was very prevalent in the USSR. Therefore, in the next decade most of the newly inserted IUDs will replace the discontinuation of IUD use caused by reasons such as termination of contraceptive effect (three to four years), expulsions, inflammatory effects and other complications, and, finally aging of IUD users past 49. To preserve IUD prevalence, it will be necessary to insert annually more than 2.5 million IUDs, which includes compensation for discontinuations, aging past 49, and the growth of the absolute number of WRA in the population. The number of terminations of IUD use (for different reasons) will increase from 1.1 million in 1990 to 2.2 million.

Taking into account the high prevalence of the IUD use, the contraceptive priority will probably change to hormonal contraception, including not only pills but also implants and injections currently not allowed in the USSR. If injections and implants are not made available, then to increase the proportion of pills up to 12 percent (20–22 percent in the contraceptive mix) it will be necessary to increase their annual sales to 81 million packets by the year 2000. Although in 1990 the import of pills increased sharply, it still did not cover half the potential demand.

Another problem will be contraception for women aged 35 and older who generally already have their desired number of children. The number of women in this age group will increase by 10 million in 1999. They will make up more than half the MWRA using contraception. The solution will be more simple than the hormonal contraception one – if sterilization for contraceptive purposes is allowed at the wish of the spouses. The fear of AIDS infection will result in a growth in the use of condoms as a family-planning method. To give a proportion of up to 10–12 percent (20 percent in the contraceptive mix), their annual sales should be about 1 billion units. According to Ministry of Health data, this occurred in 1990.

Vaginal spermicides, although less popular, are also used. If their share in the contraceptive mix remains at the level of 2.5–3 percent, it means an increase in their annual consumption by a factor of two: 10–11 million packets.

To reach the goal of decreasing the total abortion rate to 1.1, the contraceptive prevalence must increase by at least a factor of two and reach 56–60 percent, or the number of married women of reproductive age currently using contraception should increase from 13.5 million in 1987 to 28.5 million in 2000. The achievement of these goals will depend on the policy and efforts of the government.

References

Bongaarts, J., 1978. A Framework for Analyzing the Proximate Determinants of Fertility. *Population and Development Review* 4:105–132.

Bongaarts, J., and S. Kirmeyer, 1980. Estimating the Impact of Contraceptive Prevalence on Fertility: Aggregate and Age-Specific Versions of a Model. In *The Role of Surveys in Analysis of Family-Planning Programs*. Ordina, Liège.

Katkova, I., I. Manouilova, and A. Avdeev, 1989. Le comportement procreateur et la santé des femmes et des enfants en URSS. In G. Calot, A.G. Vishnevsky, and L.L. Rybakovskiy, eds., *Natalité et famille*. Editions du Progrès, Moscow.

Nortman, D., and E. Hofstatter, 1980. *Population and Family Planning Programs: Data Through 1978*. The Population Council, New York, NY.

State Committee of the USSR on Statistics (Goskomstat), 1988. *Naseleniye SSSR 1987: Statisticheskiy sbornik* (Population of the USSR 1987: Statistical Collection). Finansy i statistika, Moscow.

State Committee of the USSR on Statistics (Goskomstat), 1989. *Naseleniye SSSR 1988: Statisticheskiy ezhegodnik* (Population of the USSR 1988: Statistical Yearbook). Finansy i statistika, Moscow.

Verbenko, A.A., S.E. Ilyina, V.N. Chusovaya, and T.N. Alshevskaya, 1968. *Aborty i protivozachatochniye sredstva* (Abortions and Contraception). Meditsina, Moscow.

Vestnik statistiki (Herald of Statistics), 1990 (1):47.

Part II

Family Dynamics and Changing Attitudes

Part II

Family Dynamics and Changing Attitudes

Chapter 9

Family and Household Changes in the USSR: A Demographic Approach

Andrei Volkov

The family is an integral element of the social structure of any society. Changing in the course of social development, marital and family relations have a significant impact on population reproduction. People are members of families and the population is an aggregate of families, having a certain structure and changing with certain regularity. Factors that determine the rates of population growth include the type of family, time of formation, stability, and number of children. Thus the development of families could be regarded as an important component of population reproduction.

Family development in a demographic sense may be regarded as those changes in its composition and vital activity that have direct significance for population reproduction – the time of family formation, the number and the timing of births, and stability of the family. The process of family aggregation and disaggregation in the course of the family life cycle is also part of family development.

Unfortunately the influence of various demographic processes upon the changes in the family structure has been insufficiently studied. Traditionally population is regarded in demographic studies as a universe

149

of individuals – individuals who differ by age and sex and sometimes by marital status. The fact that most individuals are part of a family of different types and sizes has often been ignored in demographic analysis; this is mainly due to lack of information. Nevertheless even the investigation of common demographic tendencies may give us some insights into the formation of family structure and its changes.

One can see many common features in demographic changes of the family in developed countries during the last decades. Such phenomena as the decrease in age at marriage and fertility level, change in birth intervals, and change in the process of leaving parental families are common to many countries. At the same time each country has its own historical peculiarities. This paper attempts to highlight these peculiarities using recent demographic studies of the USSR.

A specific feature of the USSR is a great diversity of the demographic situation in its regions. If the demographic transition in the European part of the country is regarded as completed and republics like Azerbaijan and Moldavia are considered to be in the middle of transition, then the Central Asian republics are only at the first stage of demographic transition. All this predetermines the peculiarities of demographic tendencies as well as the evolution of the family in various parts of the country.

9.1   Nuptiality Trends

A family usually begins with marriage, and the appearance of a new family is usually connected with the formation of a married couple. So this chapter begins with an analysis of the characteristics of modern nuptial tendencies in the USSR.

The marriage legislation of the republics requires the mutual consent of each partner (parents' consent is not required). In most republics the law states that the minimum age to marry is 18 for both men and women. In the Ukraine and Uzbekistan, however, women may marry at 17. In exceptional cases the minimum age may be lowered for women but by not more than a year (in Russia not more than two years).

Not all people marry upon reaching the legal marriage age. In each generation some people marry earlier, others later, and some do not marry in the course of their whole life.

The never-married category is almost nonexistent in the USSR. According to nuptiality tables of 1980–1984, calculated by Darsky and Ilyina (1990), by age 50 only 3.2 percent of women and 2.1 percent of men in a cohort have never been married. This proportion should be regarded as moderate. Very early marriages are also relatively rare. By the same nuptiality tables 4.4 percent of women and 0.6 percent of men marry before age 18. By age 25, 66 percent of men and 79 percent of women are married; by age 30, 88 percent of men and 91 percent of women are married.

During the postwar years the age at first marriage of both men and women significantly decreased. In the 1930–1934 birth cohort, 19.1 percent of the women married before age 20; in the 1960–1964 birth cohort, 34.0 percent of women married by that age.

The calculations show that for cohorts born in 1942–1952 the median age when first married decreased between these cohorts from age 25.1 to 23.4 for men and from 22.4 to 21.5 for women (Volkov, 1986). Such decreases in only 10 years may be regarded as significant. The decline in men's age at marriage takes place for all ages up to 30, and for women it is a consequence of a sharp increase of young marriages. This tendency should not be considered favorable: it means that some people marry and stay under the supervision of their parents, which does not contribute to marriage stability.

Nuptiality trends in various nationalities in the Soviet Union are different. In the Central Asian republics, where early marriages are particularly common, women of the younger generations are now marrying at a later age. The change is connected with the increase in the educational level of women and the wearing down of the tradition of early marriages. Among populations in the European republics, where the proportion of early marriages was relatively low, there is a distinct increase in the share of women married at an early age. Thus, the proportion of women experiencing their first marriage by the age of 20 born between 1937 and 1941 varied from 14.4 percent in the Estonian population to 55.4 percent in the Kirghiz population. Of those born between 1957 and 1961, 22.0 percent of Estonian women and 33.9 percent of Kirghiz women were married by age 20. Thus in spite of significant differences in nuptiality trends and patterns still existing among particular populations, starting in the 1950s, a tendency toward a decrease of the difference in age at first marriage for various populations of the USSR was evident.

The findings of Darsky and Ilyina (1988) show that in recent years the tendency toward a younger age when married has stopped.

For the whole of the USSR, the proportion married at younger ages in recent years was increasing, and now it remains relatively high. This factor will contribute to further growth in the number of families.

9.2   Marriage Dissolution

A relatively high divorce rate is a distinctive feature of the family development in the USSR. In the republics a marriage may be dissolved on the basis of joint application from the spouses made at the civil registrar's office; however, if the couple has children under age 18 or has a dispute about property, then a decree by the People's Court is necessary for marriage dissolution. The civil registrar's office and the court must try to reconcile the couple. A marriage cannot be dissolved when the wife is pregnant or if the couple has a child under age one.

In the course of the last three decades, the annual number of divorces per 1,000 married couples (excluding the cases of actual but not registered marriage dissolution) was increasing, but in recent years the rate tends to stabilize. In 1958–1959, for every 1,000 marriages, 5.3 ended in divorce; in 1969–1970, 11.5; in 1985–1986 and 1988–1989, 14.1 ended in divorce. A sharp rise in the divorce rate in the late 1960s is explained by simplifications in the legal procedure to divorce.

According to surveys, relatively low stability is a common feature of marriages contracted for the first time by women under 20 and 25 and over. The frequency of divorces, according to Bondarskaya (1983), is the highest during the first five years of marriage.

Differences in divorce frequency exist according to territory and sociodemographic groups of population. Among Latvian women married between 1970 and 1974, 27.1 percent divorced within 10 years of marriage; among Russians, 20.1 percent; among Estonians, 20.0 percent; among Turkmens, 3.9 percent; among Azerbaijanis, 4.4 percent; and among the Tajiks, 4.6 percent. Marriages are dissolved more frequently in urban areas than in rural areas.

The analysis of divorce dynamics by marriage cohorts considering marriage duration shows that the rate of divorces was increasing up to

Table 9.1. Divorces in marriage cohorts in the USSR, cumulated rates, per 1,000 marriages.

Year of marriage	By end of 1st year	By end of 3rd year	By end of 5th year	By end of 10th year
1960	4.0	22.4	42.6	116.2
1965	5.5	43.3	86.5	176.7
1970	11.5	54.8	106.8	210.2
1975	13.2	63.1	122.3	220.8
1980	12.3	66.7	127.8	–
1984	10.8	64.3	121.6	–
1985	10.9	64.0	–	–
1986	11.6	67.4	–	–
1987	13.4	–	–	–
1988	15.4	–	–	–

Sources: For 1960–1970, estimate made by Tolchinsky (1979); for other years, estimates made using Tolchinsky's method.

the cohort of 1980 (see *Table 9.1*). Starting with the 1980 cohort one can see a decrease in divorce rates during the first three years of marriage, and there is evidence that the rate stabilizes during the next few years of marriage. However, it is too early to determine a fundamental change in divorce tendencies.

The other cause of marriage dissolution is the death of a spouse, mainly the husband. Widowhood is a more common cause of marriage dissolution than divorce. Estimates can be obtained using a decrement table which considers marriage dissolution frequencies due to a spouse's death and divorce. According to the marriage dissolution table calculated on the basis of Maison's method (Maison, 1974), in 1988–1989 the expectancy of married life in the USSR for a man married at 24 and woman married at 22 (median age at first marriage) was 30.8 years. Ultimately, 45.2 percent of marriages ended by the death of husband; 25.9 percent, by death of wife; and 28.9 percent, by divorce (see *Table 9.2*).

In late 19th-century Russia, there were practically no divorces. Married life expectancy was 28.5 years. Since then the decrease in mortality has increased marriage duration by 12 years, but the increase in divorce has shortened it by 10 years. Therefore, the total rise in the expected duration of marriage is only two years.

Table 9.2. General characteristics of hypothetical marriage cohorts: men married at age 24; women married at age 22.

Characteristic	1896–1897	1988–1989		Increase from 1896–1897 to 1988–1989	
		Divorces excluded	Divorces included	Divorces excluded	Divorces included
Life expectancy (years)					
Men at age 24	37.86	45.22		+ 7.36	
Women at age 22	39.60	51.87		+12.27	
Married life expectancy (years)	28.54	40.44	30.83	+11.90	+ 2.29
Percent of marriages in cohort ended by					
Death of husband	52.9	63.4	45.2	+10.5	– 7.7
Death of wife	47.1	36.6	25.9	–10.5	–21.2
Divorce	–	–	28.9	–	+28.9

In the course of the first 25 years of married life, when a woman is still in the reproductive age, 40 percent of marriages are dissolved: 8 percent because of the death of a husband, 3 percent because of the death of a wife, and 29 percent because of divorce. The expected duration of married life within a woman's reproductive period is decreased by more than five years. Nevertheless, with widespread fertility regulation, marriage dissolution does not significantly influence the number of children of those women who remarried after divorce or widowhood, in comparison with those whose marriage was not dissolved. But, when divorced and widowed women remain unmarried, marriage dissolution results in an irretrievable loss of population reproduction. From the point of view of family development marriage dissolution also leads to an increase in one-parent families.

Chances to marry for those who do this for the first time and for those whose marriage was terminated are quite different for men and women and for various age groups. For widowed and divorced men between ages 20 and 55, the frequency of a remarriage exceeds the frequency of a first marriage in the same age group (this is partly because those remarrying in this age group were younger at the first marriage). It is quite different for women. The probability of marrying for divorced and widowed women between 20 and 40, and in recent years in all ages, is much lower than for those marrying for the first time.

Demographic determinants of remarriages are different. Belova (1983) investigated remarriages using *ad hoc* data. On the basis of nuptiality tables for divorced and widowed women of a hypothetic cohort of 1969–1978, she found that divorced women remarry sooner after the dissolution and more often than widowed women. The probability of remarriage for women whose marriage was dissolved before 25 is 20 percent to 30 percent higher than for women whose marriage was dissolved at ages 25 to 29, and 100 percent higher than for those who dissolved their first marriage at 30 or older.

Belova and Moreva (1988) studied the impact of children on remarriages. They found that within 10 years after the first marriage dissolution about 50 percent of the childless women and only 25 percent of the women with two or more children from their first marriage remarried. After 20 years of the first marriage dissolution, these numbers are 60 percent and 34 percent, respectively. On average, within 20 years after the dissolution of the first marriage less than half of the women remarry: with 44.4 percent of those divorced and only 31.8 percent of those widowed.

9.3 Fertility Trends

The basic tendency in the evolution of fertility in the USSR during the last 70 years was the increase in the prevalence of the small families. This tendency was interrupted by short-term increases in fertility levels in the late 1930s, partly as a consequence of the legal prohibition of abortions, and during several years after World War II.

However, transition to a small family was not simultaneous in all regions and all ethnic and social groups of the population. Temporal discrepancy in the transition is connected to peculiarities in socioeconomic development and sociocultural traditions in different regions.

At the beginning of this century, the birth rate in Russia among the Orthodox population did not differ from the indigenous population in the Central Asian republics. However, in the European part of the country the transition to the small family was determined by radical, sometimes severe, destruction of socioeconomic conditions and by changes in the social functions of the family and in its role in the system of social values. Social changes had less influence on the family and the traditional lifestyle in Central Asia and, hence, less influence on demographic behavior. The relatively slow change in this behavior is partly explained by the fact that a family with a large number of children determined by cultural traditions in this region now is regarded as a national peculiarity worth preserving. Meanwhile, Belova *et al.* (1988) found that the fertility level of indigenous people in Central Asia had begun to decline in urban populations and in rural populations.

Thus in spite of significant differences in fertility level, by the late 1970s the common tendency of decreased fertility began to be noticed in all regions. More and more couples started to determine and control the number of children in their family and the time of birth. This is illustrated by the results of a 1985 sociodemographic survey (see *Table 9.3*). The table contains the data for female birth cohorts on the number of births to married women beyond childbearing ages and on total number of children (already born and still expected) for cohorts who are of reproductive age. There is an obvious tendency toward stabilization of fertility indexes in urban regions as well as in rural areas, though the

Table 9.3. Number of children born and expected by birth cohorts in the USSR in 1985 sample survey, per 100 married women.

Cohorts year of birth	Age at the survey date	Number of children								
		Total			Urban			Rural		
		Born	Expected[a]	Total[b]	Born	Expected[a]	Total[b]	Born	Expected[a]	Total[b]
1925–1929	55–59	227	–	227	186	–	186	292	–	292
1930–1934	50–54	246	–	246	193	–	193	328	–	328
1935–1939	45–49	236	–	236	189	–	189	324	–	324
1940–1944	40–44	239	8	247	193	6	199	330	11	341
1945–1949	35–39	221	18	239	188	14	202	310	29	339
1950–1954	30–34	206	40	246	178	31	209	275	62	337
1955–1959	25–29	165	82	247	146	64	210	206	119	325
1960–1964	20–24	103	146	249	94	114	208	119	202	321
1965–1966	18–19	49	210	259	48	162	210	51	275	326

[a] Expected, as reported by women interviewed.
[b] Born and expected.
Source: State Committee of the USSR on Statistics (Goskomstat), 1988.

rural cohorts expect to have more children than the urban cohorts. On average, total number of children tends to stabilize at the level of 2.5 children per married woman.

According to the 1985 sociodemographic survey, there are almost no differences between republics in the proportion of families that do not expect to have children. This proportion is extremely low (less than 1 percent, as a rule, but in Latvia it reaches 2.4 percent). More than half of the women are going to have two children in republics with established low fertility levels such as Russia, Latvia, Lithuania, and Estonia. In the Ukraine and Byelorussia more than 60 percent of the women plan to have two children. The proportion of women planning to have one or two children in these republics ranges between 76 and 83 percent.

In republics with a high level of fertility there is no pronounced trend in the preferred number of children. On the one hand, the proportion of women intending to have one child is low in these regions (from 2.3 percent in Tajikistan to 2.9 percent in Azerbaijan); on the other hand, the proportion of women planning to have four children or more is higher (from 55 percent in Kirghizia to 76 percent in Uzbekistan). The respondents in these republics were most often indigenous women, who are influenced by traditional values.

On average, the modern urban family has two children. The birth of the second child is often postponed; thus, the mean interval between the first and the second births among the Russians, Ukrainians, Byelorussians, Lithuanians, and Latvians is more than four years, whereas for the peoples of Central Asia it is less than 2.5 years (see Chapter 4).

It should be noted that, on average, every 10th child is born to women out of legal marriage. The proportion of extramarital births varies from 2.5 percent in Azerbaijan and 3.5 percent in Turkmenistan to 18 percent in Georgia and 25 percent in Estonia.

The fertility level increased in 1982–1984 as a consequence of more state support to families with children. During the first year after the law was enacted, the proportion of married women with one child who gave birth to another child increased by 12 percent; women with two children, by 14 percent; women with three children, by 10 percent.

The analysis shows that this fertility increase in the early 1980s was temporary and did not show evidence of a change in the basic tendency. One can expect the process of transition to a small family to continue. This is confirmed by the results of a survey of young couples (both

Table 9.4. Ideal and expected number of children according to the 1989 survey of young married couples.

| Number of children | Regarded as ideal by | | Expected |
	Husband	Wife	
0	3.0	2.7	4.4
1	7.3	8.0	9.6
2	51.3	53.3	51.5
3	21.6	20.4	18.5
4	7.4	7.0	7.0
5 and more[a]	9.5	8.6	8.9
Total	100.0	100.0	100.0
Mean	2.61	2.56	2.50

[a]For the calculation of mean, this group was given the weight of 6.

spouses were not older than 30) conducted in April 1989. Respondents explicitly preferred a small family: more than 80 percent of husbands and wives considered the family with no more than three children ideal (see *Table 9.4*). Nevertheless, expectations of young couples will ensure the replacement of generations if all couples in each cohort have as many children as they expect.

9.4 Separation of Young Couples from Parents

Changes in life-style create a preference not only for a small family but for a simple nuclear family. Two large surveys of young family formation and development were conducted in the USSR in 1984 and 1989. The surveys polled 45,000 to 50,000 married couples. The couples were in their first marriage, and each spouse was under 30.

More than 75 percent of the respondents who gave definite answers answered yes to the question, Do you want to separate from parents? Such an aspiration was mostly expressed by those in households with older and younger generations living together but keeping the house (completely or partly) separately. However, even among those in the same household with their parents, the proportion of those wanting to separate was significant.

Will this aspiration come true? The surveys shows a strong tendency toward earlier separation of grown-up children from parental families. According to the 1984 survey, 87 percent of men and 86 percent of women

Table 9.5. Changes in the living arrangements of young married couples according to the 1984 survey.

	Living arrangement before marriage	Arrangement after marriage (in % of column 1)		
		With parents		Without
		Own	Spouse's	parents
Men				
With parents	86.9	49.3	15.6	35.1
Without parents	13.1	3.2	17.7	79.1
Total	100.0	43.3	15.9	40.8
Women				
With parents	86.3	18.1	46.3	35.6
Without parents	13.7	1.6	24.6	73.8
Total	100.0	15.9	43.3	40.8

lived with their parents just before marriage (see *Table 9.5*). Of those, about one-third separated from parents immediately after marriage.

Using the survey data, a table based on the separation frequencies was produced, which shows the regularities in the separation of young married couples from parental families. The probability of separation is high during the first three years of married life and then slowly decreases. According to the 1984 survey, during the first 10 years of marriage, about 59 percent of the young families separate from parents and about 16 percent of the marriages are dissolved, mainly by divorce. (Estimates of divorce and widowhood frequencies were taken from the 1978–1979 marriage dissolution table.) By the 11th year, slightly more than 25 percent of the marriage cohort continues living with parents.

Similar results were obtained from the 1989 survey. Among young couples married for seven or more years, 33.9 percent lived without parents immediately after marriage and, at the time of the survey, 22.0 percent continued to live with parents and 41.1 percent separated from parents during their married life. Only 3.0 percent of couples moved back to their parents' home.

Of course, the data on separation do not characterize the process of family change completely. Particularly, they do not take into account the probability of a parent's death but the survey does consider the death of a young spouse. The survey does not investigate the process of separation depending on the number of children and the time of their birth. One cannot exclude the fact that the solution of a young family's housing problem may change the nature and the time of its separation.

According to the studies of Rouje *et al.* (1983), the process of family separation is accompanied by the appearance of so-called family groups – family unions, related in kinship, living separately, but closely connected with each other. The existence of such relations with parents living separately was confirmed by the 1984 and 1989 surveys: most young married couples living separately keep close relations with their parents.

Thus, in spite of a high degree of separation by young families there is a period in their married life, especially at the earlier stages, when they rely on support from parents, for rearing children in particular. When the parental family has several children, the older children leave the parental home first; younger children often never separate from their parents and live with them until their death.

9.5 Family Structure Dynamics

The combination of all demographic processes forms a certain structure of the population and indicates certain tendencies of its change.

In the USSR population censuses, family is defined as a group of two or more persons, related to each other by blood or marriage, who live together and have a common budget. This is similar to the definition of a household in censuses of other developed countries. The difference is that in the Soviet censuses nonrelatives are not included as part of the household (they are regarded as separate families or as living alone), and individuals living alone are not regarded as households of one person. Individuals living alone are divided into two categories: those who maintain regular financial connections with a family (such individuals are regarded as members of the family living separately) and those who do not (they are regarded as living alone).

One of the distinctive features of the USSR population family structure is a relatively large proportion of individuals living alone. The absolute number of those living alone and those regarded as the members of the family living separately is 29.4 million people or more than 10 percent of the population (1989 census). Most people living separately from their families reside in urban settlements, especially in large cities.

Table 9.6. Distribution of families by size in USSR population, in percent.

No. in family	Total population			Urban population			Rural population		
	1959	1979	1989	1959	1979	1989	1959	1979	1989
2	26.0	29.7	31.4	27.1	29.1	31.2	24.9	30.6	31.7
3	26.0	28.9	25.8	28.9	32.2	28.4	23.3	23.0	20.4
4	21.7	23.0	24.4	23.0	24.9	26.1	20.5	19.6	20.6
5	13.4	9.5	9.5	12.1	8.5	8.7	14.7	11.3	11.5
6	7.2	4.1	4.3	5.3	2.9	3.2	8.9	6.2	6.5
7+	5.7	4.8	4.6	3.6	2.4	2.4	7.7	9.3	9.3
Avg. size	3.71	3.51	3.50	3.53	3.32	3.35	3.88	3.76	3.83

In 1989, about 90 percent of the population in the country was part of a family (household). The total number of families was 73.1 million.

Essential changes took place in the distribution of families between urban and rural areas as a result of urbanization and rural–urban migration. In 1939 there were half as many families in cities as in rural areas. Today nearly 68 percent of families in the the USSR live in cities or urban-type settlements.

Currently, changes are taking place in family size (see *Table 9.6*). The number of families in the USSR is rising, but their size is decreasing. Sufficiently obvious is a general tendency toward an increase in the proportion of small families (two to four persons) and a decrease in the proportion of large families (seven or more persons). During the last decade the proportion of small families somewhat decreased; the share of large families remains practically unchanged. Due to the slight rise in fertility in the mid-1980s the proportion of medium-size families (four to five persons) increased. The proportion of two-person families also increased, which is, apparently, connected with the separation of young couples from their parents. There is a slight decrease of average family size, which is now at the level of 3.3 to 3.4 in urban regions and above 3.8 in rural areas.

Meanwhile significant ethnic-territorial differences in average family size remain unchanged. The maximum for cities took place in Tajikistan (4.74) and Turkmenistan (4.71); the minimum occurred in Estonia (3.10) and Latvia (3.09). For rural areas, the maximum is in Tajikistan (7.01) and the minimum is in Byelorussia (3.03).

As the studies show, the decrease in family size in the course of the last decades has been characterized by the following factors:

- The increase in the number of young marriages as a consequence of decreased age at marriage and general changes in population age–sex structure.
- Tendency toward young family separation.
- Tendency toward a family with a small number of children as a consequence of family planning.
- Accumulation of one-parent families in the population.

The influence of each factor is difficult to estimate quantitatively, but their total impact results in family distribution according to so-called demographic types of family (see *Table 9.7*).

At present the major (80 percent) family type in the USSR (in urban and rural areas) is a simple nuclear family, consisting of a married couple with or without children. About 20 percent of this type includes one spouse's parents and other relatives.

Families (households) with two and more married couples are not common (4.5 percent of all urban and 6.1 percent of all rural families). The percentage of such families is larger in the republics with high fertility level, where marriage at an early age is common and the tradition of an extended family remains. Relatively high is the proportion of families consisting of one parent (mainly mother) with children. Families of this type make up 12.5 percent of all urban families and 10.6 percent of all rural families.

Such family structure is typical for all regions, although the differences in demographic trends result also in family distribution by type. In *Table 9.7* family distribution by type is given not only for the entire USSR (columns 1 and 2), but also for groups of republics that are demographically similar. They differ by the proportion of extended families and one-parent families, which reflects peculiarities of marriage – family relations and reproductive behavior.

The proportion of married couples living with one or both parents is relatively small in the republics of the European USSR. But it is especially large in the republics of Transcaucasia and Central Asia, where the traditional extended family is maintained. The proportion of one-parent families is larger in Russia and the Baltic republics, where there

Table 9.7. Distribution of families by type.

Family type	% of families, USSR		Deviation from the 1989 average by groups of republics (%)					
	1979	1989	Baltic republics	Slavic republics	Moldavia	Kazakh-stan	Trans-caucasian republics	Central Asian repub-lics
Married couple[a]	66.1	65.6	1.1	1.2	8.5	-1.9	-10.2	-7.6
Married couple and spouse's parent	13.3	12.3	-1.9	-0.6	-3.5	2.5	4.2	2.8
Two or more married couples[a]	4.3	5.1	-2.6	-1.2	-1.5	0.0	7.1	9.2
One-parent family	11.8	12.2	2.7	0.5	-1.9	-1.0	-1.8	-3.6
Other types	4.5	4.8	0.7	0.1	-1.6	0.4	0.7	-0.8

[a]With or without children.

Baltic republics: Latvia, Lithuania, Estonia.

Slavic republics: Russia, Ukraine, Byelorussia.

Transcaucasian republics: Georgia, Azerbaijan, Armenia.

Central Asian republics: Kirghizia, Tajikistan, Turkmenistan, Uzbekistan.

is a high frequency of divorce and extramarital births, than in other republics. In the cities of Russia, the Ukraine, and Byelorussia this proportion is 12.6 percent and in Baltic cities it is 14.0 percent of all families; in rural areas of Central Asian republics it is only 6.9 percent.

During the last decade the distribution of families by demographic type did not change significantly. In cities and in rural areas the proportion of families consisting of two or more married couples rose, the proportion of one-parent families with children rose in cities and decreased in rural areas. For this one may conclude that, in spite of transition, a relatively stable family structure of population is forming in the USSR.

9.6   Conclusion

Demographic changes reflect a deep transformation of the society – changes in social values and priorities, changes in social psychology, and changes in life-style. Different social groups of the population are affected in different ways.

Changes of a progressive nature often conflict with shortcomings and difficulties of social development. Thus, for instance, nuptiality trends are determined by the transformation of the institution of marriage, connected with the change in woman's position in society and the relaxation of religious regulations and social morals.

The transition to low fertility was caused, on the one hand, by massive employment of women outside the home and, on the other hand, by economic difficulties and the poorly developed service sector. In many cases late separation of adult children from their parent's home is also explained by economic reasons.

Family development has been influenced by changes in demographic structures, particularly changes in the age and sex structures, caused by the famine of the 1930s, repressions, and population losses during the war. Demographic consequences of history will still have their influence on demographic processes in subsequent decades.

Thus, demographic process dynamics is a consequence of changes in the institution of family. It also has its own influence on the changing family structure. It is possible to study the causes of changes that occur. To measure their consequences or to forecast deep sociodemographic effects is beyond the framework of this paper. The underlying processes in the population family structure must be studied in more detail.

References

Belova, V.A., 1983. Povtorniie braki i rozhdaemost (Remarriages and Fertility). In *Socialno-demograficheskiye issledovaniya braka, semj'i, rozhdaemosti i reproduktuvnikh ustanovok* (Sociodemographic Studies of Marriage, Family, Fertility and Reproductive Attitudes). Armenian SSR State Planning Committee, Institute for Economics and Planning, Yerevan.

Belova, V.A., and E.M. Moreva, 1988. Povtorniie braki zhenshin: situatsiia i faktory (Women's Remarriages: Situation and Factors). In A.G. Volkov, ed. *Metodologiya demograficheskogo prognoza* (Backgrounds of Population Forecasting). Nauka, Moscow.

Belova, V.A., G.A. Bondarskaya, and L.E. Darsky, 1988. Sovzemennye problemy i perspektivy rozhdaemosti (Current Problems and the Prospects of Fertility). In A.G. Volkov, ed. *Metodologia demograficheskogo prognoza* (Backgrounds of Demographic Forecasting). Nauka, Moscow.

Bondarskaya, G.A., 1983. Analiz razvodimosti v real'nikh pokolenijakh zhenshin (Divorces in Married Cohorts of Women). In *Socialno-demograficheskie issledovaniya braka, semj'i, rozhdaemosti i reproduktuvnikh ustanovok* (Sociodemographic Studies of Marriage, Family, Fertility and Reproductive Attitudes). Armenian SSR State Planning Committee, Institute for Economics and Planning, Yerevan.

Darsky, L.E., and I.P. Ilyina, 1988. Vliyanie brachnoi structury na uroven rozhdaemosti (The Influence of Marital Structure on Fertility Level). In A.G. Volkov, ed. *Metodologiya demograficheskogo prognoza* (Backgrounds of Population Forecasting). Nauka, Moscow.

Darsky, L.E., and I.P. Ilyina, 1990. Normalizatsiya brachnosti v SSSR (Normalization of Nuptiality in the USSR). In A.G. Volkov, ed. *Demograficheskiye protsessy v SSSR* (Demographic Processes in the USSR). Nauka, Moscow.

Maison, D., 1974. Ruptures d'union par deces ou divorce. *Population* No. 2.

Rouje, V.L., I.I. Elisseyeva, and T.S. Kadibour, 1983. *Struktura i funktsii semeynich grupp* (Structure and Functions of Family Groups). Finansy i statistika, Moscow.

State Committee of the USSR on Statistics (Goskomstat), 1988. *Naseleniye SSSR, 1987: Statisticheskiy sbornik* (Population of the USSR, 1987: Statistical Collection). Finansy i statistika, Moscow.

Tolchinsky [Darsky], L.E., 1979. Ocenka urovn'ia razvodimosti v SSSR (Estimate of Divorce Level in the USSR). In A.G. Volkov, ed. *Demograficheskoie razvitie semj'i* (Family Development from Demographic Viewpoint). Statistika, Moscow.

Volkov, A.G., 1986. *Sem'ia-obiekt demografii* (The Family as the Object of Demography). Mysl, Moscow.

Chapter 10
Marital-Status Composition of the Soviet Population

Irina Ilyina

During the 20th century the marital structure of the population in the USSR was greatly influenced by socioeconomic changes caused by forced industrialization, urbanization, and the modification in rural life-styles in the early 1930s. World War I, the revolution and the years following the civil war, and World War II altered the marital structure. The marital structure of the population was affected by deformations in the sex structure of the population, the result of losses during the wars.

These deformations deprived some generations of a normal married life (Ilyina, 1977). The marital structure of the population was influenced not only by sex disproportion, but by changes in the age structure. The wife, as a rule, is younger than the husband. Therefore, the fertility decline in 1915–1920, 1933–1935, and 1941–1945 distorted the number of individuals at marriageable age (Andreev *et al.*, 1990). Increased fertility had its impact on the marital structure when cohorts born in these years reached marriageable age.

Migration flows, imbalanced by sex, played a part in the deformation of the marital structure in the USSR population. Jobs were created disregarding population structure; male and female cities appeared. In the late 1940s males predominantly migrated to cities which lessened the chances for women in rural areas to marry. The situation was reversed when intensive migration of women from rural to urban areas began in the late 1960s.

These events took place along with changes in marriage and family relations caused by changes in the society itself. Ethnographic and sociological studies show that by the early 20th century the patriarchal type of family, typical for an agrarian economy, almost disappeared (Kharchev, 1979; Zhdanko, 1990). It was replaced by a type of family with an urban life-style, in which spouses share duties and responsibilities. In the late 19th century women were mainly occupied with their household and were dependants of their father or husband. The censuses of the Soviet period registered an increase in the proportion of women employed outside the home.

Sociocultural and socioeconomic differences among nationalities of the Russian Empire predetermined the differences in the demographic situation. The demographic transition, and the family evolution connected with it, intensified population heterogeneity in the country.

Changes in legislation were significant for marital-structure trends (Nechayeva, 1980). Laws established in the first post-revolutionary decade (1917, 1920, 1926) were replaced by strict regulations of 1936 and 1944. The laws of 1956 and 1968 were more liberal. The institution of marriage changed qualitatively: marriage became more liberal; divorce became more frequent; and illegal unions and illegitimate births increased but at a much lower rate than in the West.

Of specific interest is the evolution of widowhood and widowerhood, following the increase in the length of life. With the increase in longevity the proportion of those widowed decreases. But the proportion of those widowed, as well as of those divorced, in the population may depend on the desire and possibility of remarrying and on the attitude of the society toward remarriage.

Unfortunately, the possibility of analyzing marital-status change in the population is limited. We have no continuous statistical series to follow the evolution of the marital structure. The 1937, 1939, 1959, and 1970 censuses considered only two categories of the marital status: married and unmarried. The category of unmarried was not subdivided. This limited the data on marital status. The 1897 and 1926 censuses, as well as the 1979 and 1989 censuses, considered four traditional categories of the marital status: married, never married, widowed, and divorced.

Nonetheless, data on proportion married in the censuses from 1897 to 1989 are comparable because the marital status was always determined independent of the fact of legal and illegal marriage.

10.1 Trends in Proportions Married

According to the 1897 census, in the late 19th century the population of the Russian Empire was characterized by an early age when first married and a high proportion married in each cohort. This type of nuptiality was different from the European type (Hajnal, 1965).

The evolution of the proportion married from census to census is presented in *Table 10.1*. The socioeconomic situation in 1926–1939 did not favor early marriages. Young women often worked outside the home, learned a skill, and migrated to cities. The ruling ideology did not encourage early marriages. The proportion of married women under 20 years of age was decreasing up to the 1970 census. In 1926 almost 30 percent of women were married at ages 18 and 19; in 1959 only 17 percent of women at these ages were married. The situation is analogous for men. In the mid-1960s the tendency of a decrease in the age at marriage reappeared, which is reflected in the increase of the proportion of married women under age 20 in the 1970, 1979, and 1989 censuses.

In the prewar years the proportion married decreased at various rates (excluding the age group of 40 to 44), according to comparison with the 1926, 1937, and 1939 censuses (*Figure 10.1*). In the 1940s there was a sharp decrease in the proportion of married women caused by the war. By 1970 the proportion married slightly increased. At ages under 25 and over 50 the proportion married increased steadily up to 1989. In the period between 1959 and 1989 there was a decrease in the proportion married at ages 25 to 40 and at ages 40 to 49 from 1979 until 1989. Apparently, this decrease is connected with the rise in the number of divorces not compensated by remarriages.

The chance to marry, and thus the proportion married, is closely connected with the age and sex structure of the population. At present estimates of the USSR age and sex structure are available for the period between 1926 and 1989. These estimates show that the sex ratio was not stable (*Table 10.2*). The indicators for 1946, 1956, and 1989 reflect

Table 10.1. Proportion married in the Russian Empire and the USSR, by census year, per 1,000 respondents.

Age	1897	1926	1937	1939	1959	1970	1979	1989
Men								
16–17	45[a]	8	9	4	5	4	7	9
18–19		122	63	53	41	39	38	41
20–24	549	474	381	336	274	289	384	372
25–29	886[b]	808	745	738	800	772	785	758
30–34		909	892	891	922	887	865	841
35–39	913[c]	939	931	929	953	933	891	859
40–44		944	936	940	962	946	908	863
45–49	870[d]	937	927	935	963	952	925	868
50–54		914	915	921	956	952	930	877
55–59		885	893	900	943	948	930	887
60–64	680[e]	824	847	823	922	932	918	880
65–69		742	787		889	903	898	862
70+		552	625	611	739	778	777	741
Women								
16–17	158[a]	54	59	40	29	26	27	35
18–19		289	266	250	171	186	198	230
20–24	775	687	613	614	501	559	602	623
25–29	879[b]	850	791	787	759	827	806	805
30–34		851	826	818	776	853	831	830
35–39	807[c]	818	807	800	725	839	824	815
40–44		763	763	759	623	790	806	785
45–49	660[d]	702	690	688	549	719	768	752
50–54		614	605	593	485	603	692	719
55–59		550	517	497	433	501	583	647
60–64	361[e]	438	429	363	390	408	440	540
65–69		363	337		322	326	333	408
70+		202	205	168	169	196	170	170

[a]For the age group 15–19. [c]For the age group 35–44. [e]For ages 60 and over.
[b]For the age group 25–34. [d]For the age group 45–54.

sharp disproportions in the population by sex at certain ages. This explains the decrease of the proportion of married women after the war. These sex disproportions in cohort generations over age 45 in 1946 were caused not only by World War II, but also by World War I and the civil war of 1918–1922. These events account for the lower proportions of married women in 1939 in comparison with 1926.

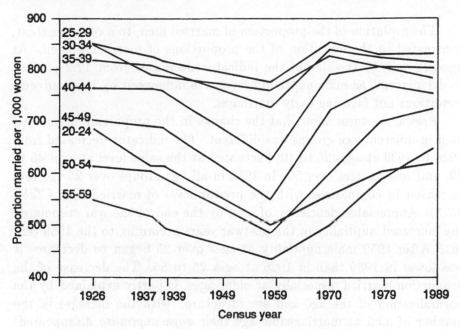

Figure 10.1. Proportion of married women in the USSR.

Table 10.2. Number of men per 100 women in selected years and selected age groups in the USSR.

Year	15–19	20–24	25–29	30–34	35–39	40–44	45–49	50–54	55–59	60–64	65–69
1926	100	97	89	93	90	86	82	84	79	82	77
1936	93	93	98	95	87	90	85	77	71	72	65
1946	92	71	66	64	66	63	61	73	68	60	53
1956	98	97	91	70	64	62	63	58	55	63	57
1989	105	101	101	100	98	96	90	88	81	65	47

Sources: Andreev *et al.*, 1990; State Committee of the USSR on Statistics, 1990.

Marriage-rate decrease from 1926 to 1959 at ages under 20 is not connected with sex disproportions. It may be explained by the influence of socioeconomic factors on the age at marriage. At ages over 20 these factors influenced the disproportions in the age and sex structure. Following the normalization of the age and sex structure in later postwar years, the proportion of married women began to increase. One must remember that the 1926 marital structure was altered by World War I and the civil war.

The evolution of the proportion of married men, to a certain extent, contrasted to the evolution of the proportions of married women. At ages 18 and 19 the size of the indicator decreased from 12.2 percent to 4.1 percent, as male nuptiality level was influenced by these adverse conditions not favoring early marriages.

Prewar censuses show that the change in the proportion of married men in different age groups was different. The indicator decreased from 1926 to 1939 at ages 25 to 40, fluctuated at the same level at ages 40 to 49, and rose at ages over 50. In 1959 in all age groups over 25 the rate increased in comparison with the prewar level of married men (*Table 10.1*). Appreciable deficiency of men by the end of the war stimulated the increased nuptiality in the postwar years according to the 1959 census. After 1959 male nuptiality at ages over 25 began to decrease: it was lower in 1989 than in 1926 at ages 25 to 55. The decrease of the proportion married, especially at older ages, is partly explained by the normalization of the age and sex structure: with the increase in the number of men at marriageable age their *supernuptiality* disappeared. This tendency is particularly explicit among men over 30 (*Figure 10.2*). First, the proportion married increased as a result of sex disproportions; the proportion then decreased as a result of normalization of sex proportion. The decrease was due to the high frequency of divorces.

Sex disproportions existed during war and postwar periods. These disproportions influenced nuptiality not only quantitatively but also qualitatively, and resulted in disruptions in marriage. Norms and values changed; illegal unions and divorces increased; the stability of marriages weakened.

The divorce procedure was simplified in 1965, and the number of divorces since 1966 began to rise. But the increase in the number of divorces was not parallel with an increase in the number of remarriages. That is why the change in the proportion married among men and women in 1959–1989 showed not only the tendency of nuptiality change, but the rise in the number of divorces as well.

The investigation of marital-structure dynamics is supplemented with a generalized indicator, that is, the number of years in marriage between ages 15 and 64 (*Table 10.3*). Usually two versions of the indicator are computed: with mortality and without mortality (Tolts, 1979). We use the indicator computed without mortality. It is computed as a sum of proportions married in five-year age intervals from 15 to 64,

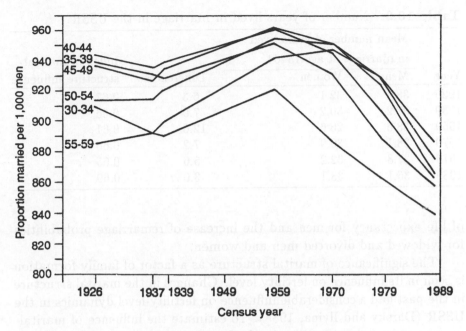

Figure 10.2. Proportion of married men in the USSR.

multiplied by five. Theoretically, within the given age interval, one can remain in marriage for 50 years; but this number is actually smaller by the number of years lived before the first marriage and by the number of years lived after a divorce or death of spouse or before a new marriage is contracted. For real cohorts the number of years lived in marriage depends on the events that affect each generation and reflects the influence of socioeconomic and demographic factors on nuptiality during a period of time. For a hypothetical cohort the indicator reflects the situation of a period of time, generalizing the nuptiality of different cohorts. One would think that the number of years lived by women in marriage must be larger, because they marry earlier. But the opposite is registered in all censuses; this is due to a higher frequency of remarriages among divorced and widowed men. In the period between 1926 and 1959 the gap in the number of years lived in marriage by men and women widened. After 1959, the sex structure normalized to three years in 1989. Between 1979 and 1989 the narrowing in the gap was particularly evident. Apparently that was the impact of two factors in the 1980s: the growth

Table 10.3. Number of years lived in marriage in the USSR.

Year	Mean number of years lived in marriage at age 15–64		Differences	Index of marital-structure influence
	Men	Women		
1926	38.4	32.1	6.3	0.67
1939	37.2	30.2	7.0	0.66
1959	38.6	26.6	12.0	0.61
1970	38.1	30.9	7.2	0.66
1979	37.8	32.2	5.6	0.65
1989	36.1	33.1	3.0	0.69

of life expectancy for men and the increase of remarriage probabilities
for widowed and divorced men and women.

The significance of marital structure as a factor of family formation
is seen in its influence on fertility level. Changes in the marital structure
in the past had a considerable influence on fertility-level dynamics in the
USSR (Darsky and Ilyina, 1988). To estimate the influence of marital-
structure changes on fertility level, the marital-structure index (Coale,
1969), independent of age structure, is usually used. In the USSR the
index was continuously decreasing until 1959 (*Table 10.3*) reflecting the
deterioration of the marital structure. After 1959 it increased again.
Along with this, marital births concentrated in younger ages and female
marital structure at older ages had less influence.

10.2   Changes in the Proportion Single

The category of single comprises never married, widowed, and divorced,
depends on different factors, and requires a differentiated approach.

10.2.1   Never married

The proportion never married among men and women, registered by
censuses is listed in *Table 10.4*. Any pronounced trend of its increase
or decrease is not evident. Evident are increasing proportions never
married among women at ages 50 to 59 in 1979 and at ages 60 to 69 in
1989, among women born between 1919 and 1928. These were the war
and prewar years, and a sharp sex disproportion was the reason for the
increased proportion of never married women in these generations.

Table 10.4. Proportion never married in the Russian Empire and the USSR, by census year, per 1,000 respondents.

Age	1897	1926	1979	1989
Men				
15–19	955[a]	950	979	974
20–24		510	599	601
25–29	444[b]	175	174	195
30–34		72	78	92
35–39	95[c]	41	45	60
40–44		29	28	42
45–49	43[d]	25	17	32
50–54		22	13	23
55–59	30	19	10	16
60–64	26[e]	18	8	12
65–69		17	8	12
70+		17	9	10
Women				
15–19	841[a]	861	903	886
20–24		277	355	330
25–29	209[b]	89	114	117
30–34		53	61	66
35–39	65[c]	39	37	49
40–44		38	35	41
45–49	46[d]	36	41	33
50–54		35	51	33
55–59	43	34	59	42
60–64	48[e]	32	54	54
65–69		32	39	64
70+		31	31	43

[a]For the age group 15–24. [c]For the age group 35–44. [e]For ages 60 and over.
[b]For the age group 25–34. [d]For the age group 45–54.

The change in proportions never married at younger ages depends on the age at first marriage and shows the tendency of change. It is usually used as marriage rate. The analysis of dynamics of age at first marriage shows that the marriage rate did not undergo any significant change among men and among women; after the increase caused by the disproportion of the sex structure, it returned to the early 20th-century level (Darsky and Ilyina, 1990a).

From 1926 until 1989 the period of being single (at ages under 50) slightly increased. The increase illustrates the change in the mean number of years lived before the first marriage in the age interval of 15 to 64: in 1926, 9.3 for men and 7.5 for women; in 1979, 9.8 for men and 8.5 for women; and in 1989, 10.2 for men and 8.3 for women. The dynamics of this indicator, for both men and women, is influenced by the increase in the age at first marriage. From 1926 to 1989 an estimated mean age at first marriage rose from 20.9 years to 21.7 years among women and from 23.4 years to 24.4 years among men.

10.2.2   Widowed and Divorced

The category of single includes individuals divorced and widowed as well. The size of these groups – their proportion in total population at each age group – depends both on intensity of growth of these groups due to divorce or death of spouse and on intensity of decrease due to death or remarriage. The relative size of these groups depends not on the general level of mortality but on the relative death rate. It is known that the mortality level is higher for those widowed and divorced than for those married (Volkov, 1986), and thus the former's share may decrease because of higher mortality. At present we cannot measure the influence of this factor, but it should be considered in data interpretation.

The rise in the number of widowed men and women in population depends, primarily, on male and female mortality levels. It is natural to expect that the proportion widowed, under 70 at any rate, must decrease with the increase of life expectancy. *Table 10.5* shows the proportion of widowed and divorced men and women. The proportion widowed significantly decreased since 1926. This decrease may be due to the increase of mean life expectancy: for men from 36.0 years in 1926 (Andreev *et al.*, 1990) to 64.6 years in 1989 (State Committee of the USSR on Statistics, 1990); for women from 41.0 years in 1926 to 74.0 years in 1989. The decrease of the proportion widowed among men is smaller than among women, evidently not only because their life expectancy increased less than that of women's, but because the gap between male and female life expectancy widened. Without remarriages the proportion widowed at each age would depend only on mortality. Actually those who are widowed often remarry, with widowers remarrying more often than widows. Thus at ages over 35, remarriage rates in 1978–1979 for widowers exceeded remarriage rates for widows by more than three times. By

Table 10.5. Proportion widowed and divorced in the Russian Empire and the USSR, by census year, per 1,000 respondents.

Age	Widowed				Divorced			
	1897	1926	1979	1989	1897	1926	1979	1989
Men								
15–19	0[a]	0	0	0	0[a]	1	0	0
20–24		3	0	1		8	12	12
25–29	6[b]	5	1	1	1[b]	8	38	38
30–34		9	3	3		6	54	59
35–39	16[c]	11	5	4	1[c]	5	58	73
40–44		19	8	8		5	55	83
45–49	41[d]	31	13	16	1[d]	4	44	81
50–54		57	20	26		4	37	70
55–59	98	89	31	42	1	4	28	51
60–64	290[e]	152	51	66	1[e]	3	22	37
65–69		235	77	95		3	17	26
70+		425	200	230		2	11	14
Women								
15–19	1[a]	1	0	1	0[a]	4	3	3
20–24		11	3	3		21	36	34
25–29	14[b]	36	9	7	1[b]	21	70	65
30–34		76	17	14		16	90	86
35–39	54[c]	127	34	26	2[c]	12	103	107
40–44		187	54	46		9	104	125
45–49	144[d]	251	82	86	2[d]	7	98	126
50–54		342	155	129		5	101	115
55–59	294	408	271	213	2	4	85	94
60–64	585[e]	523	453	319	1[e]	3	58	82
65–69		599	585	460		2	41	63
70+		761	774	750		1	21	30

[a] For the age group 15–24. [c] For the age group 35–44. [e] For ages 60 and over.
[b] For the age group 25–34. [d] For the age group 45–54.

1988–1989 the gap between the values of remarriage rates for widowers and widows, though decreased, remained large.

As a result of these processes during the last 10 years the proportion of widows in most age groups continued to decrease and the proportion of widowers over 45 years of age increased. The mean number of years lived in widowerhood or widowhood between ages 15 and 64 shows a decreasing trend: in 1926, 1.9 for men and 9.8 for women; in 1979, 0.7 for men and 5.4 for women; and in 1989, 0.8 for men and 4.2 for women.

From 1926 to 1979 the indicator decreased by almost three times for men and almost two times for women. During the last 10 years the mean number of years lived in widowerhood practically did not change for men; for women the number of years in widowhood decreased by more than a year. In addition, in 1989, as well as in 1926, the indicator for women was five times larger than for men.

The relative number of the divorced population at each age is determined by the divorce level, the frequency of remarriages, and the relative mortality level of those divorced. In 1926 the influence of divorce was insignificant: death of a spouse was the primary means for marriage dissolution. The frequency of remarriage did not compensate divorces, especially among women; and the proportion divorced increased. Data for the country as a whole are not available, but for the Ukraine the frequency of remarriages among divorced women over 25 years of age from 1926 until 1989 decreased by more than one-third (Pustokhod and Tratsevsky, 1930). Remarriage rates by age for women in 1978–1979 and 1988–1989 in the USSR were two times lower than the rates for men. We cannot estimate the influence of divorce on relative mortality. From 1926 to 1989 at ages over 25 the proportion divorced among both men and women increased many times (*Table 10.5*). The number of years lived in divorce rose as follows: in 1926, 0.2 for men and 0.5 for women; in 1979, 1.7 for men and 3.7 for women; and in 1989, 2.5 for men and 4.2 for women.

We see that during the time equal to the length of two demographic generations, from 1926 to 1989, a principal shift in the prevalence of widowhood and divorce took place. The number of years lived in widowhood in 1926 was almost 10 times longer for men and 20 times longer for women than the number of years lived in divorce. By 1989 these interrelations for women were almost identical and for men the number of years lived in widowerhood was three times less than the number of years lived in divorce.

10.3   Variation of Male and Female Marital-Status by Ethnicity

The USSR is demographically heterogeneous. Nationalities differ in their culture, family, and life traditions; they have different social and demographic characteristics of marriage and family. Ethnic nationality is

Table 10.6. Mean number of years lived by men of different nationalities in each marital status from age 15 to age 64.

Nationality[a]	Married 1979	Married 1989	Never married 1979	Never married 1989	Widowed 1979	Widowed 1989	Divorced 1979	Divorced 1989
Tajiks	39.9	40.0	8.4	8.4	1.1	1.0	0.6	0.6
Uzbeks	39.9	39.9	8.4	8.4	1.1	1.1	0.6	0.6
Kirghiz	39.4	38.7	8.9	9.3	0.8	1.0	0.9	1.0
Moldavians	39.6	38.4	8.8	9.3	0.8	1.1	0.8	1.2
Turkmen	38.9	38.2	8.7	9.2	1.9	1.9	0.5	0.7
Azerbaijanis	37.6	37.6	11.2	11.1	0.7	0.8	0.5	0.5
Ukrainians	38.6	37.2	9.4	9.9	0.6	0.7	1.4	2.2
Armenians	37.9	37.2	10.6	10.9	0.6	0.8	0.9	1.1
Kazakhs	37.8	36.9	10.3	10.9	1.1	1.2	0.8	1.0
Byelorussians	38.1	36.6	10.1	10.5	0.5	0.7	1.3	2.2
Georgians	36.7	35.9	12.0	12.6	0.7	0.8	0.6	0.7
Russians	37.5	35.7	9.6	10.3	0.7	0.8	2.2	3.2
Lithuanians	36.2	35.2	11.6	11.6	0.6	0.8	1.6	2.4
Latvians	34.3	33.4	12.0	12.0	0.7	0.8	3.0	3.8
Estonians	33.7	32.5	12.8	13.1	0.7	0.8	2.8	3.6
All	37.8	36.3	9.8	10.4	0.7	0.8	1.7	2.5

[a]Nationalities are presented in the order of decrease in the mean number of years lived in marriage in 1989.

the main differentiating factor of demographic processes in the country. Nuptiality levels of major nationalities in the republics differ greatly (Darsky and Ilyina, 1990b). The differences are explained both by age and marital status and by frequency of widowhood or widowerhood and divorce. The mean number of years lived in marriage in 1989 was 36.3 years for all the males of the USSR, but varied from 32.5 years among Estonians to 40.0 among Tajiks (*Table 10.6*). For women the differences are similar: from 30.1 years among Estonian women to 38.4 years among Tajiks; and among all women in the USSR 33.3 years (*Table 10.7*).

The maximum nuptiality level is evident in Central Asia with Moldavia close behind. The Baltic nationalities and Russians experience the minimum number of years. Nuptiality level, measured by the mean number of years lived in marriage, decreased for men in the last 10 years for most nationalities, but remained unchanged for Uzbeks, Azerbaijanis, and Tajiks. The maximum decrease took place among Russians.

Table 10.7. Mean number of years lived by women of different nationalities in each marital status from age 15 to age 64.

Nationality[a]	Married		Never married		Widowed		Divorced	
	1979	1989	1979	1989	1979	1989	1979	1989
Tajiks	38.0	38.4	5.5	6.1	5.7	4.4	0.8	1.1
Uzbeks	37.0	37.8	6.1	6.2	6.0	4.8	0.9	1.2
Turkmen	36.0	35.7	6.6	8.0	6.7	5.4	0.7	0.9
Kirghiz	35.6	35.5	6.4	7.0	6.7	5.6	1.3	1.9
Moldavians	33.8	34.9	8.1	7.3	5.7	5.1	2.4	2.7
Armenians	32.8	34.4	9.1	8.7	5.8	4.3	2.3	2.6
Ukrainians	32.9	34.2	8.3	7.8	5.2	4.0	3.6	4.0
Byelorussians	32.8	34.0	9.0	8.5	5.2	3.9	3.0	3.6
Azerbaijanis	32.6	33.8	9.0	9.4	7.2	5.5	1.2	1.3
Georgians	32.5	33.2	10.6	10.6	5.3	4.3	1.6	1.9
Kazakhs	32.9	32.7	8.0	8.9	7.9	6.7	1.2	1.7
Russians	31.8	32.7	8.4	8.2	5.4	4.1	4.4	5.0
Lithuanians	32.2	32.6	10.9	10.0	4.2	3.7	2.7	3.7
Latvians	30.0	30.6	10.9	10.2	4.3	3.6	4.8	5.6
Estonians	30.0	30.1	11.4	11.2	4.3	3.6	4.3	5.1
All	32.3	33.3	8.4	8.3	5.5	4.2	3.8	4.2

[a]Nationalities are presented in the order of decrease in the mean number of years lived in marriage in 1989.

Female nuptiality level increased for most nationalities, but slightly decreased among Kazakh, Kirghiz, and Turkmen populations. Armenian women had the largest increase in the indicator.

   National differences in never married are quite significant, but the level is relatively low. The proportion never married in 1989 was maximal among Estonians: at age 50 to 54 years it was 8.2 percent for men and 8 percent for women. It is preferable to estimate this level according to the 1989 census, for the age group of 50 to 54, not 45 to 49 as is the usual practice. Those 45 to 49 years of age in 1989 were born in the period of war, when fertility level was very low, so their nuptiality is of a unique character. The proportion never married is mainly influenced by the age at the first marriage. The mean number of years lived before marriage for men varies between ages 15 and 64 from 8.4 years among Uzbeks and Tajiks to 13.1 years among Estonians; for women from 6.1 years among Tajiks to 11.2 years among Estonians. This indicator consists of two parts: years lived by remaining single and years lived before marriage. The proportion of the former was quite small. For example, for Estonians, experiencing maximal celibacy of 13.1, only 0.4 years fall in the proportion never married.

Notable for marriage at a late age are two groups: Baltic nationalities (Lithuanians, Latvians, and Estonians) and Transcaucasian nationalities (Georgians, Armenians, and Azerbaijanis). The Tajiks and Uzbeks – nationalities of Muslim culture and traditional agriculturists – marry at the youngest ages. In spite of the fact that in recent decades early marriage almost disappeared among them, their age at marriage remains low. The change of this indicator during the last 10 years among various nationalities was different and not very significant. For Russian, Ukrainian, and Byelorussian men the mean number of years lived before the first marriage rose, and for women of those nationalities it decreased. Maximal changes of the indicator were among Turkmen: for men and women age at first marriage increased.

For men, the mean number of years lived in widowerhood is smallest among Baltic nationalities, Byelorussians, and Ukrainians and is greatest among Turkmen. For women, the indicator is smallest among Baltic people and Byelorussians and is greatest among Kazakhs. Apparently the differences of this indicator depend on national mortality differences. The mean number of years lived by widows is four to five times greater than the number of years lived by widowers. From 1979 until 1989 this indicator decreased significantly for women of all nationalities and increased for men of all nationalities, except Uzbeks and Tajiks. This increase is explained not only by the fact that male mortality during this decade decreased more than that of female mortality, but also by the fact that during this period remarriage probabilities for widowers decreased and for widows increased.

The mean number of years lived in divorce shows the largest national differences: for men this indicator varies from 0.5 years among Azerbaijanis to 3.8 years among Latvians; for women from 0.9 years among Turkmen to 5.6 years among Latvians. Divorce level, which determines these differences, is the lowest among the peoples of Central Asia and Azerbaijan and is the highest among Russians, Latvians, and Estonians. During the 1980s the proportion divorced and the mean number of years lived in divorce significantly rose for both men and women. For men the greatest absolute increase of the mean number of years lived in the divorced state could be observed among Russians and Byelorussians; for women among Baltic nationalities. The relative growth for men was highest among Ukrainians and Byelorussians; for women among Kazakhs and Kirghiz. This growth is connected with the changing divorce frequency and remarriage frequency among those divorced.

The number of years lived in divorce and widowerhood or widowhood differs among populations. For women the mean number of years lived in divorce exceeds the mean number of years lived in widowhood for Russians, Latvians, and Estonians – nationalities with the highest divorce level. Among men, besides the nationalities mentioned, Ukrainians, Byelorussians, and Lithuanians live more years in divorce than in widowerhood.

Thus, in spite of fundamental changes in the institution of marriage in the 20th century noted by ethnologists and sociologists (Zhdanko, 1990; Kharchev, 1979), there were no essential changes in population marital structures measured by traditional methods. The exception is the replacement of the category of widowed by the category of divorced, with insignificant changes in the total category of single. The traditional approach allows one to see the normalization of the marital composition and the normalization of marital structure that was altered by wars. But it does not show qualitative changes in the traditional categories of marital status and does not allow one to measure the extent of alternative forms of marriage.

References

Andreev, E.M., L.E. Darsky, and T.L. Kharkova, 1990. Istoriya naseleniya SSSR 1920–1959 gg (The History of the USSR Population, 1920–1959). In *Ekspress-informatsiya, Seriya "Istoriya statistiki"* (Express-Information, History of Statistics Series), Issue 3–5 (Part 1). Goskomstat SSSR, Moscow.

Coale, A.J., 1969. The Decline of Fertility in Europe from the French Revolution to World War II. In S.J. Behrman and L. Corsa, eds. *Fertility and Family Planning: A World View*. University of Michigan Press, Ann Arbor, MI.

Darsky, L.E., and I.P. Ilyina, 1988. Vliyanie brachnoi structury na uroven' rozhdaemosti (The Influence of Marital Structure on Fertility Level). In A.G. Volkov, ed. *Metodologiya demograficheskogo prognoza* (Backgrounds of Demographic Forecasting). Nauka, Moscow.

Darsky, L.E., and I.P. Ilyina, 1990a. Normalizatsiya brachnosti v SSSR. (Nuptiality Normalization in the USSR). In A.G. Volkov, ed. *Demograficheskiye protsessy v SSSR* (Demographic Processes in the USSR). Nauka, Moscow.

Darsky, L.E., and I.P. Ilyina, 1990b. Tablitsy brachnosti osnovnyh natsional' nostey soiuznykh republic SSSR [Nuptiality Tables for Titular Nationalities in Union Republics of the USSR (By the 1985 Sample Sociodemographic Survey Data)]. In *Ekonomika, Demografiya, Statistika: Issledovaniya i problemy* (Economics, Demography, Statistics, Studies, and Problems). Nauka, Moscow.

Hajnal, J., 1965. European Marriage Patterns in Perspective. In D.V. Glass and D.E.C. Eversley, eds. *Population in History*. London.

Ilyina, I.P., 1977. Vliyanie voin na brachnost' sovetskikh zhenshchin (The Influence of Wars on Nuptiality of Soviet Women). In A.G. Vishnevsky, ed. *Brachnost', rozhdaemost' i smertnost' v Rossii i v SSSR* (Nuptiality, Fertility, and Mortality in Russia and the USSR). Statistika, Moscow.

Kharchev, A.G., 1979. *Brak i sem'ia v SSSR* (Marriage and Family in the USSR). Mysl, Moscow.

Nechayeva, A.M., 1980. *Sem'ia i zakon* (The Family and the Law). Nauka, Moscow.

Pustokhod, P., and M. Tratsevsky, 1930. Shlubnist' na Ukraine (Nuptiality in the Ukraine). In M. Ptoukha, ed. *Demografichesky zbornik* (Demographic Articles). Kiev.

State Committee of the USSR on Statistics (Goskomstat), 1990. *Demograficheskiy ezhegodnik SSSR* (USSR Demographic Yearbook, 1990). Finansy i statistika, Moscow.

Tolts, M.S., 1979. Nekotorye obobschaushchie kharakteristiki brachnosti, prekrashcheniya i dlitel'nosti braka (Some Generalizing Characteristics of Nuptiality, Marriage Dissolution and Duration). In A.G. Volkov, ed. *Demograficheskoye razvitiye sem'i* (The Demographic Development of the Family). Statistika, Moscow.

Volkov, A.G., 1986. *Sem'ia-obiekt demografii* (The Family as the Object of Demography). Mysl, Moscow.

Zhdanko, T.A., ed. 1990. *Semeiniy byt narodov SSSR* (Family Mode of Life of the USSR Peoples). Nauka, Moscow.

Chapter 11

Marital and Fertility Experience of Soviet Women

Frans Willekens and Sergei Scherbov

In most regions of the world, social change is accompanied at some stage by a significant change in family life. The traditional functions of the family are being eroded. Today, the individual has the choice of obtaining economic, social, and even emotional support from relationships and institutions other than the family. Although most people rely on the family for many of the traditional functions, a growing number of people do not choose a family-oriented life-style. As a consequence, new trends have become manifest:

- People marry later and some do not marry at all.
- Marriages which are dissolved by age 50 are more likely to be dissolved by divorce than by the death of one partner.
- The proportion of married people who are committed to a single family for their entire lifetime is decreasing.
- Women have fewer children and give birth at later ages; the lives of women are structured less around childbearing; women increasingly derive status from activities other than raising children.
- Marriage and fertility are linked less than they used to be; the proportion of children born outside legal marriage is on the increase.

In addition, mortality decline has an effect of its own. Because of increased life expectancy, a woman may expect to live longer after she completes raising children. Consequently, the share of childbearing and childraising years in the total lifetime is decreasing, not only because of a decline in fertility, but also because of a decline in adult mortality.

These trends are manifest in many Western countries, as well as the USSR. There is, however, a very significant difference. In the Soviet Union, marriage is universal and Soviets tend to marry much younger than their Western European counterparts. The proportion of women who never marry is between 1 and 2 percent. The sociodemographic survey of 1985 revealed that 29.7 percent of the girls born in 1940–1944 first married by the age of 20 (State Committee of the USSR on Statistics, 1988); half of the 1942 cohort was married by age 22.4 (Volkov, 1986). The proportion marrying at an early age is increasing. Of those born in 1950–1954, 32.1 percent married before their 20th birthday (median age 21.5), and so did 34.0 percent of the 1960–1964 cohort. Housing shortages, prejudices against cohabitation and modern contraceptives, inadequate sexual education, and lack of effective contraceptives are the main reasons for early marriage. *The European* (August 31–September 2, 1990) recently reported that as many as half of the brides in some parts of the country are pregnant when they marry. Childlessness is very rare in the Soviet Union. The proportion of childless people remains stable between 6 and 8 percent, and childlessness is mainly due to natural infecundity which is estimated at 5 percent.

Fertility decline has been a consequence of the decline in births of high parity. The period total fertility rate (TFR) is about 2.45. The sociodemographic survey of 1985 showed that 11.8 percent of the married women between ages 18 and 59 have three or more children. Darsky and Scherbov (1990) calculated from the survey of 1985 that 95.2 percent of the women who married in 1970–1974 had at least one child. They also found that 24.6 percent of women who gave birth during the 1970–1974 period had three or more children. These general figures mask very important regional differences. The fertility transition was completed as early as 1959 in the European USSR (Coale *et al.*, 1979; Jones and Grupp, 1987). It is beginning only now in the Central Asian republics. For instance, the proportion of women who gave birth in 1970–1974, and who had three or more children in 1985, was 12.7 percent among Russians and 89.1 percent among Uzbeks (Darsky and Scherbov, 1990).

Although marriage is universal, the marriage institution is decreasing in significance. The divorce rate (number of divorces per 1,000 married couples) increased from 5.3 in 1958–1959 to 11.5 in 1969–1970 and further to 15.2 in 1978–1979 (*Demograficheskiy entsiklopeditcheskiy slovar'*, 1985). In 1984–1985, it was 14.1 percent (State Committee of the USSR on Statistics, 1988); in 1989, it was calculated to be 14.2 percent, amounting to about 1 million divorces. Experts attribute the high divorce rate to early and forced marriages. Many young couples rush into marriage when they barely know each other and are unprepared for the responsibilities. Divorces are easily obtained provided children are not involved. Couples fill in a few forms, pay a fee of 100 roubles (almost half a month's salary), and go their separate ways. Not all divorces can be attributed to dissatisfaction with the partner. In large cities such as Moscow and Leningrad, the number of divorces is inflated by so-called paper marriages. This is when residents marry people keen to move to these places who do not meet the requirements of the residential permit system, which was established to control the number of inhabitants. Remarriages are becoming more important. In 1978, remarriages contracted in a year constituted 14.3 percent of all marriages; in 1989 it was 21.5 percent (State Committee of the USSR on Statistics, 1989a, 1990).

The purpose of this chapter is to explore the changes in nuptiality and fertility in the Soviet Union, and the associated changes in women's lives. Multistate life-table analysis is introduced to generate complete marital and fertility histories (biographies) of women as they pass through the reproductive ages (16–50). The biographies are *synthetic* biographies since they are not completely observed but inferred from the available data on nuptiality and fertility. The life table is a method used to determine the biography that is consistent with a set of vital rates and to assess the impact of changes on these rates. The multistate life table has become a useful technique in family demography (see, for example, Schoen *et al.*, 1985; Zeng Yi, 1986, 1990; Bongaarts, 1987; Espenshade, 1987; Willekens, 1987; Keyfitz, 1988). In addition to the multistate life-table method, the theory of staging or sequential processes is used to describe marital and fertility careers. This theory focuses on the occurrence and timing of chains of events (Chiang, 1984; Willekens, forthcoming). Cohort data are used when available.

The first section of the chapter presents the marital biographies
women would have experienced if the rates of marital change observed
in 1989 had prevailed. Data limitations prevent the use of cohort data
and a comparative analysis over time and space. The second section
describes fertility histories. Two birth cohorts are distinguished: 1940–
1944 and 1950–1954. The combination of the marital and fertility histories to provide a picture of the complete life course of Soviet women is not
pursued because the marital careers are based on period data, whereas
cohort data are used to reconstruct the fertility careers. Prospective biographic indicators are dependent not only on patterns of marital change
and fertility, but also on mortality. Since marital and fertility careers
are estimated up to age 50, the effect of mortality is small. The USSR
life table for 1986–1987 indicates that 95.1 percent of females aged 16
survive to age 50 (State Committee of the USSR on Statistics, 1989b).

11.1   Marital Career

In 1989, the State Committee of the USSR on Statistics published the
number of marriages by age, sex, and marital status prior to marriage,
as well as the number of divorces by age (five-year age groups) and sex.
The data are shown in *Table 11.1*. They are derived from the marriage
certificates issued in 1989 (Jones and Grupp, 1987).

   To construct the marital history of women from the data, rates of
marital change must be estimated. The 1989 census provides data on
population by age, sex, and marital status as of January 1989. The
marital-status composition of the female population is shown in *Figure
11.1* and the figures are given in *Table 11.2*. The data are considered
adequate estimates of the population at risk. Occurrence–exposure rates
by specific age are estimated by dividing the number of marital events
by the population at risk. The occurrence–exposure rates of widowhood
were estimated in the following way. First it was assumed that the age
difference between bride and bridegroom is about two years. Next the
number of deaths of married men in each age category was divided by
the number of married women two years younger. These rates were later
adjusted for the ages above 45 using Gompertz's curve. The occurrence–
exposure rates are shown in *Table 11.3*.

   Two types of life-table analyses are carried out. First, the multistate,
marital-status life table is prepared. The table shows, for various ages,

Table 11.1. Number of marriages by previous marital status and number of divorces, in thousands.

Age	From never married to married	From widowed to married	From divorced to married	From married to divorced
16–17	130,697	45	61	597
18–19	652,691	450	4,196	16,844
20–24	945,735	3,537	68,060	177,034
25–29	242,141	8,078	119,451	230,529
30–34	77,993	10,472	104,498	183,466
35–39	29,675	10,419	68,873	132,095
40–44	11,740	8,810	40,142	78,842
45–49	7,014	10,389	26,742	49,539
50–54	7,956	14,943	25,913	48,632
55–59	5,926	11,018	11,474	22,611
60+	14,857	22,780	12,222	24,511

Table 11.2. Female population by marital status and age, 1989.

Age	Never married	Married	Divorced	Widowed
16–17	4,121,228	149,594	3,294	3,318
18–19	2,957,777	913,179	26,203	4,477
20–24	3,339,385	6,315,256	346,538	32,033
25–29	1,416,680	9,757,497	786,899	87,242
30–34	775,169	9,762,860	1,010,289	163,914
35–39	511,683	8,579,182	1,126,154	274,741
40–44	286,341	5,552,076	880,200	322,726
45–49	257,997	5,900,049	984,115	669,880
50–54	305,059	6,702,759	1,070,010	1,203,812
55–59	347,586	5,343,377	771,653	1,759,104
60–64	487,155	4,849,872	733,554	2,865,574
65–69	365,925	2,329,665	360,763	2,632,636
70+	543,142	2,160,856	378,789	9,541,269

the probabilities of being at a given marital status, the probabilities of marital change, and the expected sojourn time in each marital status. Four marital states are distinguished: never married, married, divorced, and widowed. Second, order-specific marital states are considered. A distinction is made between first marriage, second marriage, and higher-order marriages; first marriage dissolution, second dissolution, etc. The marital-life course is viewed as a staging process. A stage is a period

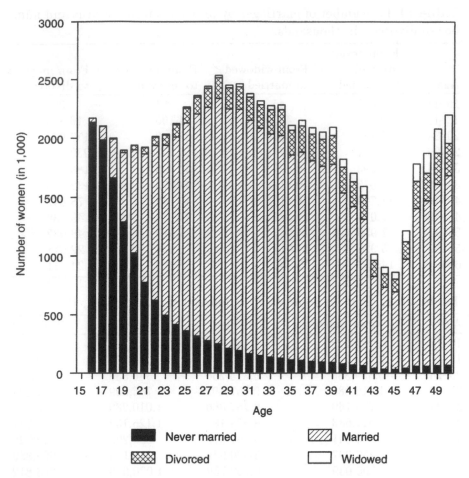

Figure 11.1. Female population by marital status.

or episode of life, characterized by the marital state occupied and the number of times one entered the marital state (Chiang, 1984; Willekens, 1988). It is stressed that the marital career is studied up to age 50. Prospective indicators, such as the expected duration of marriage, are for the period up to (but not including) age 50.

Life-table analysis shows that the probability of marrying (by age 50) is 97.9 percent. The figure is high if compared with that of other European countries. At age 25, 78.5 percent of women are married, according to the life-table analysis. The marriage of some (6.2 percent) has

Table 11.3. Occurrence–exposure rates of marital change by age for females, 1989.

Age	From never married to married	From widowed to married	From divorced to married	From married to divorced	From married to widowed
20	0.2944	0.1259	0.2034	0.0276	0.0016
21	0.3098	0.1127	0.2101	0.0293	0.0018
22	0.2835	0.1138	0.2036	0.0287	0.0020
23	0.2622	0.1076	0.1979	0.0274	0.0022
24	0.2300	0.1039	0.1812	0.0264	0.0023
25	0.2028	0.0959	0.1692	0.0255	0.0024
26	0.1828	0.1024	0.1637	0.0248	0.0025
27	0.1651	0.0929	0.1535	0.0240	0.0025
28	0.1448	0.0891	0.1427	0.0232	0.0025
29	0.1375	0.0874	0.1377	0.0223	0.0027
30	0.1213	0.0758	0.1256	0.0213	0.0027
31	0.1099	0.0740	0.1146	0.0203	0.0029
32	0.0963	0.0645	0.1025	0.0193	0.0031
33	0.0865	0.0589	0.0934	0.0183	0.0031
34	0.0779	0.0528	0.0842	0.0174	0.0033
35	0.0721	0.0503	0.0787	0.0167	0.0038
36	0.0614	0.0409	0.0653	0.0160	0.0039
37	0.0576	0.0390	0.0611	0.0156	0.0045
38	0.0498	0.0351	0.0544	0.0153	0.0044
39	0.0453	0.0300	0.0487	0.0148	0.0044
40	0.0472	0.0329	0.0528	0.0138	0.0051
41	0.0367	0.0235	0.0411	0.0130	0.0056
42	0.0408	0.0282	0.0462	0.0121	0.0061
43	0.0417	0.0297	0.0462	0.0114	0.0067
44	0.0351	0.0217	0.0386	0.0107	0.0074
45	0.0321	0.0202	0.0357	0.0100	0.0081
46	0.0236	0.0144	0.0244	0.0093	0.0089
47	0.0258	0.0153	0.0255	0.0088	0.0098
48	0.0282	0.0161	0.0279	0.0082	0.0107
49	0.0276	0.0144	0.0260	0.0077	0.0118

already been dissolved at that age, mainly because of divorce (5.9 percent). *Figure 11.2* shows probabilities of occupying given marital states for various ages. The life-table estimates are very close to the census data on those aged 25 (78.5 percent are married, 5.6 percent are divorced, and 0.5 percent are widowed). An early divorce has a significant impact

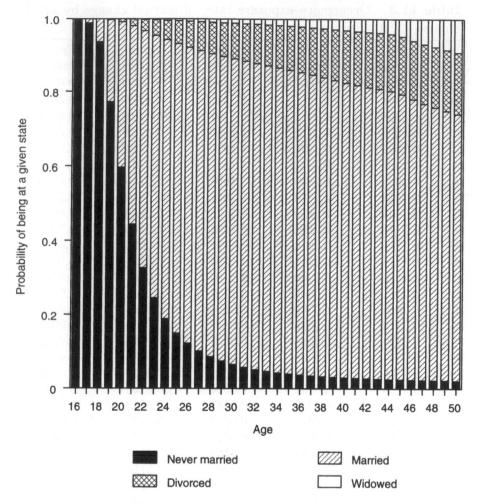

Figure 11.2. Marital-state probabilities.

on a woman's marital biography. For instance, the probability that a
20-year-old will be divorced at age 30 is 24.6 percent if that person was
already divorced at age 20, and 11.3 percent if that person was married
at age 20. In other words, a woman is much more likely to be married
at age 30 if she has not been divorced before. The difference reflects the
differences in the age profiles of marriage and remarriage. The impact
of an early divorce on marital status diminishes as the person gets older.
The marital status of a person at age 40 is not very much affected by
the marital status at age 20. The probability of being divorced at 40 is

20.3 percent for those who are divorced at 20, and 15.4 percent for those who are married at that age. The finding that a person "forgets his or her past" can be attributed to the Markovian assumption underlying the multistate life-table model. The marital-state probabilities by age and marital status at age 20 are shown in *Table 11.4*.

The probability that marriage ends in divorce before age 50 is 33.3 percent. The probability that marriage ends in widowhood is smaller, since older women are excluded from this analysis. The probability that a woman celebrates her 50th birthday in widowhood is 9.1 percent.

The probability measures of the marital-life course may be augmented by duration measures. How long does a woman spend in each marital state, provided that she experiences the rates of marital change observed in the USSR in 1989? The sojourn times are determined by two parallel processes. The first process is marital change; the second is mortality. The effect of mortality is small, except at high ages, and is not considered in this chapter.

The marital biography of a woman may be described by the timing of marital change and the sequence of marital states occupied. Each sojourn in a marital state is a stage of marital life. Multistate life-table analysis shows that women marry for the first time at age 22.4, on average. However, not all women marry. Those who marry before age 50, do so for the first time at age 21.6, on average. The multistate life-table measure is inflated because 2.1 percent of the women never marry. The mean age at first marriage observed in the 1989 population was 22.15 (calculated from single-year age data on marriages). The difference is due to the age composition of the population. Of the 34 years that separate ages 16 and 50, 23.6 years are spent in marriage. The sojourn time in marriage is influenced by the marital history. A person who is married at age 20 may expect to spend more time in marriage beyond that age than a person who is not married yet or who is already divorced (25.4 years versus 21.6 and 20.4 years). As marital change becomes less likely at higher ages, the marital status one occupies becomes a better predictor of the expected sojourn time in each marital state as age increases. For instance, the number of years a woman of age 40 may expect to spend in each marital state is determined by her marital status at that age. If she is married, she probably stays married and 9.2 years of the 10 that separates her from her 50th birthday are spent in marriage, on average. If she is divorced, however, she may look

Table 11.4a. Marital-state probabilities by age when marital status at age 20 is never married and married.

	Never married				Married		
Age	Never married	Married	Divorced	Wid-owed	Married	Divorced	Wid-owed
20	1,000	0	0	0	1,000	0	0
21	745	251	3	0	974	25	2
22	546	441	12	1	952	45	3
23	412	564	23	2	934	61	4
24	317	647	34	3	921	74	6
25	252	700	45	4	909	83	7
26	205	735	55	5	900	92	9
27	171	759	63	6	892	98	10
28	145	776	72	8	885	104	11
29	126	787	79	9	878	109	12
30	109	795	85	10	873	113	14
31	97	800	91	12	868	117	15
32	87	804	97	13	862	121	16
33	79	805	102	15	857	125	18
34	72	805	107	16	852	129	19
35	67	803	112	18	846	133	21
36	62	802	116	20	841	136	23
37	59	798	121	22	834	141	25
38	55	794	126	25	827	145	28
39	53	789	131	27	820	149	31
40	50	783	136	30	813	154	33
41	48	779	140	33	807	157	36
42	46	775	144	35	801	161	38
43	44	772	147	37	797	163	40
44	43	768	149	40	792	165	44
45	41	761	151	47	783	167	50
46	40	749	153	58	769	168	62
47	39	737	156	67	757	171	71
48	38	729	159	74	748	174	79
49	37	721	160	82	739	175	87
50	36	713	161	90	730	176	94

forward to only 1.7 years of married life. The time in marriage is even less if the woman is widowed at age 40. *Table 11.5* shows the expected sojourn time in each marital state by age and marital status at each age.

Table 11.4b. Marital-state probabilities by age when marital status at age 20 is divorced and widowed.

	Divorced			Widowed		
Age	Married	Divorced	Widowed	Married	Divorced	Widowed
20	0	1,000	0	0	0	1,000
21	181	818	0	117	2	882
22	329	670	1	206	6	788
23	442	557	1	285	11	704
24	529	469	2	350	17	633
25	592	405	3	406	23	571
26	639	356	4	450	30	520
27	677	318	5	494	36	470
28	706	288	7	528	42	430
29	727	265	8	557	48	394
30	745	246	9	583	54	363
31	757	232	10	602	60	338
32	767	222	12	620	65	316
33	773	214	13	632	70	298
34	777	209	15	642	75	283
35	779	205	16	650	80	270
36	780	202	19	657	84	259
37	778	201	21	660	89	251
38	776	201	23	662	94	245
39	772	202	26	662	99	239
40	768	203	28	662	103	235
41	765	203	31	662	107	231
42	762	205	33	661	111	228
43	760	204	35	662	114	224
44	757	204	39	663	116	221
45	751	204	45	658	119	223
46	739	204	57	650	121	229
47	728	206	66	642	124	234
48	720	207	73	636	126	238
49	713	207	80	630	128	242
50	705	207	88	624	129	246

We may study the marriage pattern by viewing the marital career as a staging process. Two events are distinguished: marriage and marriage dissolution. The occurrence of an event initiates a new stage, and a sequence of stages defines a career. *Figure 11.3* exhibits the time spent

Table 11.5a. Expected number of years spent in each marital state by age when marital status at age x is never married and married.

	Never married				Married		
Age	Never married	Married	Divorced	Widowed	Married	Divorced	Widowed
20	4.58	21.60	3.05	0.76	25.39	3.75	0.86
21	4.98	20.45	2.83	0.73	24.55	3.60	0.85
22	5.62	19.11	2.58	0.69	23.72	3.44	0.83
23	6.30	17.72	2.32	0.65	22.90	3.28	0.82
24	7.05	16.28	2.06	0.60	22.09	3.11	0.80
25	7.76	14.88	1.81	0.55	21.28	2.94	0.78
26	8.39	13.52	1.58	0.51	20.48	2.76	0.75
27	8.98	12.20	1.35	0.46	19.68	2.58	0.73
28	9.51	10.93	1.15	0.41	18.89	2.40	0.71
29	9.92	9.75	0.97	0.37	18.10	2.21	0.69
30	10.31	8.57	0.80	0.32	17.31	2.03	0.66
31	10.57	7.49	0.65	0.28	16.52	1.84	0.63
32	10.75	6.49	0.52	0.24	15.73	1.66	0.61
33	10.78	5.59	0.42	0.21	14.94	1.49	0.58
34	10.71	4.78	0.33	0.18	14.14	1.32	0.55
35	10.54	4.06	0.25	0.15	13.33	1.16	0.51
36	10.29	3.39	0.19	0.12	12.52	1.01	0.48
37	9.92	2.84	0.14	0.10	11.69	0.86	0.44
38	9.47	2.34	0.11	0.08	10.87	0.73	0.40
39	8.93	1.92	0.08	0.07	10.03	0.60	0.36
40	8.33	1.57	0.06	0.05	9.18	0.49	0.33
41	7.70	1.22	0.04	0.04	8.32	0.39	0.29
42	6.97	0.97	0.03	0.03	7.44	0.30	0.27
43	6.24	0.72	0.02	0.02	6.53	0.22	0.24
44	5.49	0.49	0.01	0.01	5.63	0.16	0.21
45	4.67	0.32	0.00	0.01	4.73	0.11	0.16
46	3.80	0.19	0.00	0.00	3.84	0.07	0.10
47	2.88	0.11	0.00	0.00	2.91	0.04	0.05
48	1.95	0.05	0.00	0.00	1.96	0.02	0.02
49	0.99	0.01	0.00	0.00	0.99	0.00	0.01

in each stage by women with different marital careers. A woman who experiences only one event by the age of 50 is married at that age. Her career is shown at the left. As the number of events increases, the time spent in each stage becomes smaller.

Table 11.5b. Expected number of years spent in each marital state by age when marital status at age x is divorced and widowed.

	Divorced			Widowed		
Age	Married	Divorced	Wid-owed	Married	Divorced	Wid-owed
20	20.36	8.92	0.73	16.61	2.22	11.17
21	19.32	8.98	0.70	15.49	2.02	11.49
22	18.19	9.15	0.66	14.46	1.84	11.70
23	17.03	9.34	0.63	13.39	1.66	11.95
24	15.83	9.58	0.59	12.35	1.48	12.16
25	14.66	9.79	0.55	11.33	1.32	12.36
26	13.51	9.98	0.51	10.37	1.16	12.47
27	12.34	10.20	0.47	9.32	1.00	12.68
28	11.18	10.39	0.42	8.35	0.85	12.80
29	10.06	10.56	0.38	7.40	0.72	12.88
30	8.93	10.73	0.34	6.46	0.59	12.95
31	7.86	10.85	0.30	5.63	0.48	12.89
32	6.85	10.89	0.26	4.83	0.38	12.79
33	5.93	10.85	0.22	4.12	0.30	12.57
34	5.09	10.72	0.19	3.49	0.24	12.27
35	4.33	10.51	0.16	2.94	0.18	11.88
36	3.62	10.25	0.13	2.42	0.14	11.44
37	3.05	9.84	0.11	2.02	0.10	10.88
38	2.53	9.38	0.09	1.65	0.07	10.28
39	2.09	8.84	0.07	1.33	0.05	9.61
40	1.71	8.23	0.06	1.08	0.04	8.88
41	1.33	7.63	0.04	0.82	0.03	8.15
42	1.05	6.91	0.03	0.66	0.02	7.33
43	0.77	6.20	0.02	0.47	0.01	6.51
44	0.52	5.46	0.01	0.31	0.01	5.69
45	0.34	4.66	0.01	0.20	0.00	4.80
46	0.19	3.80	0.00	0.11	0.00	3.88
47	0.11	2.89	0.00	0.07	0.00	2.93
48	0.05	1.95	0.00	0.03	0.00	1.97
49	0.01	0.99	0.00	0.01	0.00	0.99

The data do not permit an investigation of how cultural and economic changes are affecting the marital biographies of women. We know that the divorce rate started to increase in the 1960s from a low of 5.3 percent in the late 1950s to 11.5 percent in the late 1960s. Remarriages have become much more common. The proportion of marriages that are

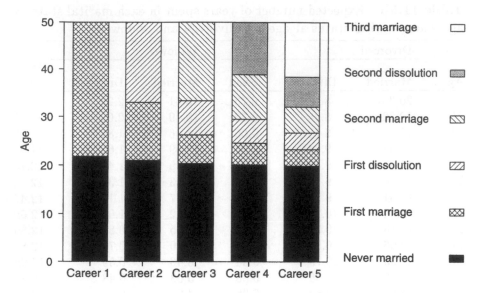

Figure 11.3. Time spent in marital stages for different marital careers.

remarriages increased from 14 percent in the late 1970s to 23 percent in the late 1980s. Part of the increase can be attributed to the increased prevalence of divorces. The rate at which divorced women remarry in a year increased too. The specific reasons for these changes remain unknown although they are likely to be related to the changing status of women.

11.2   Fertility Career

The period TFR in the Soviet Union is about 2.45, which means that, on average, a woman has two to three children during her lifetime. The TFR remains stable for many years. Not all women have a fertility career. Between 6 and 8 percent of the women remain childless. Natural sterility amounts to about 5 percent. Marriage is not a limiting factor since the proportion of women who do not marry is very small. Almost all married women who are able to have children do so.

The fertility careers of women will be studied using cohort data. For this purpose, we used data on the number of children born to women of different ages and parity (see Chapter 2). The data were reconstructed by cohort. To study the fertility careers of women by birth cohort, the data were transformed into occurrence–exposure rates. The transformation consists of two steps. First, the number of children born in order of birth to women of a given cohort are estimated for single years of age from five-year age data. Second, rates at which women of a given parity and age have an additional child (occurrence–exposure rates) are estimated from the data on the number of children born, assuming that fertility is the outcome of a Markov process. The occurrence–exposure rates are the parameters of staging processes, underlying the fertility careers. The rates are given in *Tables 11.6* and *11.7*. The rates serve as an input to estimate the density distribution of children born by birth order. The densities for the 1940–1944 and the 1950–1954 cohorts are shown in *Figure 11.4*. The figure demonstrates the increase in first and second births, and the decline in higher-order births.

In this section, the fertility career is studied irrespective of the marital career. The marital career described in Section 11.1 is based on period rates of marital change observed in 1988. The fertility rates used are cohort rates for the 1940–1944 and the 1950–1954 birth cohorts.

The fertility careers of women start relatively early. Fertility before age 20 accounts for about 10 percent of total fertility and the mean age at first childbirth is about 23. These figures do not change much as fertility declines, due to continued early marriage. Fertility decline in the Soviet Union is therefore predominantly affected by a decline in higher-order births.

Figures 11.5 and *11.6* show the parity distribution of women, born in the 1940–1944 and 1950–1954 cohorts. Women born in 1940–1944 had on average 2.05 children by the end of their reproductive career. Women with children had 2.34 children on average. Analysis of childbearing as a staging process shows that about 12.3 percent remained childless, 23.2 percent had one child, 39 percent had two children, and more then 25 percent had three or more children. For women born in 1950–1954 about 7.5 percent remained childless, 27.3 percent had one child, 42.5 percent had two children, and about 23 percent had three or more children. The proportion with six or more children was as high as 6 percent. Large families are in Asian republics with high-fertility nationalities [TFRs: Tajiks 6.9; Turkmen 6.5; Kirghiz 6.5; Uzbeks 6.4; Kazakhs 5.6 (Darsky and Scherbov, 1990)]. Women of more recent cohorts are as likely to have children as women of older generations. The age at which women have

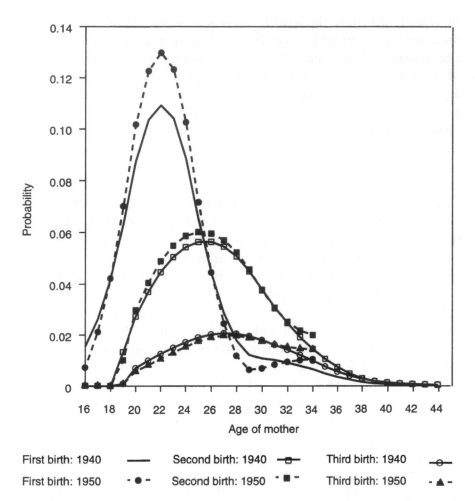

Figure 11.4. Probability densities of births by order for 1940–1944 and 1950–1954 cohorts.

their first child is decreasing, probably due to increased sexual activity at younger ages and the lack of adequate knowledge on and availability of contraceptives.

Childbearing is generally completed by age 37. The TFR of the 1940–1944 cohort at age 37 is 2.00. The percentage of children borne by women before age 20 is 8 percent, 46 percent before 25, 77 percent before 30, and 94 percent before 35. The fertility careers of most women extend over a period of 15 years between the ages of 20 and 35. At age

Table 11.6. Parity occurrence–exposure rates, 1940–1944 cohort.

Age	Parity					
	0	1	2	3	4	5
16	0.016	0.000	0.000	0.000	0.000	0.000
17	0.027	0.000	0.000	0.000	0.000	0.000
18	0.046	0.000	0.000	0.000	0.000	0.000
19	0.073	0.152	0.000	0.000	0.000	0.000
20	0.110	0.209	0.823	1.056	0.000	0.000
21	0.147	0.199	0.359	0.207	0.000	0.000
22	0.183	0.180	0.235	0.184	0.000	0.000
23	0.213	0.163	0.178	0.192	0.000	0.000
24	0.226	0.150	0.144	0.209	0.468	0.000
25	0.206	0.142	0.123	0.228	0.463	0.689
26	0.169	0.139	0.105	0.229	0.442	0.381
27	0.126	0.139	0.091	0.216	0.393	0.304
28	0.088	0.137	0.078	0.193	0.343	0.282
29	0.065	0.133	0.067	0.163	0.299	0.285
30	0.061	0.122	0.056	0.130	0.262	0.301
31	0.061	0.107	0.048	0.103	0.233	0.309
32	0.058	0.091	0.040	0.081	0.209	0.302
33	0.053	0.075	0.033	0.063	0.184	0.280
34	0.046	0.059	0.026	0.049	0.155	0.238
35	0.036	0.044	0.020	0.038	0.124	0.181
36	0.026	0.031	0.014	0.029	0.095	0.130
37	0.018	0.020	0.010	0.022	0.071	0.091
38	0.012	0.013	0.007	0.016	0.052	0.062
39	0.008	0.008	0.005	0.012	0.038	0.042
40	0.006	0.005	0.004	0.008	0.027	0.031
41	0.005	0.004	0.003	0.006	0.019	0.022
42	0.003	0.003	0.002	0.004	0.013	0.016
43	0.003	0.002	0.001	0.003	0.009	0.012
44	0.002	0.002	0.001	0.002	0.007	0.010

20, 15 percent of the women have at least one child. The percentage of women with at least one child at higher ages are: 55 percent at age 24, 65 percent at age 25, 82 percent at age 30, and 86 percent at age 35. At age 30, one-third of the women have one child and another third have two children. The proportion of childless women at that age is 18 percent. There are 16 percent who have three or more children.

The mean age of mothers at childbirth is given in *Tables 11.8* and *11.9*. The mean ages shown are the ages at which women had their

Table 11.7.   Parity-specific occurrence–exposure rates, 1950–1954 cohort.

Age	Parity 0	1	2	3	4	5
16	0.007	0.000	0.000	0.000	0.000	0.000
17	0.022	0.000	0.000	0.000	0.000	0.000
18	0.044	0.000	0.000	0.000	0.000	0.000
19	0.078	0.153	0.000	0.000	0.000	0.000
20	0.126	0.258	0.934	0.746	0.000	0.000
21	0.177	0.222	0.279	0.190	0.000	0.000
22	0.229	0.188	0.175	0.181	0.000	0.000
23	0.279	0.162	0.133	0.193	0.136	0.000
24	0.313	0.145	0.111	0.212	0.386	0.000
25	0.296	0.134	0.097	0.230	0.425	0.718
26	0.241	0.130	0.085	0.227	0.389	0.392
27	0.162	0.128	0.075	0.209	0.334	0.290
28	0.089	0.126	0.066	0.183	0.283	0.241
29	0.051	0.120	0.057	0.152	0.240	0.219
30	0.058	0.109	0.049	0.119	0.205	0.210
31	0.076	0.096	0.043	0.095	0.183	0.205
32	0.093	0.085	0.039	0.078	0.170	0.201
33	0.111	0.077	0.036	0.066	0.163	0.195
34	0.129	0.073	0.035	0.059	0.159	0.188

first, second, etc., child. For instance, a woman born in 1940–1944 with at least one child at age 30 has the first child at age 22.7, on average; women with at least two children at age 30 have the second child at age 25.2, etc. The mean age at which women who have reached the end of their fertility had their first child is 23.4 years. The second child is born at age 27.2.

Tables 11.10 and *11.11* provide estimates of average birth intervals. For calculating birth intervals we should not take the difference between mean ages, because not all women of a given parity experience an additional birth (see Feichtinger, 1987).

The interval between two consecutive births depends on the completed parity. Let $x_{i.j}$ denote the mean age at the i-th birth of those women whose completed parity is $j(j = i, i + 1, i + 2, \ldots)$. The $x_{i.j}$ values are not observed, but estimated from the parameters of the fertility process, i.e., the age- and parity-specific fertility intensities (occurrence–exposure rates). The estimates are approximations and should be treated

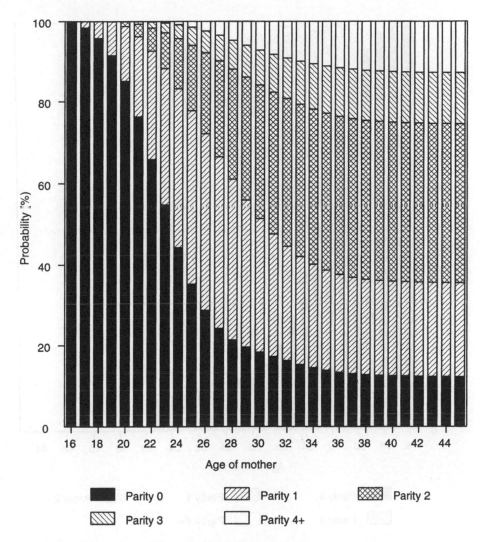

Figure 11.5. Stage probabilities for the 1940–1944 cohort.

with caution. For instance, the values of $x_{1.2}$ are obtained by summation of the densities of second birth weighted by the difference between the age at second birth and the *average* age at which *these* women had their first child. It is equivalent to the sum of the densities of the second child weighted by the average age of the first child at the time of birth of the second child. The method is an improvement on the technique suggested by Ryder (Wunsch and Termote, 1978). Ryder approximates

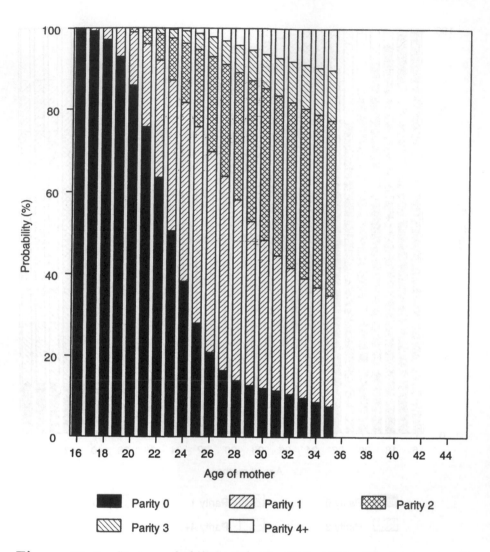

Figure 11.6. Stage probabilities for the 1950–1954 cohort.

the birth interval by the difference between the mean ages at two consecutive childbirths, divided by the parity progression ratio. The Ryder method overestimates the birth interval considerably if the parity progression ratio is small. The *average* birth intervals in *Tables 11.10* and *11.11* are the weighted sum of the intervals by completed parity (by age 45), the weights being the probability that a woman in parity i will end up with j children (Chiang and Van den Berg, 1982; Feichtinger, 1987).

Table 11.8. Mean age of mother at childbirth, 1940–1944 cohort.

Age	Birth order					
	1	2	3	4	5	6
16	0.0	0.0	0.0	0.0	0.0	0.0
17	16.5	0.0	0.0	0.0	0.0	0.0
18	17.1	0.0	0.0	0.0	0.0	0.0
19	17.8	0.0	0.0	0.0	0.0	0.0
20	18.5	19.5	0.0	0.0	0.0	0.0
21	19.3	20.2	20.5	0.0	0.0	0.0
22	20.0	20.8	21.1	21.5	0.0	0.0
23	20.6	21.5	21.7	22.2	0.0	0.0
24	21.1	22.1	22.3	22.9	0.0	0.0
25	21.6	22.7	22.9	23.6	24.5	0.0
26	22.0	23.2	23.5	24.3	25.1	25.5
27	22.2	23.8	24.1	25.0	25.7	26.1
28	22.4	24.3	24.7	25.7	26.3	26.7
29	22.6	24.8	25.3	26.3	26.9	27.4
30	22.7	25.2	25.8	26.8	27.5	28.1
31	22.8	25.6	26.3	27.3	28.1	28.8
32	22.9	25.9	26.7	27.7	28.6	29.5
33	23.0	26.2	27.1	28.1	29.1	30.1
34	23.1	26.5	27.5	28.4	29.5	30.6
35	23.2	26.6	27.8	28.7	29.9	31.1
36	23.2	26.8	28.0	29.0	30.2	31.5
37	23.3	26.9	28.2	29.2	30.5	31.8
38	23.3	27.0	28.4	29.4	30.8	32.1
39	23.3	27.0	28.5	29.5	30.9	32.3
40	23.4	27.1	28.6	29.6	31.1	32.4
41	23.4	27.1	28.6	29.7	31.2	32.5
42	23.4	27.1	28.7	29.8	31.3	32.6
43	23.4	27.1	28.7	29.8	31.3	32.7
44	23.4	27.2	28.7	29.9	31.4	32.7
45	23.4	27.2	28.8	29.9	31.4	32.8

A final measure characterizing the fertility career is the probability that a child borne by a woman of a given age is the first, second, third, or higher-order child. Of all the children born to the 1940–1944 cohort, 43 percent are first children, 31 percent second children, 12 percent third children, and 14 percent fourth or higher-order children. Of the children born to the 1950–1954 cohort (for mothers up to age 35), 46 percent are first children, 33 percent second children, 11 percent third children, and

Table 11.9. Mean age of mother at childbirth, 1950–1954 cohort.

Age	Birth order					
	1	2	3	4	5	6
16	0.0	0.0	0.0	0.0	0.0	0.0
17	16.5	0.0	0.0	0.0	0.0	0.0
18	17.2	0.0	0.0	0.0	0.0	0.0
19	18.0	0.0	0.0	0.0	0.0	0.0
20	18.7	19.5	0.0	0.0	0.0	0.0
21	19.5	20.3	20.5	0.0	0.0	0.0
22	20.2	20.9	21.1	21.5	0.0	0.0
23	20.8	21.5	21.7	22.2	0.0	0.0
24	21.3	22.1	22.3	22.9	23.5	0.0
25	21.8	22.7	22.9	23.6	24.3	0.0
26	22.1	23.2	23.6	24.4	25.0	25.5
27	22.3	23.8	24.2	25.0	25.7	26.1
28	22.5	24.3	24.8	25.7	26.3	26.7
29	22.6	24.8	25.4	26.3	26.9	27.3
30	22.6	25.2	25.9	26.8	27.5	27.9
31	22.7	25.5	26.4	27.3	28.0	28.6
32	22.8	25.8	26.8	27.7	28.5	29.2
33	22.9	26.1	27.3	28.1	29.0	29.9
34	23.0	26.4	27.7	28.5	29.5	30.5
35	23.1	26.6	28.1	28.9	30.0	31.1

10 percent fourth or higher-order children. The analysis confirms the observation that fertility change in the Soviet Union is characterized by an increase in the proportion of women with children, but a decrease in the number of large families.

The fertility career of Soviet women is characterized by an early start, mainly associated with early marriage. Women with several children at any given age start, on average, earlier than women with one or two children. The model replicates the observed relation between age at first birth and level of completed fertility. The relation may not be represented fully by the model since no micro-data are used and heterogeneity between women is not taken into account. Most women have at least one child and two-thirds have two or more children. For mothers aged 40, the first child is 16 years old on average. It is 18 when more children are present.

The fertility career study of women is incomplete without consideration of the means to control fertility. Most women rely on abortion to

Table 11.10. Mean age of mother at first child and average birth interval, 1940–1944 cohort.

	Birth order					
Age	1	2	3	4	5	6
16	0.0	0.0	0.0	0.0	0.0	0.0
17	16.5	0.0	0.0	0.0	0.0	0.0
18	17.1	0.0	0.0	0.0	0.0	0.0
19	17.8	0.0	0.0	0.0	0.0	0.0
20	18.5	1.0	0.0	0.0	0.0	0.0
21	19.2	1.1	0.3	0.0	0.0	0.0
22	19.8	1.3	0.5	0.4	0.0	0.0
23	20.4	1.5	0.7	0.7	0.0	0.0
24	20.8	1.7	1.0	1.0	0.0	0.0
25	21.2	2.0	1.2	1.2	0.9	0.0
26	21.5	2.2	1.4	1.5	1.1	0.4
27	21.6	2.5	1.6	1.7	1.2	0.6
28	21.7	2.7	1.8	1.9	1.4	0.9
29	21.8	3.0	1.9	2.1	1.6	1.1
30	21.8	3.2	2.1	2.2	1.8	1.4
31	21.8	3.4	2.3	2.4	2.0	1.7
32	21.9	3.6	2.4	2.5	2.1	2.0
33	21.9	3.8	2.5	2.6	2.3	2.3
34	21.9	3.9	2.7	2.7	2.5	2.6
35	21.9	4.0	2.8	2.7	2.6	2.8
36	21.9	4.1	2.9	2.8	2.7	3.0
37	22.0	4.1	2.9	2.9	2.8	3.2
38	22.0	4.2	3.0	2.9	2.9	3.4
39	22.0	4.2	3.0	2.9	3.0	3.5
40	22.0	4.2	3.1	3.0	3.0	3.6
41	22.0	4.2	3.1	3.0	3.0	3.7
42	22.0	4.2	3.1	3.0	3.1	3.7
43	22.0	4.3	3.1	3.0	3.1	3.8
44	22.0	4.3	3.1	3.0	3.1	3.9
45	22.0	4.3	3.1	3.0	3.1	3.9

Birth intervals of less than nine months are due to multiple births.

control their fertility. In the Soviet Union, about 6 million abortions are registered each year compared with 5 million births. Abortions are concentrated in the European republics, mainly Russia and Ukraine. These republics registered 5.2 million abortions compared with 3.3 million live births (State Committee of the USSR on Statistics, 1989a). Population

Table 11.11. Mean age of mother at first child and average birth interval, 1950–1954 cohort.

| | Birth order | | | | | |
Age	1	2	3	4	5	6
16	0.0	0.0	0.0	0.0	0.0	0.0
17	16.5	0.0	0.0	0.0	0.0	0.0
18	17.2	0.0	0.0	0.0	0.0	0.0
19	18.0	0.0	0.0	0.0	0.0	0.0
20	18.7	0.8	0.0	0.0	0.0	0.0
21	19.5	1.0	0.3	0.0	0.0	0.0
22	20.1	1.1	0.5	0.4	0.0	0.0
23	20.6	1.3	0.7	0.7	0.0	0.0
24	21.0	1.6	0.9	1.0	0.6	0.0
25	21.4	1.8	1.1	1.2	0.8	0.0
26	21.6	2.1	1.4	1.5	1.0	0.5
27	21.8	2.3	1.6	1.7	1.2	0.7
28	21.9	2.6	1.8	1.9	1.4	0.9
29	21.9	2.9	2.0	2.1	1.6	1.1
30	21.9	3.1	2.2	2.2	1.8	1.4
31	21.9	3.3	2.3	2.4	1.9	1.7
32	21.9	3.5	2.5	2.5	2.1	2.0
33	22.0	3.7	2.7	2.6	2.3	2.3
34	22.0	3.8	2.8	2.7	2.5	2.5
35	22.0	4.0	3.0	2.8	2.7	2.8

Birth intervals of less than nine months are due to multiple births.

experts believe that women who undergo abortion have two abortions for each live birth. That means six pregnancies in a lifetime, four of which are aborted. The abortions that occur in the Central Asian republics are concentrated in those republics which have a high proportion of European nationalities (Russians, Ukrainians, and Germans) such as Kazakhstan. Contraceptives are not popular in the Soviet Union. Statistical data on contraceptive use are lacking. In the sociodemographic survey of 1985, no information was collected on contraceptive use. Contraception was not an issue to be discussed before perestroika. Information collected in some special studies may have been published in medical journals. Experts state that the general public and doctors have prejudices against hormonal contraceptives, due to expected health hazards. Other contraceptives are not generally available. Information on the availability and use of contraceptives is not published.

11.3   Conclusion

Nearly all Soviet women marry and have children. They marry and have children at young ages. In some parts of the country, half the brides are pregnant when they marry. Early marriage is both a determinant and a consequence of early fertility. Soviet women differ greatly from women in Western Europe in the early start of their marital and fertility careers. One-third of the women marry before their 20th birthday. There are 15 percent with at least one child at age 20; at age 25, 65 percent have one or more children. Recently published data and life-table analysis can be used to describe marital and fertility careers. The model underlying the career paths is a Markov model. Data limitations prevent the estimation of more complicated models. The parameters of the career processes, the occurrence–exposure rates of marital change and fertility, are estimated. Once estimated, they permit a reconstruction of entire biographies. The biographic indicators include probability and duration measures.

No attempt has been made to link marital and fertility careers. The marital career is estimated from period data, whereas cohort data are used to generate the fertility career. Changes in marriage patterns may greatly affect period data, which should therefore be interpreted with the greatest care. An integration of both careers into a single staging process would benefit the assessment of the impact of changes in age when married on fertility. In the Soviet Union, age when married is not the major factor in fertility, except in Central Asian republics.

The analysis of marriage and fertility can only provide a first impression of the processes that determine observed patterns. Because of the ethnic composition, great regional differences exist. Marital and fertility change in the Soviet Union cannot be understood without the regional and/or nationality component.

References

Bongaarts, J., 1987. The Projection of Family Composition Over the Life Course with Family Status Life Tables. In J. Bongaarts, T. Burch, and K.W. Wachter, eds. *Family Demography: Methods and Their Applications*, Clarendon Press, Oxford.

Chiang, C.L., 1984, *The Life Table and its Applications*. Krieger Publishing Co., Malabar, FL.

Chiang, C.L., and B.J. van den Berg, 1982. A Fertility Table for the Analysis of Human Reproduction. *Mathematical Biosciences* **62**:237–251.

Coale, A.J., A. Anderson, and E. Härm, 1979. *Human Fertility in Russia Since the Nineteenth Century*. Princeton University Press, Princeton, NJ.

Darsky, L., and S. Scherbov, 1990. *Parity-Progression Fertility Tables for the Nationalities of the USSR*. WP-90-53. IIASA, Laxenburg.

Demograficheskiy entsiklopeditcheskiy slovar' (Demographic Encyclopedia), 1985. Sovietskaya Entsiclopediya, Moscow.

Espenshade, T.J., 1987. Marital Careers of American Women: A Cohort Life Table Analysis. In J. Bongaarts, T.K. Burch, and K.W. Wachter, eds., *Family Demography: Methods and Their Applications*. Clarendon Press, Oxford.

Feichtinger, G., 1987. The Statistical Measurement of the Family Life Cycle. In J. Bongaarts, T.K. Burch, and K.W. Wachter, eds. *Family Demography: Methods and Their Applications*. Clarendon Press, Oxford.

Jones, E., and F.W. Grupp, 1987. *Modernization, Value Change and Fertility in the Soviet Union*. Cambridge University Press, Cambridge.

Keyfitz, N., 1988. A Markov Chain for Calculating the Durability of Marriage. *Mathematical Population Studies* 1(1):101–121.

Schoen, R., W. Urton, K. Woodrow, and J. Baj, 1985. Marriage and Divorce in Twentieth-century American Cohorts. *Demography* 22:101–114.

State Committee of the USSR on Statistics (Goskomstat), 1988. *Naseleniye SSSR, 1987: Statisticheskiy sbornik* (Population of the USSR, 1987: Statistical Collection). Finasy i statistika, Moscow.

State Committee of the USSR on Statistics (Goskomstat), 1989a. *Naseleniye SSSR, 1988: Statisticheskiy ezhegodnik* (Population of the USSR, 1988: Statistical Yearbook). Finasy i statistika, Moscow.

State Committee of the USSR on Statistics (Goskomstat), 1989b. *Tablitsy smertnosti i ozhidaemoy prodolzhitel'nosti zhizni naseleniya* (Tables of Mortality and Life Expectation of the Population). Finasy i statistika, Moscow.

State Committee of the USSR on Statistics (Goskomstat), 1990. *Demograficheskiy ezhegodnik SSSR, 1990* (USSR Demographic Yearbook, 1990). Finasy i statistika, Moscow.

Volkov, A.G., 1986. *Sem'ia-obiekt demografic* (The Family as the Object of Demography). Mysl, Moscow.

Willekens, F.J., 1987. The Marital Status Life Table. In J. Bongaarts, T.K. Burch, and K.W. Wachter, eds. *Family Demography: Methods and Their Applications*. Clarendon Press, Oxford.

Willekens, F.J., 1988. A Life Course Perspective on Household Dynamics. In N. Keilman, A. Kuijsten, and A. Vossen, eds. *Modelling Household Formation and Dissolution*. Clarendon Press, Oxford.

Willekens, F.J., forthcoming. *Life Course Analysis: Stochastic Process Models*.

Wunsch, G.J., and M.G. Termote, 1978. *Introduction to Demographic Analysis: Principles and Methods*. Plenum Press, New York, NY.

Zeng Yi., 1986. Changes in Family Structure in China: A Simulation Study. *Population and Development Review* 12(4):675–703.

Zeng Yi., 1990. *Family Dynamics in China: A Life Table Analysis*. University of Wisconsin Press, Madison, WI.

Chapter 12

Teenage Marriages in Latvia

Andis Lapinš

Great interest in the study of marriages in Latvia was aroused when Latvia recorded the highest divorce rate in the USSR, and one of the highest divorce rates in the world in the early 1960s. In 1979 the divorce rate peaked at 5.5 divorces per 1,000 inhabitants, in the following years it remained level at 4.7 to 4.8, and only started to decline in 1985. In 1989 it was 4.2. The divorce rate in Latvia is higher than in most European countries: France, 1.9 (1987); Federal Republic of Germany, 2.1 (1988); Belgium, 2.0 (1987).

It is often assumed that one factor in the breakup of a marriage is the age of the partners. For many social and psychological reasons teenage marriages are of special concern, especially since almost all of them have at least one child.

The mean age at marriage was decreasing in Latvia from the mid-1950s until the early 1980s. Then age when first married started to increase. Despite this increase, the number of married couples under age 18 also increased (*Table 12.1*). This study focuses on this special group.

This investigation studies the demographic, economic, and social problems of married couples under age 18 and the formation of their families. The study was organized in Riga, Latvia. The sample was

Table 12.1. Teenage marriages in Latvia as percent of the total number of first marriages.

Year	Males	Females
1987	0.7	2.5
1988	0.9	3.1
1989	1.0	3.6

drawn from couples who were permitted to marry under the age of 18, the minimum age to marry determined by the Latvian Marriage and Family Code. All the marriages were registered between 1987 and 1988. Questionnaires were sent to 308 couples, and 121 completed forms were processed. It is difficult to evaluate the representativeness of this sample because no structural information is available for the whole population of those who married below age 18.

12.1   Sociodemographic Structure of the Couples Surveyed

Of the females surveyed, 20 percent were married at age 16 and 80 percent at age 17. The majority, 71 percent, was married between the ages of 17.5 and 18. Of the males studied, 33 percent were married between ages 17 and 17.5, and 67 percent were between ages 17.5 and 18. The 16-year-old brides predominantly married 18-year-old bridegrooms. In general 17-year-old brides chose bridegrooms who were between ages 21 and 24; only 10 percent married men of the same age. The 17-year-old bridegrooms for the most part married 17- and 18-year-old brides. There were only a few cases when a bridegroom or a bride under age 18 married a partner who was five or more years older.

In the majority of cases the minors married partners of the same nationality. At the time of the survey, 3 percent of the women were divorced and 5 percent acknowledged that their marriage was in trouble. Among men 3 percent were divorced; one was a widower.

12.2   Pre-nuptial Relationship

The couples met under a variety of circumstances. Of those answering the survey, 35 percent met through friends and acquaintances, 21 percent met at parties and discotheques, 10 percent lived in the same neighborhood, and 34 percent met in other situations.

The average duration of the pre-nuptial relationship was one year and three months. About half the courtships did not exceed a year: in 40 percent of the cases the relationship lasted from half a year up to a year; in 10 percent, less than half a year; but in 20 percent, less than three months. A short pre-nuptial relationship is often considered a risk factor of marriage stability.

The first sexual activity took place, on average, five months into the relationship; however, 45 percent of those questioned acknowledged that they were sexually active in the first three months. All of the respondents had a child. Women became pregnant 10.5 months after their first sexual relation, on average. Registration of marriage followed, on average, after five months.

Some of the reasons for marriage given by women included: "I was pregnant" (29 percent); "We loved each other" (24 percent); "I wanted to be independent" (19 percent).

The men surveyed gave the following motives: "mutual love" (23 percent); "pregnancy" (23 percent); "love toward the partner" (17 percent); "the need for independence from parents" (16 percent).

12.3   Children in Families of Teenagers

All of the couples surveyed had children: one child in 72 percent of families; two, in 28 percent. Thirteen percent of the wives had become mothers at the age of 16; 44 percent, at the age of 17; 31 percent, between the ages 18 and 19. Eleven percent of the husbands had become fathers at the age of 17; one became a father at the age of 16.

The results of the survey show that the women expected to have 2.1 children; men expected to have 2.4 children. A family with two children was preferred by 62 percent of women and 44 percent of men. A family with three children was preferred by 22 percent of women and 44 percent of men. One child was preferred by 14 percent of women and 8 percent of men.

The ideal number of children in a family was 2.6 for women and 2.8 for men. A family with two children was considered ideal by 44 percent of women and 31 percent of men. A family with three children was considered ideal by 39 percent of women and 40 percent of men.

12.4   Living Conditions

In the first year of married life 34 percent of the minors received some financial help from parents or relatives: 18 percent of these families were fully dependent on parents, but 16 percent had a common budget with parents' families. Thus, in this respect 34 percent of the families were not independent, autonomous social formations in the first year of marriage. In the second year, the situation changed slightly: 16 percent of families were fully dependent on the parents; 14.2 percent of these families had a mutual budget with parents.

In the first year of marriage, regular pecuniary aid from the wife's or husband's parents was received by 30 percent; in the second year, by 24 percent. Irregular aid was received by about 20 percent. Four percent of families of minors could not depend on pecuniary aid in the first year of cohabitation; 9 percent, in the second year.

Only 12 percent of those questioned stated that their families could do without pecuniary aid in the first year of cohabitation; 17 percent, in the second year.

Of the husbands answering, 47 percent said that the family's financial resources were used by both mates equally; 35 percent said, mainly by the wife; 14 percent said, mainly by themselves. Of the wives responding, 40 percent said that these resources were used by both mates equally; 38 percent said, mainly by themselves; and 16 percent said, by husbands.

Housekeeping is done either by the wife or by both mates. The husband participates in housekeeping to a much smaller extent, as compared to the wife. Child care is the responsibility of the wife; however, 33 percent of those questioned acknowledge that the child is cared for by both parents in turn. Cooking is done by the wife (70 percent), by both mates (20 percent), or by wife's (husband's) parents (8 percent). The daily tidying up is done mainly by the wife (60 percent). The general tidying up is done in most cases by both mates.

12.5   The Minimum and Optimum
Age of Marriage

Only some of those questioned were of the opinion that the minimum age for registration of marriage in Latvia should be 17 or even 16. A minimum age for marriage for males below 18 was preferred by 6 percent of women and 5 percent of men interviewed. A minimum age below 18 for females was preferred by 11 percent of women and 8 percent of men.

The women surveyed were of the opinion that the optimum age of marriage for women is between 18 and 20. The optimum age of marriage for men, in women's opinion, is between 20 and 22. In men's opinion, the optimum age of marriage for women is between 18 and 20; for themselves, 20 and 25.

12.6   Sex and Morality Attitudes

An opinion is often expressed that marriage of minors is a consequence of early puberty and liberal sexual morality. Of the men questioned 25 percent considered casual pre-nuptial sexual relationships permissible. However, considerably more men (46 percent) considered pre-nuptial sexual relationships only in a serious relationship. Of the men questioned 28 percent were of the opinion that pre-nuptial sexual activity was permissible only with the future wife. One male respondent considered all pre-nuptial sexual activity impermissible.

Women have more restrictive views on sex and morality. Only 14 percent of those questioned considered casual pre-nuptial sexual activity permissible. Pre-nuptial sexual activity in a serious relationship was considered permissible by 34 percent of women, i.e., by 12 percent less than men. Pre-nuptial sexual relationships with the future husband were acknowledged by 39 percent of women, which fully concurs with men's point of view regarding relations with the future wife.

Men placed more restrictions upon women than upon themselves. Of those questioned 18 percent considered pre-nuptial sexual relationships for women impermissible; 43 percent of men were of the opinion that such relationships may take place only with the future husband, which concurs with the women's point of view. Thirty-six percent of men approved of pre-nuptial sexual activity in a serious relationship; yet only 4 percent approved of it in a casual relationship.

Women's views regarding men's sexual activity are more liberal: 21 percent of them had the opinion that a man may have pre-nuptial sexual relations in a casual relationship. Thirty-four percent of women accepted such relations in a serious relationship. Thirty-nine percent felt it was acceptable for a man to engage in pre-nuptial sexual relationships, which concurs approximately with men's opinion.

Of the men questioned 16 percent revealed a double-standard view on sex morality: "What is permissible for a man, is impermissible for a woman." None of the women supported this double standard. This double standard was not accepted by 35 percent of the men questioned and 51 percent of the women questioned whose attitude to pre-nuptial sexual activity was very tolerant.

It is also interesting to study the attitude to consensual unions. Twenty-one percent of the men questioned and 20 percent of the women questioned considered permissible the fact that couples having formed a family sometimes do not register their marriage. Thirty-three percent of men and 48 percent of women acknowledged that in specific cases a marriage may not be registered. Seven percent of men and 8 percent of women consider unregistered marriage impermissible.

12.7   Satisfaction with Family Life

Eighty-two percent of husbands declared that their satisfaction with family life had risen with the birth of a child. None of the husbands expressed dissatisfaction after having become a parent.

Unfortunately, the birth of a child worsened the material welfare for some of the new families (45 percent of men and 55 percent of women). Forty-three percent of young fathers and 33 percent of young mothers were of the opinion that their material welfare had remained the same, and 12 percent pointed out that their material welfare after the birth of the child had risen.

Forty-four percent of husbands and 31 percent of wives declared that after the birth of the child their relationship had improved; 46 percent and 52 percent, respectively, noticed no change. After the birth of the child, relations with the wife worsened for 9 percent of husbands, and those with the husband worsened for 17 percent of the wives.

After the birth of a child, relationships of young mothers improved both with own parents (42 percent) and with those of their husband's (16 percent); they remained the same in 42 percent and 61 percent, respectively. In 23 percent of cases the birth of a child worsened relationships with the husband's parents; and in 16 percent of cases, it worsened with those of their own parents.

The husband's relationships to parents and parents-in-law also improved after the birth of a child, yet not as much: with parents-in-law the relationships improved in 22 percent of cases; with own parents the relationship improved in 11 percent of cases. Seventeen percent of young fathers declared that relationships had worsened with own parents, and in 12 percent of cases they also worsened with the wife's parents.

Satisfaction with family life after the birth of a child was expressed more by husbands (55 percent) than by wives (42 percent). It did not change with 43 percent of men and 45 percent of women. Thirteen percent of women declared that their satisfaction with family life after the birth of the child had lowered (only one husband gave a similar response).

On the whole, 69 percent of husbands and 58 percent of wives were satisfied with their family life; 25 percent of husbands and wives were not satisfied. Six percent of husbands and 12 percent of wives were dissatisfied with their family life.

If relationships between parents were unsatisfactory, 31 percent of men and 33 percent of women answered that it would be better for a child to stay with one parent. Thirty-one percent of husbands and 25 percent of wives had the opinion that it is better for a child to live with both parents, notwithstanding their bad relationship.

12.8   Summary

The survey of families of teenagers who married below age 18, who had registered their marriages in 1987–1988, was carried out in Riga, Latvia. All women questioned were pregnant before marriage. When the motives of marriage were arranged according to their significance "mutual love" prevailed, followed by "pregnancy." One-third of families of minors were not independent, autonomous social formations, and received some subsistence from their parents. Sixteen percent of the men questioned supported a double standard on sexual morality: "What is permissible for a man, is impermissible for a women." The average desired family size was 2.1 for women and 2.4 for men. On the whole, 69 percent of husbands and 58 percent of wives were satisfied with their family life, but 6 percent of husbands and 12 percent of wives were dissatisfied.

Chapter 13

Large Families in Low-Fertility Regions: A Social Portrait

Lidia Prokophieva

The topic of large families has recently come under public scrutiny. One reason for this is the desire to protect children from the present economic transformation, which often implies a lower standard of living for families.

The republics of the Soviet Union show two very different fertility patterns: one in the Central Asian republics and the other in the European republics. A large family is the prevailing type of family in the Central Asian republics. The populations in these regions practice old traditions of having many children in a family. Due to the increase in the population of Central Asia in absolute and relative terms this phenomenon gains additional importance.

The concept of a large family is different in different republics: a family with five children in the Central Asian republics is a medium-size family; in the European republics this would be considered a large family.

This chapter studies and describes the phenomenon of large families, especially in the regions where nuclear families with one or two children prevail. Studies of this kind have been made in several centers, particularly in the Institute of Sociology of the Academy of Sciences in Moscow (Antonov and Dudchenko, 1988). However, families in Moscow do not seem to be fully representative even of European Russia; in terms of educational background, skill, and interests, families in Moscow would be nearer to the characteristics of the populations in the Baltic states.

In this study we have tried to answer the following questions: What are the causes of this phenomenon in modern Russia? What impels families to behave contrary to standard demographic behavior in the conservative environment of a small town? To what extent may the opinion that a large family is the result of the immoral life-style of parents be justified? Such a negative attitude toward large families in the USSR has been based on socioeconomic factors: with a generally low standard of living annoyance is felt toward the families who qualify for privileges and assistance from the society because of their large size.

From the studies we have tried to provide a social portrait of large families. We have studied the motivation for a large family, the socioeconomic status of the parents, and the degree of their responsibility for the well-being and upbringing of their children. In addition, the essential differences in the way of life have been considered.

We use data from the socioeconomic survey carried out in Taganrog, a medium-sized city in Russia, in early 1990. Taganrog is a typical industrial center in European Russia, with a population of 250,000 to 300,000. It is also typical with respect to industry structure, sex and age structure of the population, family structure, level of income, housing conditions, urban social infrastructure, and so on. On the basis of a 5 percent sample selective aggregate of urban families, a survey of demographic structures and standards of living of the population had been carried out for the city where one- or two-child families were prevalent. In parallel, a specific large-family subproject had been executed with targeted excerpts covering all large families in the area.

Here we should define a large family. In demography often one- or two-child families are viewed as being small; three- or four-child families belong to the medium-size category; large families are those with five or more children. A family is entitled to receive a family allowance for large families starting with the fourth child. The threshold to large-family status may be subject to the region of the country or the pattern of population reproduction prevailing there. As for Moscow and other

Table 13.1. Demographic structure of small (one or two children) and large (four or more children) families, in percent.

	Two parents			Single parent		
		Of them			Of them	
	Total	2 genera-tions	> 2 genera-tions	Total	2 genera-tions	> 2 genera-tions
Small	82.8	72.6	27.4	17.2	57.6	42.4
Large	85.1	90.2	9.8	14.9	84.0	16.0

large cities in European Russia a three-child family may seem to be a deviation from "the social standard" (Achildieva, 1990).

In medium-size cities of European Russia the traditional approach to family size seems to be somewhat different; therefore we use a locally accepted principle for defining large families. The poll studied all of the families with four or more children who had been receiving or were receiving large-family allowances. Thus, we had an opportunity for comparing the standard of living of families which have usually been characterized as medium-size (four children) and proper large-size families (five or more children). Further analysis has shown that such a demarcation is useful also from the viewpoint of a social profile of those family groups. A total of 250 families were polled which amounted to less than 1 percent of the families in the city.

13.1 Demographic Structure of Families

Families with four children make up 45.2 percent of the sample of large families; families with five children, 36.9 percent; families with six or more (maximum of 8), 17.9 percent. Two-parent families make up the overwhelming majority of large and small families (*Table 13.1*). Families with one or two children used to be considered small. Among large families the proportion of one-parent families is small.

The fact that the overwhelming majority of children in large families have two parents is a positive factor of the demographic structure of these families. As for large families in which children are brought up only by their mother, the analysis shows that they significantly differ from single-parent families with one or two children. Only one-third of large one-parent families were formed as a result of a divorce or a parent's

Table 13.2. Age of mothers in small (one or two children) and large (four or more children) families, in percent.

	Under 25	25–30	30–35	35–40	40–45	45+	Average age
Small	12.0	16.8	19.2	24.4	11.3	16.3	35.5
Large	1.2	9.0	34.1	39.5	10.8	5.4	35.8

death. Most single-parent families comprise a mother with children from several unions, mostly unregistered.

The evolution of a modern family has been accompanied with changing attitudes to traditional marital status (Goldtorp, 1987); however, in this study it implies an irresponsible attitude toward family and children. In one-quarter of one-parent families children do not know their father. Nuclear families (with two generations present) prevail among large families; the proportion of extended families (which include grandparents or other relatives) is much smaller than among families with one or two children.

One could conjuncture that parents of small families are young and have not separated from their parents' household. However, the analysis of the age structure of mothers shows that subgroups of families with one or two children contain a significantly large number of both young and mature families (wife age 45 or older); the average age of women is nearly the same for small and large families: 35.5 years in families with one or two children and 35.8 years in large families (*Table 13.2*).

13.2   Social Structure of Families

Another aspect to study is the educational background and professional status of parents in large families. The following groups have been singled out: service personnel, low-skilled workers, medium-skilled workers, high-skilled workers, employees with special education, employees with high education, high- and medium-level managers, and others. In this study a family's social type is determined by the member of the family with the highest qualifications; in most cases it is the husband's qualification and in extended families it is the eldest family member's qualification.

Table 13.3. Per capita monthly incomes of small and large families, in percent.

| | Roubles | | | | | | Mean per capita income in roubles |
	Up to 50	50–75	75–100	100–125	125–150	150 and more	
Small	2.9	9.9	19.6	22.7	19.0	25.9	118.2
Large	33.3	41.7	17.8	5.4	0.6	1.2	63.0

We find that two-thirds of those polled are workers' families of high- and middle-class status, with the high-class group prevailing. This does not mean that there are no so-called asocial families among the large ones, where neither husbands nor wives have qualifications or work. The share of these families is 4 percent. In addition, we find that in 12 percent of families the parents lack qualification for securing a required standard of living for their children. It is this last group of families that stirs up anxiety in society and gives reason for a negative attitude on the part of the population to large-family benefits. But these benefits are needed because three-fourths of these families are below the poverty line, equal to 85 roubles per person per month in the beginning of 1990 (Mozhina, 1991).

What are the components of a large-family income? Is the size of the family the only factor which may lower a family below the subsistence level? Analysis shows that the total per capita income in large families is lower than that in small families. Per capita income distribution of small and large families is given in *Table 13.3*. A woman with many children would not be capable of contributing to the total family budget; in 40 percent of cases she either does not work or only carries out unskilled low-paying jobs. As for males, they would be striving to make up for the inability of mothers to perform production-related activities.

The data reflect a situation that is peculiar on average for all families, while for some families the opportunities for a wage increase and a husband's ability to get it may differ. Until recently the possibilities of increasing a family's income through additional employment have been limited. But when asked, Would you like to moonlight for additional pay?, nearly 40 percent of men having five or more children replied negatively. It is interesting to note that in families with four children the men's feeling of responsibility for the children's well-being will be higher;

Table 13.4. Classification of husband and wife by type of career and size of family, in percent.

	Descending	Horizontal	Weakly growing	Ascending
Husbands				
Small	12.3	21.3	30.5	35.9
Large	19.0	32.4	25.4	23.2
Wives				
Small	18.8	45.5	25.6	10.1
Large	48.9	36.7	10.0	4.4

among them nearly three-quarters would like to have perquisites, mainly medium-skilled and white-collar workers with high and secondary special educational background.

An essential characteristic of a working person's social mobility is the type of a career during one's working lifetime. We have determined four groups of working careers: ascending (a significant improvement of educational and qualification level, promotion), weakly growing (insignificant shifts toward better skills), horizontal (lack of promotion, but increased wages due to transverse movements such as a change in occupation or job), descending (socio-occupational mobility due to a deteriorating skill, demotions, and so on). The data obtained show that there are differences in the types of careers pursued by a husband and wife in relation to the number of children in the family (*Table 13.4*).

The data confirm that a large family would mostly be a result of deliberate choices of the way of living, mode of behavior in private life and at the place of work, aimed at the family and upbringing of children. This may explain why a woman in a large family will be less successful in her career than a woman in a family with one or two children. However, it is amazing to see a similar regularity with men. To a certain degree it may be explained by the search for better paying jobs which would not necessarily be linked with upgraded skills; it may be a transfer of workers to hazardous jobs with better pay or of technical staff to nontechnical positions because salaried employees' remuneration is still less than that of nontechnical workers.

Women may experience problems trying to realize their professional status irrespective of the number of children; nearly half of mothers with one or two children may experience under present conditions only

Table 13.5. Women's educational level in the small and large families, in percent.

	Educational background at time of marriage				Educational status	
	Higher	Special secondary	Secondary	Below secondary	Improved	Same
Small	14.2	27.5	38.5	19.8	31.2	68.8
Large	3.7	12.9	41.4	42.0	18.3	81.7

horizontal transfers (change of occupation or place of work without skill development or promotion). At the first stages of a family life cycle, when a woman's attention would mostly be concentrated on rearing small children, the lack of an adequately developed social infrastructure (services, good preschools, and so on) deprives her of the opportunity of being promoted.

In the case of mothers with many children a deteriorating occupational status becomes a decisive factor. More than a half of the mothers polled in large families would prefer to look after their children provided material provisions are available. Only 30 percent of women would pursue a professional career. From the analysis of the major types of female careers in families of different sizes, we conclude that in large families the choice will generally be in favor of the family: employment in most cases remains a means of earning income for the family, not an aspiration to realize oneself in a certain occupation.

The results of the survey provide an opportunity for keeping track of the interdependence of the dynamics of the mother's educational growth and the number of children in a family. Contrary to the assertions of some sociologists, conventional wisdom has been verified: women without specific education or skills would pattern their lives on having many children. Less than 20 percent of mothers with many children had higher or secondary training (*Table 13.5*).

Young families burdened with the need for acquiring necessities should not be compared with older families who have similar incomes; the latter do not face the problems of furnishing a home, such as buying furniture and durables. Large families may also be differentiated on the basis of the same terms: there are families whose property would be quite meager (beds, a table, sometimes a refrigerator and TV set); on the other hand, 45 percent of large families do have color TV sets,

Table 13.6. Housing conditions for small and large families, in percent.

	Private flat, house	Room in parents' flat	Room in shared flat	Dormitory or rented flat
At time of marriage				
Small	18.9	42.9	10.1	28.1
Large	12.5	33.9	13.1	40.5
At time of survey				
Small	72.5	5.9	16.5	5.5
Large	72.8	4.4	18.4	4.4

and 15 percent have cars. However, a comparative analysis of large and small families shows that the families with one or two children have, on average, more possessions.

We studied the correlation between the number of children in a family and housing conditions when we examined the population status in the city of Taganrog in 1980 (longitudinal analysis). On the basis of past studies two conclusions were made on the effects of housing conditions on the number of children in a family in the regions with a low birth rate: on the one hand, a good house or flat would promote the desire of a family to have one more child; on the other hand, families would have one or more children in the hope of getting an individual dwelling much faster (Prokophieva, 1991).

In the past would-be large families lived in less favorable conditions than the families who limited themselves to one or two children (*Table 13.6*). At the time of this poll, the housing conditions of these two groups had become similar. One of the essential factors in this was that large families were given flats with modern amenities out of turn. Of those families who indicated the reason for their size was their desire to obtain a bigger flat, 70 percent had initiated their family life in a rented room and 30 percent had a room in their parents' flat or in a flat shared with several families. Currently the families have been given spacious municipal flats, but 40 percent of them had to go through one more stage, that is, several rooms in a shared flat or in a hostel.

13.3 Reasons for Large Families

We asked married couples how many children they planned to have, what size of family they regard as ideal, and if they planned to have more children. The results show that the share of families who have as many children as they planned for is rather small, only 12 percent. A considerable proportion of families (46 percent) did not give any thought to this question, while 42 percent wanted to have a small family. When asked about the ideal number of children in the family, 44 percent of large families said four or more; the remaining families preferred a family with two or three children.

These figures indicate that the intended size of a family and the final size of a family (18 percent of families would like to have more children) may be very different. There is also a difference between the actual large size of a family and preference to have two or three children in a family. Here, a decisive role is played by the traditions in society.

Hence, the question arises, how many couples plan to have a large family? The data in Section 13.2 characterize the objective socio-economic parameters of the families polled, but do not provide an answer to two questions: What accounts for their decision to have many children? Was this decision a deliberate one? The reply to these questions can be used to assess the respondents' decisions, their views on the value of children, their attitudes in the society toward large families, and so on. In addition, a hypothesis has been tested regarding the notion that large families would reproduce themselves. Accordingly the couples were asked questions about the number of children their parents had.

The last question asked was: Why did you have a large family? The replies to this question have been used to classify motivations for large families. The answers give a "social profile" of each group to be outlined more explicitly. The profile was used by social security officers to appraise the atmosphere for raising children when family conditions were suspected of being unfavorable because of the parents' abnormal behavior.

We single out three groups: 1) families with a deliberate number of children; 2) married couples who did not intend to have a large family; 3) families with parents who do not take much interest in raising their children. The first group of families, 44.6 percent of those polled, intend to have a large family. This group contains the largest families. The reasons for having many children given by the families in this group include: love for children (20.8 percent), religion (13.7 percent), and family traditions (10.1 percent).

The second group (39.3 percent) comprises families in which many children are to a large extent an implication of fortuitous factors: the birth of twins or triplets (3 percent), children from second marriages (13.7 percent), or unwanted children born as a result of medical errors in defining pregnancy (9.5 percent). In this group we also include the families who tried to have one more child of a different sex (5.9 percent); this reason was given often by the families with four or five children. Some families increased the number of children to acquire better housing within a shorter period of time (7.2 percent). This phenomenon seems to be inherent to the USSR where the government's controlled distribution of flats is based on living space for each person in the family.

In most cases the behavior of the first and second groups of families cannot be considered asocial, immoral behavior. This type of behavior is evident in the third group, fortunately small in size (16.1 percent). Educational opportunities for these families are scarce and often, as experts point out, would have a negative character. Any increase in allowances will not help the situation. Children in such families would need specific help from family centers at the local level.

The survey results made it possible to verify the hypothesis about the roots of large families, that is, to what extent the parents' example had been followed. Among large families only 48 percent of married couples had grown up in families with many children; however the situation in certain groups, as to why they had many children, was different. The groups having many children were often motivated by the number of children in their parents' families (60 percent); in the remaining groups the share of families where the married couples had grown up in large families did not reach 40 percent.

Some of the families in the first group had not completed their families at the time of the poll; 30 percent of these families intend to have more children. As for the ideal number of children, only 15 percent of families in the first group of large families regret having a large family and believe that it would have been better to have a smaller family (one to two children). In the second group of large families, nearly 50 percent of parents regret having a large family; of those who believe that a family with three children is the ideal size, 65 percent would be unhappy about their big size.

The comparative analysis of socioeconomic characteristics of families in each group shows that there is no uniformity. Thus, the educational

Table 13.7. Mother's educational level in ramilies by group, in percent.

	Higher	Special secondary	Secondary	Below secondary
Group I – deliberate large size	9.5	22.2	46.1	22.2
Group II – incidental large size	7.5	25.0	45.0	22.5
Group III – careless large size	–	5.3	57.9	36.8

Table 13.8. Working career of husbands and wives in large families by group, in percent.

	Descending	Horizontal	Weakly growing	Ascending
Husbands				
Group I	21.1	30.2	22.4	26.3
Group II	14.9	27.7	31.9	25.5
Group III	21.1	52.6	21.0	5.3
Wives				
Group I	50.7	29.0	14.5	5.8
Group II	38.1	45.2	11.9	4.8
Group III	66.7	13.3	20.0	–

level of mothers in the families from the third group is the lowest of the three groups (*Table 13.7*).

The third group features the lowest mean per capita income, though the number of children in these families is less than in the the first group of families. Many in the third group are less active in looking for a job (*Table 13.8*). This may explain the low career profile of both men and women in this group.

Thus the investigation of large families in a typical Russian medium-size town shows they are very difficult to categorize. This heterogeneity calls for the necessity of having a variety of social policies that address children's interests in all families.

How does one help large families in the cities of Russia? It is necessary to make the society, on the whole, and each family more prosperous; funds can then be earmarked for assisting families in dire straits. But the solution is complex. The shift from parasitic smugness to self-reliance seems to be a complicated process and everybody has to go through it.

Assistance from the state in this case will be necessary for the creation of more opportunities.

But so far in the respondents' replies one can discern an evident tendency to parasitic smugness: it manifests itself in demands to increase assistance from the state for large families to provide for all their needs.

The value of children for both the family and society should be recognized. Their education and welfare at both the family and state levels should be addressed.

References

Achildieva, E., 1990. Gorodskaja mnogodetnaja sem'ia (Large Urban Family). *Soziologicheskie issledovaniya* (Sociological Studies) 9:72–79.

Antonov, A., and O. Dudchenko, 1988. O demograficheskoj polityke v otnochenii mnogodetnych semey (Demographic Policy for Large Families). In *Sozialno-ekonomicheskie issledovaniya braka, sem'i, reproduktivnich ustanovok i voprosi prognosirovaniya naseleniya* (Socioeconomic Research of Marriage and Family). Institut soziologii AN SSSR, Erevan.

Goldtorp, J., 1987. *Family Life in Western Societies: A Sociology of Family Relationships in Britain*. Cambridge University Press, Cambridge.

Mozhina, M., 1991. O suschnosti i metodach opredeleniya minimalnogo potrebitelskogo budjeta (Essence and Methods of Defining the Minimal Cost of Living). In A.G. Vishnevsky, ed. *Demografija i sociologija: Sem'ia i semeynaya politika* (Demography and Sociology: Family and Family Policy). Institute for Socioeconomic Studies of Population, Moscow.

Prokophieva, L., 1991. Sozialno-ekonomicheskij status semey na otdelnich etapakh jiznennogo tsikla (Socioeconomic Status of the Families in Different Stages of Life Cycle). In *Narodnoe blagosostojanie: tendntsii i perspectivy* (Welfare of the Population: Trends and Prospects). Nauka, Moscow.

Chapter 14

Public Opinion on Fertility and Population Problems: Results of a 1990 Survey

Valentina Bodrova

A well-known feature of the demographic situation in the Soviet Union is that in many of the European republics fertility rates are not high enough to ensure full replacement of generations. Problems related to the low fertility rates are high levels of infant mortality, high divorce rates, and an aging population.

Many politicians and scholars realize the necessity for a new approach to government policies concerning the demographic development as part of general social policy. For a better understanding it is necessary not only to have aggregate statistics on demographic trends but also to know the views of the population on issues relating to childbearing and the family. The findings of the survey described in this chapter provide an integral picture of the most urgent – in people's view – issues related to demographic developments in the context of economic and environmental crises, and of the trends of demographic policy in the country.

Table 14.1. Distribution of respondents by socioeconomic factors, in percent.

Category	Share in survey	Share in population
Males	45.8	45.4
Age		
Under 25	21.1	18.7
25–54	53.6	54.6
55 and over	25.3	26.7
Education		
Complete or some university	17.0	12.0
Secondary and technical education	57.7	47.0
Primary education	25.3	41.0
Monthly per capita income		
Income up to 100 roubles	33.8	34.0
Income from 101 to 150 roubles	31.8	32.0
Income more than 150 roubles	34.4	34.0

In June 1990 the All-Union Center for Public Opinion and Market Research (VCIOM)[1] carried out an opinion poll entitled "The Demographic Situation as Mirrored in the Public Opinion." The purpose of the research was as follows:

• To elicit views on the most acute demographic issues in the country.
• To elicit views on the demographic problems that might face the country in the near future.

A sample of the population of the USSR above age 16 was surveyed. The poll was carried out among urban and rural residents in 28 towns and 13 villages in Russia, the Ukraine, Estonia, Georgia, Kazakhstan, Uzbekistan, and Tajikistan. A total of 2,708 people were polled. The samples are representative for urban and rural populations of the USSR. *Table 14.1* shows that the discrepancy between the selected quotas of respondents and the distribution derived from official statistics data by sociodemographic and socio-occupational groups was rather low (1.6 percent, on the average). The biggest relative deviation was among people with incomplete secondary education.

14.1　Views on Ideal Family Size

Fertility is considered one of the most important components of population dynamics. Five questions on this subject were posed in this poll.

One question (How many children should a family have today?) was asked to probe the views on the ideal number of children in a family (see *Table 14.2*). The most popular answer was a family with two or three children; 41 percent favored a two-child family and 34 percent a three-child family, in both towns and villages. Thirteen percent of the respondents preferred four or more children, twice as many as compared with the single-child-family supporters and six times more as compared with those favoring childless families.

Some substantial differences may be observed when analyzing the respondents' replies with regard to socio-demographic characteristics. The most interesting differences in the responses to this question were evident in the following categories: place of residence (republic, rural or urban area), age, marital status, family size, and nationality.

The poll showed that in Estonia, where fertility is now slightly above replacement level, a preference toward the three-child family had developed. Nearly half (48 percent) of the respondents thought a three-child family was the ideal size in a contemporary family. The percentage in Estonia was higher than in Tajikistan (46 percent) and Uzbekistan (41 percent), where in contrast with Estonia the existing pattern of reproduction is in line with the opinion expressed. The Estonian phenomenon can probably be explained by the fact that the Estonians consider low fertility a threat to their national existence; however, they are less willing to put this belief into practice.

In the Ukraine (43 percent) and in Russia (41 percent), the concept of a two-child family prevailed. In the Ukraine, in contrast with other republics, 13 percent of the respondents expressed the view that a family should consist of only one child. In Georgia 17 percent of the respondents were of the opinion that a family should include six or more children; in Russia only 6 percent of the respondents felt the same.

As the size of the community decreases, the number of those who favored a three-child family increases, which is one reason for the high fertility in those areas. Thus, as compared with Moscow (18 percent) in a district center, the percentage of respondents in favor of a three-child family was doubled (39 percent), and in the villages the percentage of respondents was even higher (41 percent).

Table 14.2. Replies to the question, How many children should a family have today? (in percent of number interviewed): 1, No obligation to have children; 2, One child; 3, Two children; 4, Three children; 5, Four or five children; 6, Six or more children; 7, No opinion.

Category	1	2	3	4	5	6	7
Total	2	5	41	34	8	5	5
Area							
Urban	2	5	41	34	8	5	5
Rural	2	5	41	32	9	7	4
Region							
Russia	2	6	41	31	10	6	4
Ukraine	2	13	43	30	3	0	9
Estonia	2	4	27	48	13	3	3
Georgia	0	3	36	24	14	17	6
Kazakhstan	2	5	41	34	12	1	5
Uzbekistan	3	4	46	41	1	2	3
Tajikistan	1	2	41	46	5	2	3
Type of community							
Moscow	11	9	53	18	3	1	5
Capital city	1	4	49	28	8	7	3
Regional, autonomous republic center	1	12	47	28	3	1	8
District, autonomous region center	1	4	41	39	6	5	4
Other town	1	8	53	29	6	1	2
Village	1	4	23	41	16	10	5
Education							
University degree	2	8	40	33	7	5	5
University undergraduate	1	7	38	33	14	3	4
Secondary technical	3	6	44	32	8	3	4
Vocational school	3	9	42	27	4	11	4
Secondary school	2	7	43	33	6	4	5
Grade 8 or 9	0	6	37	33	17	5	2
Grade 7 or under	1	4	29	43	9	6	8
Nationality							
Russians	2	8	49	31	3	2	5
Indigenous population	1	4	26	36	16	12	5
Other	1	6	36	37	15	3	2

There were substantial differences in the views of respondents belonging to various generations on the necessity of having three children in a family: among youths up to 20 years of age, 24 percent favored three-child families; among those from 50 to 54, 38 percent; among those age 60 and older, 47 percent. The educational factor had less influence than the age structure on the respondents' view of the ideal number of children in a family.

The views of the respondents either without children or with one child below 15 were similar: 29 percent of the respondents without children stated that a contemporary family should include three children, while 30 percent of those with one child under 15 stated the same. Those respondents who actually have families with two children or three or more children below the age of 15 were more inclined to say the ideal size would be three children (35 percent and 34 percent, respectively).

Among the Russian population in all republics the idea that two children would be ideal in a family was twice as widely accepted as compared with the mostly Central Asian indigenous populations. The views on a three-child family were not very different among the representatives of the various nationalities (from 31 percent to 37 percent), but they were sharply different with regard to four or more children in a family. In this case only 3 percent of the Russians and 16 percent of the indigenous population and other nationalities thought that a family should have four or five children. The opinions of men and women did not differ concerning families with four or more children. But the two child family was favored more by women than by men.

Among the factors that substantially affected the views on the number of children in present-day families were housing conditions, family monthly per capita income, and occupation. The respondents who lived in a house of their own or in a flat with four or more rooms (41 percent and 38 percent, respectively) were more likely to favor a three-child family; those who lived in a single- or double-room flat favored the two-child family. Those who rent an apartment or room were three times less likely to favor a three-child family than those who lived in a house of their own.

There was not much difference in opinions on families with three children between public sector employees (35 percent) and collective farmers or pensioners (39 percent); the opinions of students and military personnel coincided. Regarding families of four or five children the diversity of opinions was enormous. Public sector employees were 5.5

times less inclined than collective farmers to consider a four- or five-child family as desirable.

The respondents who had a high monthly income were less inclined to favor a three-child family than those with a low income (26 percent of those with more than 250 roubles per capita a month, and 39 percent of those with 51 to 75 roubles per capita a month).

14.2   Size of Parents' Family

The fact that the demographic behavior of the population has changed over the last generation may be substantiated by replies to the following question: How many children did your parents have?

Comparing the respondents' views on ideal family size to the number of children in their parents' family, one can see the changing trend in the reproductive behavior of the population (*Table 14.3*). In Russia 63 percent of the respondents' parents had three or more children, while 47 percent of the respondents said that a family today should have three or more children. In the Ukraine, the corresponding figures are 53 percent and 33 percent; in Estonia, 72 percent and 64 percent; in Georgia, 63 percent and 55 percent; in Kazakhstan, 60 percent and 47 percent; in Uzbekistan, 69 percent and 44 percent; and in Tajikistan, 60 percent and 53 percent.[2]

Some Soviet demographers find that Soviet Central Asia is rapidly moving toward demographic transition. This finding is confirmed by the information in our poll.

Many nationalities live in Russia. Undoubtedly, it is interesting to know the opinion of the Russian respondents who represent the majority of the population of the republic and whose fertility level is low. The share of Russian respondents living in Russia whose parents had three or more children and the share of those Russians who view the ideal family as including three or more children were lower than the share of respondents of other nationalities living in Russia. Of the Russian respondents, 55 percent were raised in families with three or more children, and 36 percent viewed a family with three or more children as ideal. The corresponding figures for the Russia as a whole were 63 percent and 47 percent.

Table 14.3. Replies to question, How many children did your parents have? (in percent of those interviewed): 1, Just myself; 2, Two children; 3, Three children; 4, Four or five children; 5, Six or more.

Category	1	2	3	4	5
Total	9	29	23	22	17
Area					
Urban	9	28	24	22	17
Rural	10	30	21	23	16
Region					
Russia	9	28	21	25	17
Ukraine	12	35	24	18	11
Estonia	5	23	38	22	12
Georgia	9	28	18	15	30
Kazakhstan	12	28	25	21	14
Uzbekistan	11	20	23	26	20
Tajikistan	7	33	24	19	17
Type of community					
Moscow	24	37	22	11	6
Capital city	17	27	25	18	13
Regional, autonomous					
republic center	13	36	23	17	11
District, autonomous					
region center	6	29	23	22	20
Other town	13	29	21	20	17
Village	2	21	23	33	21
Education					
University degree	9	38	25	15	13
University					
undergraduate	20	42	17	12	9
Secondary technical	12	33	23	20	12
Vocational school	9	30	24	28	9
Secondary school	11	28	20	22	19
Grade 8 or 9	7	21	26	32	14
Grade 7 or under	3	12	25	24	36
Nationality					
Russians	12	33	22	18	15
Indigenous population	7	24	24	29	16
Other	7	14	23	28	28

14.3   Views on Population Policies

Among Soviet demographers there is still discussion concerning the need for population policy both at the national level and at the regional level.

To study if the public thinks it is necessary to pursue a population policy with regard to fertility, the following questions were asked: In your republic, is there a need for urgent measures aimed at increasing the fertility? and In your republic, is there a need for urgent measures aimed at decreasing the fertility? The responses are presented in *Tables 14.4* and *14.5*.

Under conditions of economic dislocation and environmental crises more than half of the respondents favored urgent measures aimed at increasing fertility, while one-fourth (24 percent) expressed the opinion that such measures were not needed. The opinions of those interviewed differed markedly depending on nationality, family size, housing conditions, occupation, and region of the country.

The respondents most strongly supported urgent measures to increase fertility both in a republic with low birth rate (Russia, 56 percent) and in a republic with high birth rate (Tajikistan, 57 percent).

Respondents from other republics (the Ukraine, Georgia, and Uzbekistan) expressed similar notions about the necessity of pursuing a population-oriented policy (49 percent, 50 percent, and 49 percent of the respondents, respectively) in spite of great differences of demographic situations in these republics. Kazakhstan and Estonia had the highest proportion of respondents who were against population policy among the republics where polling was carried out. From the viewpoint of demographic development, Kazakhstan may be classified as belonging to a group of republics where the demographic transition from high to low fertility has started. According to the poll, the public opinion in this republic was divided nearly equally: 39 percent said yes and 40 percent said no to the need for urgent measures aimed at increasing fertility. Estonia has a steady low level of fertility, although not the lowest in the USSR. The results of the poll showed that, similar to Kazakhstan, the opinions in Estonia on the need for urgent measures aimed at increasing the fertility were divided into two nearly equal parts: that is, 46 percent in favor and 42 percent opposed.

The replies to the question on the need for decreasing the fertility level (*Table 14.5*) testify to the fact that in the USSR, on the whole, 8 percent of the urban and rural populations subscribed to this view – three times less than the number of those who are against population policies aimed at increasing fertility (24 percent). At the same time 5

Table 14.4. Replies to question, In your republic is there a need for urgent measures aimed at increasing fertility? (in percent of those interviewed).

Category	Yes	No	No opinion
Total	52	24	24
Area			
Urban	52	22	26
Rural	51	28	21
Region			
Russia	56	20	24
Ukraine	49	20	31
Estonia	46	42	12
Georgia	50	28	22
Kazakhstan	39	40	21
Uzbekistan	49	28	23
Tajikistan	57	19	24
Type of community			
Moscow	44	18	38
Capital city	52	32	16
Regional, autonomous republic center	55	23	22
District, autonomous region center	55	23	22
Other town	36	37	27
Village	63	16	21
Education			
University degree	53	29	18
University undergraduate	38	34	28
Secondary technical	55	24	21
Vocational school	53	25	22
Secondary school	48	28	24
Grade 8 or 9	46	21	33
Grade 7 or under	58	12	30
Nationality			
Russians	47	26	27
Indigenous population	64	20	16
Other	40	29	31

percent of respondents in Russia and the Ukraine, 2 percent in Tajikistan, 20 percent in Kazakhstan, and 15 percent in Estonia expressed the need for urgent measures to decrease the fertility level. The opinion that the fertility level should be decreased was expressed equally by males and females, living in cities and villages, irrespective of their marital status

Table 14.5. Replies to question, In your republic is there a need for urgent measures aimed at decreasing fertility? (in percent of those interviewed).

Category	Yes	No	No opinion
Total	8	63	29
Area			
Urban	7	62	31
Rural	8	66	26
Region			
Russia	5	65	30
Ukraine	5	61	34
Estonia	15	50	35
Georgia	10	68	22
Kazakhstan	20	54	26
Uzbekistan	6	65	29
Tajikistan	2	78	20
Type of community			
Moscow	10	49	41
Capital city	13	65	22
Regional, autonomous republic center	3	64	33
District, autonomous region center	5	72	23
Other town	13	55	32
Village	4	65	31
Education			
University degree	9	69	22
University undergraduate	14	53	33
Secondary technical	7	68	25
Vocational school	4	69	27
Secondary school	9	61	30
Grade 8 or 9	8	51	41
Grade 7 or under	2	60	38
Nationality			
Russians	8	60	32
Indigenous population	7	70	23
Other	7	56	37

and nationality. A difference of opinion was evident among people with different educational backgrounds, occupations, average incomes, and family sizes.

How can these figures be interpreted? Public opinion does not approve unconditionally of the necessity of a population policy in the country; one-quarter of the respondents were against it.

The populations in Estonia and Kazakhstan, more often than in other republics, favored a policy to lower the level of fertility in the regions where it is high. Apparently the problem of carrying out or not carrying out the population policy is not only the subject of scientific discussions. It is becoming a sociopolitical matter, since there is a threat of postponing some social programs in several regions due to a redistribution of funds to other regions.

14.4   Motives for Having a Large Family

One precondition for implementing population policies in relation to fertility is to learn the reasons for having large families.[3] A survey of the public's view on large families is essential because of the widespread view that people oppose assisting families that allegedly make themselves large just to obtain more allowances and privileges from the state.

A question was asked to discover people's motivation for having large families: Why do some people have many children? The majority of those interviewed (55 percent) said that it was because they see it as their purpose in life (*Table 14.6*). This idea is shared by urban and rural populations and males and females alike.

Judging from the results, the socio-demographic characteristics of the respondents seem to exert influence upon the differentiation of views on this matter, such as the type of community, region, nationality, family size, marital status, and occupation. The respondent's educational background and family income seemed to be less important.

Nearly four-fifths (77 percent) of the respondents who have three and more children up to age 15 stated that raising children is their purpose in life. People of indigenous nationalities point to it more often than Russians. The notion that rearing children is of paramount importance was supported by a clear majority in four nationalities: Uzbeks, Estonians, Georgians, and Tajiks.[4] In comparison this response received support from 49 percent of Russians living in other republics and 50 percent of Ukrainians.

National traditions and religious dogmas were mentioned as the second important reason for having many children. This view was expressed

Table 14.6. Replies to question, Why do some people have many children? (in percent of those interviewed, multiple responses possible): 1, To fulfill a purpose in life; 2, To obtain various privileges; 3, To follow national traditions or religious beliefs; 4, Irresponsible; 5, Unable to practice birth control; 6, Other reasons; 7, No opinion.

Category	1	2	3	4	5	6	7
Total	55	14	39	28	31	3	8
Area							
Urban	56	14	38	28	30	2	8
Rural	55	14	42	27	34	4	7
Region							
Russia	57	15	41	27	28	4	7
Ukraine	50	10	29	31	33	2	12
Estonia	56	18	50	26	27	2	2
Georgia	62	14	49	23	41	0	7
Kazakhstan	50	17	40	35	44	0	7
Uzbekistan	53	11	29	27	27	1	10
Tajikistan	59	11	40	29	29	4	8
Type of community							
Moscow	42	21	35	34	25	1	8
Capital city	56	14	47	31	36	4	3
Regional, autonomous republic center	52	10	40	31	34	4	8
District, autonomous region center	56	11	37	27	30	1	9
Other town	51	14	44	32	42	2	4
Village	63	17	36	21	24	4	11
Education							
University degree	53	11	53	37	36	4	4
University undergraduate	52	21	48	31	38	3	7
Secondary technical	56	11	41	28	34	2	6
Vocational school	60	18	22	33	28	8	5
Secondary school	60	13	38	29	29	2	7
Grade 8 or 9	57	21	42	26	29	2	10
Grade 7 or under	44	13	30	15	26	1	20
Nationality							
Russians	51	14	39	31	32	2	8
Indigenous population	62	14	40	22	28	5	8
Other	61	13	38	30	34	2	7

about equally by males and females (40 percent and 39 percent, respectively) in cities and villages; but it was expressed more often by those who were divorced (47 percent) than by those who were married (38 percent), and more often by university graduates (53 percent) than by those with a low educational background (vocational school, 22 percent; seven years of education or less, 30 percent).

Agreement with the fourth and fifth reasons was more or less the same. One-third of those interviewed (31 percent) stated that large families are due to the inability to practice birth control; this view was expressed more often in the countryside (34 percent) than in towns (30 percent) and more by women (33 percent) than by men (29 percent).

The respondents 30 years of age or older adhered to more or less similar views on birth control. The lack of birth control was mentioned more often by respondents in the most reproductive age groups: 20 to 24 (45 percent) and 25 to 29 (38 percent). This response was also given by university graduates (36 percent), university undergraduates (38 percent), and technical school graduates (34 percent), and by those who have one child or two children younger than age 15 (34 percent and 33 percent, respectively).

In Moscow every fourth respondent was of the opinion that the lack of birth control leads to large families; in large townships this view was expressed nearly one and a half times as often. A planned family requires that parents have the means to control a child's birth, which is more available in a major city than in a small township.

Divergences in replies by nationality were not significant, but they were rather substantial when the republic of domicile was considered. In Estonia *only* 27 percent of the respondents stated that the reason for having many children in a family was the inability to use contraceptives properly, but in Georgia and Kazakhstan the percentages of respondents who stated this were 41 percent and 44 percent, respectively. In the case of Estonia we put the word *only* before the figure of 27 percent of the respondents who think that inaccessibility to birth control accounts for large families, but in the case of Uzbekistan and Tajikistan the figures of 27 percent and 29 percent, respectively, are not equipollent to the Estonian 27 percent. The response to this question in Uzbekistan and Tajikistan means something different; namely, in the republics with strong religious and national traditions of having large families, nearly every third respondent expressed a significant change in the demographic perceptions of the population, which may imply that in the near future these republics may experience a decline in the birth rate.

Irresponsibility of parents was mentioned as the next reason for large families. This view was expressed by 28 percent of the respondents, both males (28 percent) and females (27 percent), in towns and in villages. Most concern was expressed by respondents in Kazakhstan (35 percent), respondents in the Ukraine (31 percent), respondents in Moscow (34 percent), those age 25 to 29 (39 percent), university graduates (37 percent), people with high incomes (29 percent), and military personnel (53 percent).

Results from the poll indicated that the view that large families were desired to obtain more privileges was not widely accepted. Only 14 percent of those questioned gave this response. The number of people in Uzbekistan, Tajikistan, and the Ukraine who believe that the reason for large families lies in the desire for more privileges was six to seven percentage points less than in Estonia and Kazakhstan.

14.5   Views on Serious Demographic and Socioeconomic Problems

Respondents were asked, Among the problems listed, which are the most serious in the USSR? Two major problems were expressed by more than half of the respondents (*Table 14.7*): low income of young families (54 percent) and double work load of the female population (55 percent). The high infant mortality rate was considered the second most important issue by 32 percent, and 30 percent of the respondents suggested that the growth in divorce rate was important. In the past Estonians were more concerned about divorce, but in this poll Georgians (39 percent) and Russians (31 percent) were more concerned about it. One-fourth of the respondents pointed to the low life expectancy in the country. Most concern about this was expressed by people living in Russia and in the Ukraine (both 29 percent), and the least concern was expressed by people in Uzbekistan (14 percent). Aging was generally not considered a problem except by the elderly.

Table 14.7. Replies to question, Among the problems mentioned below, which are the most serious? (in percent of those interviewed, multiple responses possible): 1, Low income of young families; 2, Decrease in fertility rate; 3, High level of infant mortality; 4, Growth in percentage of aging population; 5, High mortality level; 6, High mortality in the male population; 7, Growth in divorce rate; 8, Double work load of female population (employment and housekeeping); 9, Other problems; 10, No opinion.

Category	1	2	3	4	5	6	7	8	9	10
Total	54	14	32	9	25	13	30	55	4	7
Area										
Urban	54	14	32	9	26	14	30	55	4	7
Rural	53	14	33	8	23	12	28	53	5	7
Region										
Russia	54	14	33	10	29	13	31	54	4	6
Ukraine	53	13	24	9	29	13	29	60	8	8
Estonia	63	19	37	11	19	12	20	39	1	5
Georgia	53	14	48	3	16	5	39	54	0	8
Kazakhstan	50	12	44	10	19	15	25	45	6	7
Uzbekistan	49	14	20	5	14	22	27	66	4	8
Tajikistan	53	14	21	4	21	13	28	59	6	10
Type of community										
Moscow	51	6	23	10	23	10	24	52	4	13
Capital city	62	14	39	10	22	12	29	52	12	3
Regional, autonomous republic center	52	13	31	17	32	12	24	54	6	8
District, autonomous region center	51	12	25	4	18	17	33	58	5	8
Other town	55	13	40	10	28	13	30	55	2	4
Village	53	19	34	8	28	10	31	53	2	6
Education										
University degree	53	14	33	11	22	12	27	58	9	4
University undergraduate	58	18	35	9	25	10	27	52	8	7
Secondary technical	53	16	28	9	26	14	27	54	5	8
Vocational school	56	10	36	3	25	10	29	57	5	2
Secondary school	64	11	36	7	26	13	31	52	3	5
Grade 8 or 9	40	14	34	10	30	12	37	55	3	8
Grade 7 or under	46	16	26	13	21	18	28	55	1	12

14.6   Summary

The majority of the population seems to be well informed of numerical changes in the population structure. The most-informed respondents proved to be males, university graduates, Estonians, and Kazakhstans; less-informed respondents were in Uzbekistan and Russia. Russians who live in Russia seemed to be the most pessimistic about the country's population growth.

A family with two to three children seemed to be the ideal size. Factors that affected the responses to the question on the ideal size of a family were the republic of residence, type of community, age, marital status, family size, and nationality.

More than a half of the population favored urgent measures aimed at increasing the fertility level; one-fourth believed that such measures are not needed. The necessity for pursuing a population policy would be supported more often by Russians than by the respondents of the indigenous nationalities. Public opinion favored urgent measures aimed at increasing the fertility level both in the republics where it is low and in the republics where it is high.

The majority of the respondents were of the opinion that people want to have many children because they believe that rearing them is their purpose in life. The second reason for having large families was to follow national traditions and religious requirements. The third and fourth reasons for having large families were the respondents' notion that people cannot control childbearing and people are irresponsible. The opinion that large families are desired to receive various benefits is not widely spread.

The poll showed that people were well informed about differences in the life expectancies of males and females. But people were less informed about life expectation in the USSR as compared to other Western countries. The respondents with university degrees, high incomes, and military experience proved to be the most informed on this matter as compared with collective farmers, pensioners, and rural residents.

A matter not discussed in this chapter is the attitude toward refugees. A negligible number of people would welcome refugees in their vicinity, less than half would not mind refugees, and one-third would not like refugees to settle in their area. The most ardent opponents to refugees from other regions were Muscovites and rural residents. It seems essential to take into consideration the local population's views on preferred nationalities of refugees. For the majority of the respondents, nationality seemed to be of no importance; the preference to refugees of their own nationality and of any nationality akin to their own was widespread among youths from age 20 to 24. The most intolerant toward refugees of other nationalities were students of technical schools and vocational training centers as well as collective farmers and those who have been out of work for more than six months.

Three problems may become crucial for a republic (territory, district, town, or village): effects of environmental degradation on the population, overpopulation in the place of residence at present and in the future, and population aging. The last problem is considered extremely important by people over age 60. They believe that an increase in the share of elderly people may bring about serious social problems.

One-third of the population was apprehensive of the employment problem and unemployment. The problem of unemployment was expressed by respondents between ages 30 and 39, as well as by those who have three and more children younger than 15. Less concerned about unemployment in the future were those employed in the public sector and the armed forces. Collective farmers, cooperative workers, and those who have been unemployed for more than six months were more concerned.

Notes

[1] From 1 January 1992 the Russian Center for Public Opinion and Market Research (VCIOM).

[2] When doing such comparisons across generations one has to consider that subjects from large families are always overrepresented. In the sample, a family with five children has a five times greater chance of having a child than a family with one child.

[3] The nationwide policy is based on the notion that a family with three or four children is considered a large family. In the poll a large family was not specifically defined because this notion may vary in different regions.

[4] The respondents could give several replies, therefore the total figure is more than 100 percent.

Chapter 15
Public Opinion on Family Policies in Lithuania

Vlada Stankuniene and Vida Kanopiene

Over the past few years family issues such as birth-control methods and the role of women in family life and in society were given special attention among demographers and public administrators. These problems were also discussed by members of political parties, movements, unions, and various sections of society. One of their goals was to set up the main directions for family policy (FP). This chapter reviews the FP carried out in Lithuania, establishes the directions of its development, and analyzes public opinion on the methods for supporting families with children.

The chapter emphasizes those aspects of FP that cause the most anxiety and controversy: the upbringing of small children, the opportunities of mothers with small children to combine domestic and professional duties, and the demand of families for state assistance.

The article is based on two surveys carried out by the Institute of Economics of the Lithuanian Academy of Sciences. The public opinion on the effect of existing FP and family preferences has been investigated along with other problems by these surveys.

15.1   Procedures and Characteristics

The surveys were conducted in 1988 and 1990. The 1988 survey was
completed in the spring, on the eve of rapid political and socioeconomic
changes. The last few years have been distinguished by the impetuous
development of all spheres of public life and keen discussions of the socioeconomic problems including demographic ones. The 1990 survey was
also conducted in the spring. At the time of this survey the republic's
goals seemed to become more attainable. A comparison of the surveys'
results reveals the influence of changes in the attitudes toward various
aspects including demographic ones. General questions were asked to
evaluate the needs of families in the 1988 survey. More detailed questions on family needs were asked in 1990 to reveal the preferable forms
of improving family life, the problems of mothers with small children,
and the preferred ways of raising preschool children.

The 1988 survey was based on a national sample of economically
active urban and rural populations. Four regional factors were considered: the type of the settlement (from a large city to a country village),
its location, its level of development, and its major industry. The survey comprised 81 questions on four topics: labor, family, children, and
leisure. The questions were divided into five research areas: reproductive
behavior and attitudes; migration behavior; the conditions of combining
domestic and professional activities; opinions and attitudes toward state
demographic policy; and life-style. Family and employment histories, financial status, living conditions, and sociodemographic characteristics
of the respondents also were subjects of the survey.

Respondents returned the questionnaires directly to the research
group or mailed them back. Interviewing was mainly used in the rural regions to prevent any confusion in answering rather complicated
questions. Some 3,800 questionnaires were distributed by personnel de-
partments and 1,200 were delivered by mail. Correspondingly, 2,467
and 413 of the questionnaires were suitable for analysis. The survey was
representative according to age–sex composition, socioprofessional structure, and education (the discrepancy between the sample and general
characteristics was never more than ± 12 percent).

Respondents to the 1990 survey were asked 10 additional questions.
The questions were in the program of Lithuania's population questionnaire survey, carried out by the Public Opinion Research Center of the

Lithuanian Academy of Sciences. The survey was distributed to a random sample of adults (above age 18), using electoral lists for the selection. A total of 1,583 respondents answered the survey.

Though the samples of the two surveys differed, the results were representative enough to be comparable.

15.2 Evolution of Lithuanian Family Policy

Until 1988 the All-Union population policy was carried out in Lithuania. Certain elements were introduced by means of radio, TV broadcasts, and newspapers; these elements were specific for Lithuania. However, these elements also supported the All-Union population policy goals. The government decree of 1989 concerning the intensification of state assistance to families with children should be considered as the first step to Lithuania's independent family policy. The elaboration of the decree started in 1988, before the idea of a sovereign republic had real support. Under this decree the Lithuanian government started the formation of an independent population policy. It was intended to help couples plan their family and raise children. The measures increased the chances of raising preschool children at home and made it possible to improve conditions in preschool establishments (PSE, nurseries and kindergartens for children under age 6). The need to decrease the number of children in each preschool establishment (up to 10 in groups of children under age 3; and 15, above age 3) had been anticipated. Partially paid maternity leave was prolonged for mothers with children under 18 months by increasing the amount to 50 roubles or about 20 percent of the average monthly salary until one year and 35 roubles or about 15 percent for the following six months. The paid maternity leave for one year (35 roubles) and the unpaid leave for 18 months was foreseen by the All-Union regulation at that time. Unpaid maternity leave was available for three years. The provisions gave women the opportunity of choosing between employment and motherhood or of combining them. Until recently labor laws did not address the needs of families concerning women's employment.

The organization and administration of labor activities were based on the principle of full employment. The policy aimed at employing all women and was based on the social security system. These measures and the system of remuneration led the family to rely on the husband's and wife's incomes. It resulted in high employment of women. About 92 percent of working-age females were employees or students in Lithuania in the 1980s. The right of mothers with children under age 8 to work part-time, provided by the labor law in 1987, had not been exercised.

The 1990 survey asked the question, Do you know of the law giving mothers with children under age 8 the right to work part-time? Some 50 percent answered no.

The employment policy had only stimulated women's economic activities; the principle of providing favorable conditions to combine professional and family duties remained only a declaration. Poorly developed preschools did not meet the needs of the population.

The measures provided by the Lithuanian government decree of 1989 corresponded to the conditions of a centrally planned management and did not consider the peculiarities of the transitional period to a market, demand economy. Administrators had not foreseen the possible changes in employment and the state security system. However, they had eliminated juridical restrictions concerning the behavior of families raising small children. They abolished some contradictions among society, family, and individual interests and broadened women's opportunities.

As the republic moves toward a market economy, it is developing measures to encourage family self-provision. It is also reorganizing the employment system.

The credits for families expand according to their economic situation. During the turbulent summer of 1990, the partially paid maternity leave (until the child is 18 months) increased to 60 roubles by the special decree of the Lithuanian government. A decree increasing support for poor families (the per capita income of which was under 90 percent of subsistence level) was passed at the end of the year.

One important new goal for family policy is the essential change of preschool activities. It is hoped that new family policy will promote individuality and the development of every child. Further assistance is oriented toward support for young, poor, single-parent, or large families.

However, the new family policy must deal with the unstable socioeconomic and political situation in the republic. Financial resources for family support are limited. The administrators' decisions are influenced by various statements. They are often based on denying the past and advancing the opposite actions. For example, instead of promoting full employment for women, which was popular for many years, the administration is encouraging mothers to remain at home. Therefore some

Table 15.1. Preference of family policy, 1988 survey, in percent.

				Age group			
Trends of FP	Male	Female	Total	20–29	30–39	40–49	50+
Improve working conditions	32.7	36.1	35.0	32.7	34.6	37.6	36.4
Improve financial support for families	25.3	23.2	23.8	26.3	23.4	22.2	24.2
Better housing	20.1	18.5	19.1	19.5	18.1	18.7	20.5
Improve PSE	13.5	12.9	13.1	12.9	14.0	12.5	12.5
Other	8.4	9.3	9.0	8.6	9.9	9.0	6.4

PSEs are being closed, especially in rural areas. To enact effective family policy, administrators must thoroughly understand family interests and demands. The 1988 and 1990 surveys may be used as a basis for this understanding.

15.3 Attitudes Toward Family Policy

According to the 1988 survey results, the All-Union policy for families of the early 1980s was not recognized by the public. However, 65.2 percent of the respondents were sure that the government's actions could positively affect the reproductive behavior of families.

Participants were asked: Which demographic policies would most effectively influence and improve the welfare of families with children? According to *Table 15.1* the respondents preferred the development of different forms and regimes of women's employment.

The respondents based their opinions on their own experience and living conditions. The survey results showed that young respondents, especially between the ages 20 and 24, supported an increase in financial assistance and an increase in access to better housing. Housing problems are most acute among young families. However, 20.5 percent of the respondents over 50 expressed concern about the housing problem because they are aware of the problems facing their grown-up children.

The requirement for improving PSEs depends on the educational level of the respondents: the higher it is, the more often the respondents are for better PSEs. It was mostly the concern of women with higher education and employed in public health, administration, or science fields.

When asked about concrete ways of solving financial problems of families (What is the best form of financial support for families with children? 1990 survey) the respondents gave surprising answers. The ideal means of improving a family's financial situation turned out to be the possibility of enabling men to earn enough to support the entire family. This comes despite the long-held public policy that the family budget should result from a balance of two incomes (the wife's and the husband's). Paradoxically, at the same time the increase of grants for families with children was not popular. Some 8 to 9 percent of the respondents were for increased grants as well as for the proposed expansion of credit.

Urban inhabitants with a high level of education were most likely to prefer self-supporting family. For example, 72.8 percent of educated urban men were for the establishment of conditions that allow workers to earn as much as possible.

The number of children in the family had an influence on the respondents' opinion of financial support from the state. Only 49.6 percent of the respondents with four or more children were sure that they could handle the family's financial difficulties by themselves (*Figure 15.1*). Respondents having large families were more likely to prefer state grants. Some 15 percent of the respondents in this category were for monthly grants for nonworking mothers compared with 9.1 percent among those with one child; 10.3 percent, with two children; and 9.2 percent, with three children. Often young families were more likely to prefer state support: 14.7 percent of the parents under 30 were for larger grants for nonworking mothers, compared with 9.6 percent on average.

15.4   The Care for Small Children

The results of the surveys showed that people were not satisfied with the current system for raising small children: 30 percent of the respondents pointed out that they did not see preschool establishment attendance as a positive experience for children, and 27 percent found it to be a negative experience for children. But until recently most families had no alternative. At present, 45.3 percent of preschool children attend preschool establishments. Respondents were less inclined to prefer the practice of sending children under 3 to preschool establishments (*Table 15.2*). Most respondents preferred a nonworking parent's care for

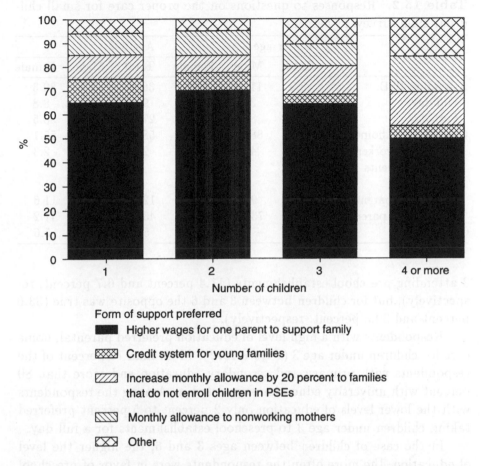

Figure 15.1. Forms of financial support by number of children, 1990 survey.

children under age 3. Less than 10 percent of the respondents were in favor of enrolling children under 3 in a preschool establishment. The cooperation of working parents and grandparents was preferred by 10.3 percent.

The respondents were more often in favor of enrolling children between ages 3 and 6 in the preschool establishments (36.4 percent: 9.3 percent favored full day and 27.1 preferred part of the day). Nonworking parents were the preferred caretakers for children in this age group.

Mothers of older children were inclined to consider their own professional interests. More men than women were in favor of children under

Table 15.2. Responses to questions on the proper care for small children, 1990 survey, in percent.

	Under age 3			Age 3–6		
	Total	Male	Female	Total	Male	Female
Attending PSE:	8.9	11.4	6.7	36.4	33.0	39.3
Full day	2.4	3.1	1.8	9.3	9.9	8.8
Part day	6.5	8.3	4.9	27.1	23.1	30.5
Care given at home by:	91.1	88.6	93.3	58.0	61.3	55.1
Child-care worker	4.8	4.9	4.7	2.4	2.5	2.3
Working parents cooperating						
with grandparents	10.3	9.9	10.6	12.2	12.8	11.6
Nonworking parent	76.0	73.8	78.0	43.4	46.0	41.2
Other	–	–	–	5.6	5.7	5.6

3 attending preschool establishments (11.4 percent and 6.7 percent, respectively), but for children between 3 and 6 the opposite was true (33.0 percent and 39.3 percent, respectively).

Respondents with a high level of education preferred parental, home care for children under age 3 (*Figure 15.2*). More than 70 percent of the respondents with primary and secondary education and more than 80 percent with university education supported it. Among the respondents with the lower levels of education only 2 percent to 5 percent preferred taking children under age 3 to preschool establishments for a full day.

In the case of children between ages 3 and 6, the higher the level of education, the more often the respondents were in favor of preschool establishments (*Figure 15.2*). Only 21 percent of the respondents with primary education and 40 percent with secondary education were for preschool establishments for children over age 3. However, most of those considering this way preferable (in the groups with higher education the absolute majority) were not for full-day preschool establishments. The higher the level of education of the parent, the less likely the respondent was in favor of raising children over age 3 only at home. This clearly was seen when the questions were answered by parents facing the problems. For example, only 29 percent of the parents under 35 with higher education supported caring for children over age 3 only at home; and 55.1 percent of parents were for enrolling children in preschool establishments (only 2.9 percent for the full day). Preferences are given to the more

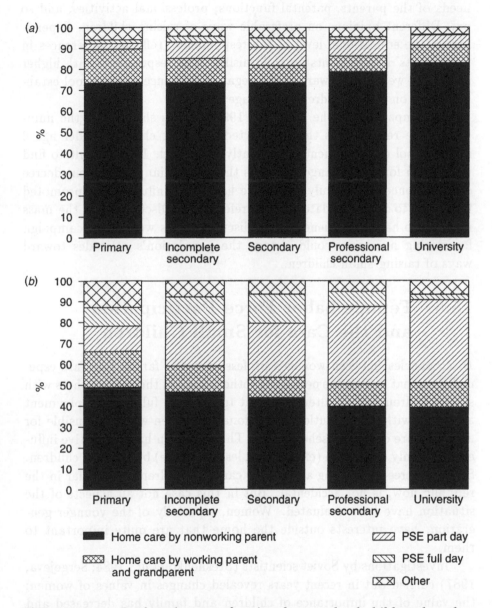

Figure 15.2. Preferred ways of caring for preschool children by educational level of respondents: (*a*) children under age 3 and (*b*) children between ages 3 and 6.

flexible preschool establishments that take into account the interests and needs of the parents, parental functions, professional activities, and so on. Different opinions on preferable ways of raising children, depending on the educational level of the respondents, reflected differences in the parents' requirements for childraising. The respondents with higher education were more aware of the negative influence of preschool establishments on small children (under age 3).

A comparison of the 1988 and 1990 surveys shows that the number of the respondents that supported enrolling children under age 3 in preschool establishments has greatly decreased. It is difficult to find the reason for such change. Perhaps the Lithuanian government decree of 1989 concerning family assistance had some influence. It promoted flexibility to accommodate family preferences. Discussions in the mass media also had an influence. Such discussions, as well as the campaign for closing nurseries, could change the population's attitudes toward ways of raising small children.

15.5   Female Labor Force Participation and the Care for Small Children

Discrepancies between women's professional and family roles are especially evident during the period of motherhood. In the past mothers with small children had no alternative but to combine full-time employment activities with family duties. Traditionally, women were responsible for a large share of the household work. This situation had a negative influence not only on women (exhaustion, less free time) but also on children. So the interest in staying at home to care for children, so popular in the republic now, is not accidental. But in this case not all aspects of the situation have been evaluated. Women, especially of the younger generation, have interests outside the home that are quite important to them.

Investigations by Soviet scientists (Steshenko *et al.*, 1984; Sergejeva, 1987) carried out in recent years revealed changes in values of women: the value of the importance of children and family has decreased and the importance of professional activities has increased. This is the result of the doctrine that all groups of the population, including women with small children, are obliged to participate in the economic sector. Was such socioeconomic policy effective in Lithuania? Two interrelated aspects of the problem should be examined to answer this question: the peculiarities of the value orientations of women and their attitudes toward ways of caring for small children.

The results of the 1988 survey showed that a woman's professional activities are not as important as family and children. Some 47 percent of the women responding to the survey placed family as most important in their life. Work and professional activities were most important to 16.9 percent of the respondents; welfare, to 15.0 percent; physical and spiritual development, to 11.9 percent; and social activities, to 9.1 percent. Family values were most important to married women with a low level of education, while work and professional careers were most important to unmarried women with high education. According to the results of the survey, only a little more than 20 percent of all the female respondents indicated that married women should take care of children and the household rather than work. Such ideas were more frequent in the group of those with primary or incomplete secondary education (37.3 percent) and less popular among professional women with higher education (13.9 percent).

Men were far more conservative in their views on the role of women in society and family. Some 35 percent of the respondents preferred women to stay at home as housewives.

Considerable importance was given to the problem of women with small children. One question in the 1988 survey was: How can a mother improve her family's welfare? The aim of the 1990 survey was to reveal the preferable behavior under current conditions. The question in 1990 was: What is the preferred activity for mothers with preschool children? The results are valuable for determining future family policy.

One-third of those responding in 1988 preferred part-time employment for mothers with preschool children. In addition, 22.3 percent considered part-time employment desirable for their entire life. The remainder felt it was a temporary necessity (while children are small).

The results of the 1990 survey were similar. Of the female respondents 38.1 percent considered part-time work most appropriate for mothers with preschool children. This opinion was also expressed by women with higher education.

The attitudes of mothers with small children toward professional activities were defined by changes during various periods of life. Young parents (especially mothers) were more keenly aware of the problems.

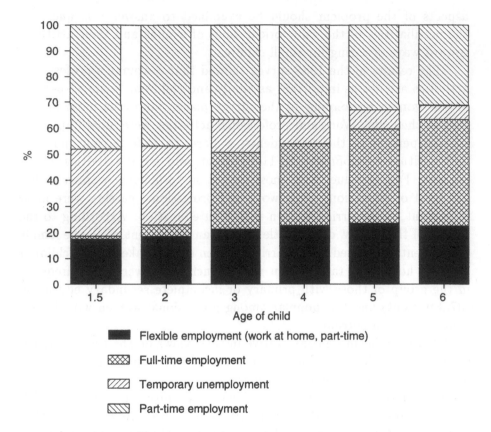

Figure 15.3. Preferred working situation of mothers by age of child, 1988 survey.

The answers of the respondents younger than 30 (with children) were analyzed. Women of this group often preferred part-time employment as compared with the sample. More of the young respondents preferred the variant "to give up the job temporarily."

The main differentiating factor which influenced the behavior of mothers was the age of the child. *Figure 15.3* shows that only 33 percent of the respondents thought that it is necessary to give up a job temporarily until the child is 3.

The attitude of those preferring part-time employment when children are small changed little with the age of a child.

15.6 Conclusions

The 1988 and 1990 surveys indicate the following:

- Most families prefer to support themselves; men should be allowed to earn as much as possible to support their families.
- The preferable way of caring for preschool children is at home by a nonworking parent. Attendance in preschool establishments is acceptable only for part of the day when the child is over age 3.
- Women with preschool children prefer to give up their job or to work under special conditions (part-time, homework). Most women with children under age 3 wish to give up their job. Mothers with children over age 3 prefer to work part-time or under other special conditions of employment.

The intensification of the state assistance for families, along with discussions in the society in the course of recent years, has greatly influenced the attitude of the population concerning the ways of caring for children. It is clearly reflected in the decrease of those in favor of PSE attendance when the child is under 3.

The results of the surveys along with the experience of other countries that are implementing family policy are an important base for promoting family policy in Lithuania. They also lead to a better evaluation of demographic factors and possible changes in reproductive behavior of the families. At the same time they may help to improve the assumptions made in population projections.

References

Sergejeva, G.P., 1987. *Zaniatost zhenshin: problemi i perspektivi* (Women's Employment: The Problems and Prospects). Economics, Moscow.

Steshenko V.S., V.P. Piskunov, and L.V. Chuiko, eds., 1984. *Trudovaja aktivnost zhenshin* (Women's Labor Activity). Naukova Dumka, Kiev.

Chapter 16

The Social Roles and Status of Women in the USSR

Natalia Rimashevskaya

The aim of this study is to assess the social status of women in Russia and the other republics of the USSR. The underlying factors of the ongoing political, social, and economic restructuring processes are studied. The existing social and professional mobility of spouses are analyzed, and the theoretical and methodological base for the solution of the so-called women's question is considered.

16.1 Major Issues Concerning Women

The 20th century has been marked by a radical change in the status of women, by a more complete and effective recognition of equality of the sexes in many countries. History has shown that genuine economic growth and social progress are impossible without the involvement of women in social life. However, the achievement of *practical* equality is one of the most complicated social problems. We have not fully recognized it and therefore we underestimate it.

But where and how do these problems manifest themselves in the USSR? Women encounter the following problems:

- While entering the sphere of production on initially equal terms with men (health, professional and educational levels, social and cultural activities), women perform jobs that require less skills; encounter a pyramidal employment structure, i.e., the higher the social status of a job, the less likely women are employed in it; mainly occupy positions without prospects for advancement.
- Women are poorly represented in managerial positions (among the chief executive officers of enterprises, women constitute less than 6 percent). In addition, women perform hazardous, manual work.
- Men's wages are higher than women's, notwithstanding that at the second stage of their career (after completion of their reproductive period) women seek hazardous, manual jobs to receive better pay and benefits.
- After marriage two-thirds of working women do not improve their professional skill and continue working at their previously attained level. In the period of scientific-technical revolution, when there is need for retraining, this inevitably has negative economic consequences for the women.
- Women often combine raising children and working outside the home; in fact, they work two shifts – one at the workplace, the other in the home. Average weekly working hours, including professional and family duties, are 73.6 hours for women and 59.4 hours for men.
- The health status of women is lower than that of men. Women become ill more frequently, spend more days in hospital, and have more chronic diseases.

What are the consequences of such a situation? They are negative both for the society and for each woman. On the one hand, existing traditions result in underutilization of a great intellectual and emotional potential: there are some activities where women are more efficient than men. On the other hand, women are highly dissatisfied with their employment, social, and family status. Today divorces are initiated more often by women than by men. Women tend to give up their social positions and retreat from political life in times of a general politicization. Evidence of this can be found in the results of the recent elections: women's representation has substantially declined. In addition, deviant behavior among women is increasing.

16.2   Social and Professional Mobility of Women

Results of a longitudinal study of families reveal the socioprofessional characteristics of married women and men and their wages as objective measures reflecting social consciousness and human behavior.

In the late 1960s a team of researchers at the Central Economic Mathematical Institute of the Academy of Sciences of the USSR conducted a comprehensive socioeconomic study of living standards, the Taganrog-II Project.[1] The Taganrog-II Project studied five topics: living standards and socioeconomic problems; life-style and consumption of cultural goods; demand and supply; urban family and its development trends; and health status of the population. In 1987–1988 the Taganrog-III Project was conducted; it also studied five topics. The results of these studies give a systematic picture of social changes during a period of transition to a market economy.

Information was gathered through questionnaires. Respondents were asked about their educational level and professional status and about their family situation. All demographic and socioeconomic indicators have been specified with regard to the year when the event occurred.[2]

The study finds that at the beginning of married life the differences between the level of education and qualification of spouses are insignificant; sometimes the educational level of wives is even higher. But the professional training of the wives begins to lag. Many wives begin their job without special training and that affects their careers (see *Tables 16.1* and *16.2*). A comparison between different cohorts shows some positive shifts in the relationship of professional levels of the spouses. In the junior cohort the share of families in which the wife's qualification was higher than that of her husband's at the time of marriage contract is twice as great as in the senior cohort. The qualifications of the husband and the wife were equal in 49.1 percent of the junior cohort at the time of marriage (*Table 16.3*).

Significant changes in the levels of qualification of the husband and wife during the cycle of the family life indicate that an optimal combination of the woman's roles until now has not been achieved; in most cases women forgo their employment career. Sometimes they are forced to give up their career because there are no possibilities of combining duties. Combining a woman's role functions can be deemed optimal

Table 16.1. Educational level of spouse at the time of marriage (senior cohort, marriage duration 20 to 25 years), in percent.

	Prim. school	Gen. second. school	Tech. school	Spec. second. school	Univ. comp. & incom.	Continuing education[a]
Blue collar						
Husband	51.6	14.7	27.4	6.3	–	20.9
Wife	53.4	25.3	18.7	1.3	1.3	13.5
White collar						
Husband	4.2	4.3	8.5	48.9	34.1	66.7
Wife	16.1	11.3	11.3	43.6	17.7	21.7
Total						
Husband	35.9	11.3	21.2	20.4	11.2	34.0
Wife	36.5	19.0	15.4	20.4	8.7	17.0

[a] Persons with higher education are excluded.

Table 16.2. Educational level of spouses at the time of marriage (junior cohort, marriage duration 5 to 10 years), in percent.

	Primary school	Gen. second. school	Tech. school	Spec. second. school	Univ. comp. & incomp.	Continuing education[a]
Blue collar						
Husband	17.8	30.9	34.2	15.8	1.3	19.9
Wife	16.3	44.6	21.7	17.4	–	10.9
White collar						
Husband	3.4	23.7	6.8	35.6	30.5	45.2
Wife	4.5	37.9	3.6	32.4	21.6	14.6
Total						
Husband	13.7	28.9	26.6	21.3	9.5	25.4
Wife	9.9	40.9	11.8	25.6	11.8	12.7

[a] Persons with higher education are excluded.

only when she freely chooses her life path. But social consciousness and public attitudes differ substantially from actual behavior.

There is widespread opinion that a woman seeks employment solely for pecuniary considerations; if the pay of the husband were substantially raised, she would remain at home to look after her children. But typically the woman not only does not leave her job, but often, during the second stage of her career, tries to get a harder job to receive higher pay.

Table 16.3. Level of qualification of spouse at the time of marriage and at the time of questionnaire, in percent.

	Senior cohort		Junior cohort	
	Time of marriage	Time of question.	Time of marriage	Time of question.
Qualifications of husband and wife are equal	45.3	19.1	49.1	23.2
Qualification of husband is higher than that of wife	49.6	68.4	40.0	60.3
Qualification of wife is higher than that of husband	5.1	12.5	10.9	16.5

The attitudes of women toward employment are illustrated in the following example. In a study at KamAS (KamAS is a leading producer of commercial vehicles in Naberejnye Chelny on the Kama) women were asked: If your husband were given a substantial pay rise, would you continue to work? Some 20 percent of white-collar workers and 26 percent of blue-collar workers would like to retain their jobs; 64 percent of white-collar workers and 32 percent of blue-collar workers would like to work part time; 12 percent of white-collar workers and 30 percent of blue-collar workers would like to have a more pleasant job; only 4 percent of white-collar workers and 3 percent of blue-collar workers would leave employment. These answers clearly indicate that women would prefer to continue working notwithstanding the fact that combining the duties of a mother and working woman amounts to doubling their everyday work load. The majority of women (68 percent) considered the family necessary, but some 15 percent of the respondents at KamAS were of the opinion that today it is unnecessary to have a family.

At the present a woman's employment is motivated not only by the necessity of having a second wage earner in the family, but also by a desire to realize one's potential, interests, and skills; to work in a collective; and to be useful outside the home. The problems of female employment are to a great extent determined by the attitudes prevailing at a given stage of the family's life cycle.

A family life cycle is characterized by several stages; each stage has specific demographic and socioeconomic features, functions, and prevailing types of activity of a family undergoing a change. Most families during their life cycles go through the same natural stages, repeating the same processes of growth, stabilization, and reduction of the number of their members.

The first child is born when the family is, on average, in the second or third year of marriage; the second, in most cases, in the sixth or seventh year. These years are also the most important years for career development. This problem of starting a family while establishing a career is solved by each spouse differently; the solution influences the employment career and accordingly the changes in income. Meanwhile the difference between the average wage of husbands and wives remains fairly stable: at the time of the marriage contract the wife's pay was 61 percent of that of her husband's pay in the senior cohort and 67 percent in the junior cohort.

Over a period of 20 years, the rate of growth of each spouse's pay differs insignificantly. But, as the analysis shows, during the first 10 years of matrimony, women fall behind in their career path; they begin to catch up only after the tenth year of marriage. In contrast, husbands' wages grow faster at the initial stages of family life. During the second decade the growth slows down.

The first 10 years of family life is a period of childbirth and child-rearing. Therefore the rate of wage growth of wives is substantially slower. During the second decade their wages grow more rapidly than that of their husbands. Considerations of the family's financial well-being demand that a woman, after the socialization of children, accelerates her professional activity. She tries to compensate for relative employment passivity in the preceding period and thus tries to raise her income; in addition, the woman attempts to achieve an adequate pension. After the tenth year of marriage the share of women who manage to raise their wages is greater than that of men. But these increases are to a great degree connected with the fact that women change their professions or move to other industries rather than raise their educational or qualificational level.

Trends in the average wage of spouses in the various cohorts differ insignificantly; the junior cohort is characterized by a more pronounced fall in the wife's wages at the initial stage of family life as compared with the senior cohorts. This can be explained by the reduction of proto- and inter-birth intervals, and by high rates of childbirth precisely at this stage which hamper a woman's professional activity.

Table 16.4. Wage change of spouse during a period of 20 to 25 years of married life, in percent.

	No change	Only decrease	Both increase and decrease	Only increase
All husbands	7.7	7.7	22.6	62.0
Blue collar	7.7	8.8	17.6	65.9
White collar	7.8	5.9	31.4	54.9
All wives	7.7	2.8	31.0	58.5
Blue collar	9.4	4.7	31.2	54.7
White collar	6.4	1.3	30.8	61.5
Total	7.7	5.3	26.8	60.2

The method of cohort analysis allows one to monitor not only the dynamics of average wages but also the changes in the distribution of pay and its differentiation over time.

With women, most of whom are employed in the service industries, the wages of different socioprofessional groups appear to be dependent on the type of industry rather than on the level of qualification and education. Women are mostly employed in industries where wages are substantially lower because of stereotypes that view women as the second breadwinner.

Studies of specific paths of wage changes of spouses during the family life cycle have shown that wages are growing for the majority of workers. As can be seen in *Table 16.4*, wages have been growing steadily for some two-thirds of all workers.

Most decreases in pay occur at the first stage of family life. At this time the appearance of the first child automatically reduces the disposable family income. An additional income reduction due to changes in the woman's work status aggravates the family's material standing. Why do couples undertake such steps? For women the most important reason is the child (44 percent of all respondents, 76 percent of respondents at the initial stage of family life); the second reason is search for more favorable working conditions; the third reason is socioprofessional changes due to poor health; and the last reason is retirement. These data show how the division of role functions influences the social status of women: combining both functions (that of employee and mother) leads to marginalization of the female labor force and uncompetitiveness in the labor market.

It should be noted that a woman faces the problem of combining employment and motherhood in the most acute form at the first stage of family life. As the study has shown, the most preferable combination is achieved through more convenient working hours (for example, part-time work, flexible schedule). But management is often unable to meet the needs of women. Therefore a woman often seeks a job with flexible working hours, even when it is of a different professional status, or takes up a position at a preschool to be with her children. Such behavior is more characteristic of blue-collar employees. In the junior cohort sharp changes in professional status are under way not only by workers of low qualification, as in the senior cohort, but also by skilled blue-collar workers and professional employees. Furthermore, almost 16 percent of women temporarily leave employment for an average period of four years (not counting the one-year leave after childbirth).[3]

The roles of family and society are different at various stages of a child's growth. These differences are determined by the specific needs of children at different ages. Recent sociopsychological studies have indicated that children experience emotional and psychological deviations caused by disruption of natural ties between child and mother. Officials must recognize the importance of parental care during the first years of life. Therefore there must be policy measures to help parents care for their children. This means that parents looking after young children must receive a percentage of their previous wages. However, such measures are not harmless from a social point of view because they widen the gap between men and women in the realization of their creative potential. The first child is born just when a woman takes her first steps in her professional career. That is why a substantial increase of child-care leave may be undesirable, and its duration should not exceed two years. As an alternative the USSR law provides that such leave should be available to both spouses or to other family members.

Returning to the topic of women's employment, it must be noted that in 13 percent to 15 percent of cases, reduction of the wife's pay is caused by the search for more favorable working conditions. However, in the remaining cases such employment changes are associated with a search for easier work.

The main reason for the drop in a husband's pay during the family life cycle is a search for more favorable working conditions (47.7 percent of cases). The second reason is the transfer from a blue-collar job to a white-collar job (since the 1970s blue-collar wages have often been higher than white-collar wages). The third reason (14 percent of all cases) is the necessity for improving housing conditions. For example, in the first years of marriage many husbands seek a job at a construction site or another enterprise with prospects of obtaining a flat, even when this entails a decrease in pay.

16.3   Typology of Employment Careers of Spouses

Analyses of the changes in social and professional status of spouses during the family life cycles reveal four types of employment careers. The first type of mobility may be called descending career. This type is experienced by workers whose professional level decreases over time. With women the main reason for such mobility is related to the necessity for a more flexible job. Some women whose status was lowered during a certain period of family life never return to their previous level. With men a descending career is related to poor health and to the search for better housing and living conditions.

The second type is called horizontal career. Workers in this category have taken jobs in another industry but without raising their level of education or qualification.

The third type of employment careers can be defined as weak growing; the worker has achieved some progress with regard to education and qualification, but the growth has been insignificant. This is the most promising group since it can expect to raise its level beyond the period under consideration.

The fourth group includes workers who have fully realized their creative potential. Such an employment can be defined as an ascending career. With blue-collar workers this presupposes the achievement of the highest level on the employment scale. Professional growth of white-collar employees is measured by their career advancement and educational level.

Table 16.5 illustrates the socioprofessional mobility relationship of spouses. More than 50 percent of husbands have followed an ascending

Table 16.5. Types of employment career of spouses during 20 to 25 years of family life in the senior cohort, in percent.

Wives	Husbands				
	I	II	III Weakly	IV	
	Descending	Horizontal	growing	Ascending	Total
I Descending	–	3.7	4.4	8.8	16.9
II Horizontal	3.0	3.6	3.7	12.5	22.8
III Weakly growing	5.1	5.9	5.9	16.2	33.1
IV Ascending	4.4	2.2	2.2	18.4	27.2
Total	12.5	15.4	16.2	55.9	100.0

employment path. The most common type for women is the weakly growing career (one-third of women). This group includes mothers who have changed jobs due to childbirth and have taken positions demanding less skill, but later have achieved a higher status in a different profession.

There is a substantial difference between the percentages of men and women in the descending career (12.5 percent for men, 16.9 percent for women). It is important to note the qualitative features of the respective paths; while women with relatively high skills at the time of marriage often work in unskilled jobs, men seldom experience such sharp changes. And if they do work in unskilled jobs it is mostly due to poor health conditions.

With regard to the careers of spouses, the study shows the predominance of families in which both spouses achieve the highest possible qualification status, though at different stages of family cycle (*Table 16.5*). Less frequent are families in which husbands have an ascending career and wives a relatively less favorable career: a weakly growing career in 16.2 percent, a horizontal career in 12.5 percent, and descending career in 8.8 percent of cases. All other combinations of employment paths are met considerably less frequently, not exceeding 5 percent.

The analysis of socioprofessional mobility of spouses has revealed some key problems of the labor potential of both men and women. In the latter case mobility is determined by the specific social consciousness. At first sight it looks as if the married couple freely chooses its life paradigm. In reality patriarchal relations predetermine all developments before they start.

16.4 From Patriarchy to Equality

Economic transformation of society presupposes not only structural changes (due to social reorientation, conversion of armament production, reduction of military spending); significant changes are taking place in property relations, with all forms of economic activity being legalized and supported. The system of self-management and self-financing means that the prosperity of the enterprise is directly related to its economic performance. This leads to changes in women's position: the more they are protected as mothers, the more they are marginalized in the employment sphere.

As a result of transition to a market economy, the problem of female employment has been aggravated; the economic reform has caused a sharp reduction in the female labor force. Transition to market mechanisms of regulation and diversity of property forms encourage the activity and responsibility of workers. The inevitable reduction in the number of available jobs due to privatization, structural changes, conversion, and demilitarization increases competition for jobs. Less skilled women cannot compete in the labor market; in addition, the situation is complicated by social factors. Therefore it is not accidental that women are the first to lose their jobs. During recent years of administrative staff reductions, 80 percent of those who lost their jobs were women. The situation is such that women are the first to lose their jobs, but are not the first to be rehired. That is why women's share in the labor force dropped to 48 percent in 1988.

The female labor market is characterized by low wages and skills, uncreative work, intermittent employment, limited range of professions, bad working conditions, horizontal professional mobility, and part-time work. Being a socially vulnerable category of the population, women to a greater degree run the risk of being forced to accept unregulated working hours, high intensity of work, and hazardous working conditions. One can expect an increase in the number of women below the level of poverty.

Do women have a chance to take up executive positions under the current situation? To answer this question it is necessary to analyze the ongoing processes of social restructuring of the society as a whole and its influence on social relations. Furthermore, one must take into account the dynamics of a woman's social status and the degree to which those structures are open to her.

Political and economic democratization, decentralization of government, social reorientation of the economic development and demilitarization, and open and regular publication of information allow every person to reveal and realize his or her potential. Perestroika removed the barrier of rigid hierarchy related to the administrative command system. But at the same time a hitherto hidden barrier has been revealed, that of gender. And this barrier is reproduced by the ideology of social protectionism with regard to women which puts them in a position of socially handicapped persons.

The model of a woman's position and activity entrenched in public opinion may be described as follows:

- It is said that by nature women as mothers represent a special work force.
- The distinguishing feature of the female work force is believed to be such that society must give women the opportunity of combining employment and motherhood through various advantages.
- Increased benefits to mothers are said to give women a choice of either pursuing a career or devoting more time to the family.
- This view emphasizes that the USSR has *over-emancipated* women. This is considered to lie at the root of many evils.

To solve these problems it is necessary to reduce women's participation in social production.

For a woman to reach an executive position under these circumstances is quite difficult. And the ongoing social and economic transformations do not contribute to the advancement of a woman. She becomes less competitive in the labor market.

What are the deep causes of the existing situation with regard to women's position in the USSR? The main cause is the existence, at the intellectual and behavioral levels, of a sex division of social roles and patriarchal relations between man and woman. The essence of patriarchy is that the world rests on some natural foundations and to undermine them would be very dangerous because it would sooner or later lead to the destruction of the society itself. The division of social functions between man and woman is precisely one of such natural foundations. By nature woman has been assigned the role of mother. At the same time man is by nature a breadwinner, a public figure, a link between the small community (the family) and the large community (society). In accordance with patriarchal attitudes the role of subordination is placed above all other things. He or she must perform his or her role even when it makes a person miserable, even when it contradicts common sense.

Against the background of the ongoing socioeconomic transformations in the USSR, patriarchal attitudes and relations have brought about a false stereotype of woman's *over-emancipation* as the cause of many of today's economic, demographic, and social problems.

The main conclusions to be drawn are as follows: first, such an attitude has revealed a barrier related to gender; second, the transition to market relations leads to a serious marginalization of women in the labor force. That is why the decisive precondition of putting an end to the existing situation is to destroy the patriarchal stereotypes and to establish a new paradigm based on granting men and women equal opportunities. The egalitarian concept is based on the conviction that the so-called natural division of labor between the sexes is of social origin. Changing this division requires destruction of old social attitudes. The new egalitarian types of relations are based on complementarity in society and family, as opposed to patriarchal relations based on domination.

The transition from a patriarchal system to an egalitarian one is of great importance for social relations.

Notes

[1] Taganrog is a typical Russian industrial city of about 300,000 inhabitants 1,000 kilometers south of Moscow. Each survey polled about 10,000 households.

[2] Three cohorts of families have been singled out: senior – duration of marriage 20 to 25 years; middle – 10 to 15 years; junior – not more than 10 years (5 to 10 years). Cohorts under 5 years of marriage have been excluded, since such a short duration could not reveal trends in the career of the couple.

[3] The study was conducted before the introduction of a 1.5-year leave.

Bibliography

Mukomel, V.I., ed., 1990. *SSSR: Demograficheskiy diagnoz* (USSR: Demographic Diagnosis). Progress, Moscow.

Perestroika – sem'e, semia – perestroike (Perestroika for the Family, the Family for Perestroika), 1990. Mysl, Moscow.

Rimashevskaya, N., 1987. Aktualnije problemi polozhenia zhenschini: po materialam economico-soziologicheskogo issledovanija (Current Problems of the Women's Situation: On the Basis of an Economic-sociological Study). *Socialisticheskiy trud* (Socialistic Labor) **7**.

Rimashevskaya, N., and N. Zacharova, 1989. *The Diversification of Women's Training and Employment: The Case of the USSR*. Training Policies Branch of ILO, Geneva.

Rimashevskaya, N., and L. Onikov, 1990. *Narodnoe blagosostoyaniye: Tendentsii i perspectivi* (People Welfare: Trends and Prospects). Nauka, Moscow.

Rimashevskaya, N., *et al.*, 1977. *Sem'ia, trud, dokhody, potrebleniye: Taganrogskie issledovaniya* (Family, Labor, Income, Consumption: Taganrog Studies). Nauka, Moscow.

Zacharova, N., A. Posadskaya, and N. Rimashevskaya, 1989. Kak my reshayem zhensky vopros (How We Solve the Woman Question). *Kommunist* **4**.

Part III

Components of Mortality Trends

Part III

Components of Mortality Trends

Chapter 17

Life Expectancy and Causes of Death in the USSR

Evgeny M. Andreev

This chapter analyzes mortality in the USSR in the context of world mortality trends. Numerical regularities and the mortality evolution of the country based on demographic theories of mortality are presented.

The theory of demographic transition in the field of mortality is based on accumulated factual material; it is therefore one of the most developed fields of demographic analysis. In this work peculiarities of the USSR population transition are presented in terms of the demographic transition theory.

17.1 Demographic Transition in Mortality Rates

17.1.1 Dichotomy of mortality factors

The theory of demographic transition is based on the division of mortality into exogenous and endogenous components assuming the dichotomy of influences on the level of mortality (Bourgeois-Pichat, 1952; Vishnevsky, 1976). As far as a mortality-factor division into exogenous effects (from outside the human organism such as environment and society) and endogenous effects (from inside the human organisms such as heredity and aging) is concerned such classifications have not met any objections. For example, Urlanis (1978) has identified internal and external mortality factors. But such a dichotomy of causes of death is not so easy.

The level of mortality in a cohort during a rather short time span is determined by the specific life-style and health factors (vitality level) of the cohort. In this case cohorts of different vitality levels living under the same conditions will have different mortality levels. The reverse is also true: cohorts of the same vitality level but living under different conditions will have different mortality levels. Further, cohort vitality is determined by biological characteristics of its members and by specific conditions of life from the moment of birth until death. In the course of life the cohort is affected by different external (exogenous) impacts. Some impacts serve as a cause of death irrespective of the vitality level; others result in mortality only for persons with poor vitality. The decline of vitality with age is the result of both endogenous (natural) aging and exogenous impacts. Mortality as a result of accumulated external impacts is similar in many aspects (inter alia, by causes of death) to endogenous mortality; this category of causes of death we define as *quasi-endogenous mortality* (Vishnevsky and Volkov, 1983).

Thus general mortality can be classified into three components: exogenous mortality, quasi-endogenous mortality, and pure endogenous mortality. One can admit that endogenous mortality in older adults is the result of natural aging, while quasi-endogenous mortality reflects premature, pathological aging.

Mortality levels from each cause of death are determined by impacts of endogenous and exogenous factors of mortality (exogenous factors of both immediate and accumulated impacts). Due to this aspect the division of causes of death into mostly endogenous and mostly exogenous factors is to a certain extent conditional. Moreover, the current classification of diseases and causes of death is not fully compatible with such groupings. It may be possible to develop a classification system that allows for a thorough study of mortality factors.

Despite these difficulties, attempts at classifying death causes into mostly endogenous and mostly exogenous categories have been made. Thus in some works on the analysis of the prospects of further mortality decline in the countries with low mortality levels, such mortality causes as respiratory diseases, infections, and violence are considered exogenous (Bourgeois-Pichat, 1952, 1978). All other causes of death are considered endogenous.

Another classification of a more universal character is presented by Preston *et al.* (1972). The authors do not use the term endogenous causes but speak of internal diseases. They include cardiovascular diseases, neoplasms, nephritis, cirrhosis of liver, diabetes, certain infant diseases, and mortality causes not classified elsewhere. External causes are classified as inorganic (deaths from accident or injury) and organic (infections, diarrhea, influenza, pneumonia, bronchitis, and pregnancy complications). The results of their calculations and Bourgeois-Pichat demonstrate the proximity of the two causes in countries with relatively low mortality levels. However, when mortality levels are high, it is advisable to exclude some digestive organ diseases from the group of endogenous causes. To simplify the matter it is possible to consider the whole class of diseases as exogenous.

The decision regarding the classification of different diseases as either endogenous or exogenous depends on many factors including the living conditions of the population. For example, quite recently rheumatism (a disease of a quasi-endogenous nature) played a considerable role in the class of circulatory system diseases. There are many reasons for considering malignant neoplasms as a quasi-endogenous cause of death; however, the ratio of endogenous and exogenous components in cancer mortality is still to be investigated. It is almost impossible to separate endogenous and quasi-endogenous causes of mortality considering only etiological considerations or using statistical methods.

17.1.2 Stages of demographic transition

The demographic transition in mortality rates began in Western Europe and gradually spread to other countries. In the second half of the 20th century the transition and the increase of life expectancy determined by it covered practically all countries of the world. In the countries with relatively low mortality levels, including the USSR, life expectancy increased by a factor of three in the transitional period. The development of demographic transition in the field of mortality proves the theory and demonstrates that transitional processes are not confined to one-dimensional primitive patterns.

The main content of demographic transition is based on a decrease in the influence of exogenous factors upon mortality level. As a result the role of endogenous factors increases, and the potentials of human longevity are manifested more fully. It leads to radical reconstruction of cause-of-death structures: causes specified by mostly exogenous factors are substituted for mostly endogenous causes. Thus in the pre-transitional period most deaths were caused by infections and other acute diseases, violence, starvation, and so on. In the process of transition the growing proportion of deaths from cardiovascular diseases and other chronic diseases which are considered endogenous and quasi-endogenous are observed.

This process is determined not only by the relative growth of quasi-endogenous mortality as a result of more intensive restriction of immediate exogenous factors. It is not always possible in the process of exogenous mortality restriction to liquidate a certain class of exogenous effects completely. Some exogenous factors are weakened or compensated so that they are no longer the immediate cause of death; only then does it become possible to eliminate their negative impact on the vitality level. (That is the case when the lethality of a certain disease is liquidated, or considerably decreased, but the morbidity still exists.)

Demographic transition is substantially linked to the basic improvement of human environment, and this linkage introduces new exogenous factors (such as environmental pollution). The structure of immediate environmental factors is changing. When global exogenous factors are eliminated, some minor factors can in certain cases survive and become even more significant; among these are factors connected with specific aspects of family life, individual behavior, and so on. Injury factors also increase as new technology is implemented.

Thus, the process of limiting the influence of exogenous mortality factors and their restructuring (and the resulting restructuring of causes of death) depends on the country, region, or population group under study. Period mortality levels reflect both cohort biographies and living conditions. These levels explain the variety of observed mortality trajectories, and most display specific interchanges of relatively fast decline and stability and in some cases mortality growth in specific age groups. A slowdown of mortality decrease was observed in some European countries in the late 1950s and early 1960s. Stable age-specific mortality growth was registered during this period in 19 out of 27 countries with relatively low mortality levels (United Nations, 1982). Today the decline of mortality levels is slower compared with the decline in the 19th century.

According to recent calculations (Andreev and Dobrovol'skaya, 1979) the restructuring of cause of death accounts for about 50 percent of life-expectancy increase in England and Wales. At the same time, if the structure of cause of death had not changed but the mean cause-specific age at death had changed (as was actually observed), then the expectation of life would have increased by only 75 percent of the actual observed growth. This increase can be explained by the substantial growth of mean age at death from exogenous causes determined by the growth of the mean age of living and by the growth of the endogenous mortality component of the cause which was classified as exogenous. English cause-of-death statistics covers the longest period of time; time series for other countries are much shorter. Nevertheless, analysis of less detailed data proves the universality of these regularities.

Changes in mortality structures by cause of death were observed in countries with low mortality levels up to the early 1960s. However, mortality decline in the last period was connected not with changes of mortality structures but with the delay of death for each cause of death. The analysis of mortality data in 30 countries with low mortality levels performed by Andreev and Shaburov shows that the increase of life expectancy after 1970 was mainly due to the growth of the mean age at death (*Table 17.1*).

The increase of mean age at death was observed not only for endogenous and quasi-endogenous causes but also for exogenous causes. Thus the mean age at death from respiratory diseases for males exceeded 70 years and for females 75 years in many of the countries studied; this increase demonstrates the growth of endogenous mortality components. The increase of mean age at death is manifested first of all for such acute respiratory diseases as pneumonia and influenza.

Analysis of mortality trends by age and cause of death indicates four principal stages of demographic transition. The first stage is the

Table 17.1. Components of life-expectancy increase, probability and mean age at death, by cause of death.

Cause of death	Males		Females	
	Probability of death	Mean age at death	Probability of death	Mean age at death
Infectious and parasitic diseases	0.208	0.213	0.517	0.721
Neoplasms	0.436	0.680	0.263	0.366
Diseases of circulatory system	0.274	0.795	0.302	0.434
Diseases of respiratory system	0.143	0.529	0.185	0.464
Diseases of digestive system	0.361	0.422	0.324	0.453
Accidents, poisonings, and violence	0.220	0.381	0.203	0.625

Source: Andreev, 1987.

restriction of extraordinary periodic increases of mortality by particularly dangerous infections (cholera, smallpox, and typhus) and famine.

The second stage of demographic transition is characterized by extension and intensification of social control over immediate exogenous factors; as a result the mortality level of principal infections (child infections, diarrhea), respiratory diseases (influenza, pneumonia), and some other diseases decreases drastically or is eliminated. The second stage is the result of economic growth and industrialization. The mortality level at this stage is substantially influenced by the negative consequences of the industrial revolution such as environmental pollution; as a result the quasi-endogenous mortality (caused by circulatory system diseases and malignant diseases in younger ages), as well as mortality caused by accidents, increases.

The third stage is the gradual elimination of negative industrialization consequences and quasi-endogenous mortality along with further eradication of purely exogenous components. These processes are based on environmental-protection measures, improved labor and domestic conditions, and promotion of healthy life-styles. This stage results in increases of mean age at death from endogenous and quasi-endogenous

Table 17.2. Basic indexes of life tables by cause of death in the USSR, urban population in 1938–1939 and 1966.

Cause of death	Probability of death		Mean age at death	
	1938–1939	1966	1938–1939	1966
Males				
All causes	1,000	1,000	41.2	65.4
Diseases of circulatory system	203	483	66.8	73.8
Neoplasms	82	208	65.2	66.1
Accidents, poisonings, and violence	60	104	42.1	44.5
Diseases of respiratory system	155	82	31.4	61.0
Infectious and parasitic diseases and diseases of digestive system	346	68	24.9	56.1
Other and unknown	154	55	41.1	46.5
Females				
All causes	1,000	1,000	48.2	73.5
Diseases of circulatory system	268	655	72.6	79.4
Neoplasms	95	167	68.7	67.6
Accidents, poisonings, and violence	24	36	46.3	54.0
Diseases of respiratory system	140	59	35.9	64.3
Infectious and parasitic diseases and diseases of digestive system	293	39	22.7	59.2
Other and unknown	180	44	52.4	47.9

Source: Indexes are based on state statistics.

causes; increases of endogenous components (and as the result the increase of mean age at death) of former purely exogenous causes (such as respiratory diseases); and decreases of injuries.

The elements of the fourth stage are only beginning to manifest themselves in the countries with low mortality levels. The major features of this stage are the further decrease of infant mortality due to the preventive measures, efficient treatment of hereditary and congenital diseases, and the nursing of children born prematurely. Other features include the decrease in mortality rates of the aged. At present the fourth stage can be considered a result of a highly efficient and developed health-care system.

Table 17.3. Basic indexes of life tables by cause of death in the USSR and in five developed countries in the mid-1960s.

Cause of death	Probability of death		Mean age at death	
	USSR	Five countries	USSR	Five countries
Males				
All causes	1,000	1,000	65.9	67.9
Diseases of circulatory system	473	450	75.4	72.3
Neoplasms	173	192	65.3	68.5
Accidents, poisonings, and violence	107	73	43.1	48.8
Diseases of respiratory system	105	76	62.1	71.1
Infectious and parasitic diseases and diseases of digestive system	75	71	56.2	65.1
Other and unknown	67	138	52.8	62.5
Females				
All causes	1,000	1,000	74.0	74.3
Diseases of circulatory system	643	502	80.3	78.4
Neoplasms	133	170	66.4	70.1
Accidents, poisonings, and violence	33	46	49.8	63.4
Diseases of respiratory system	84	60	66.6	75.7
Infectious and parasitic diseases and diseases of digestive system	42	57	56.0	70.5
Other and unknown	65	165	60.7	70.0

Sources: USSR calculations based on state statistics; calculations of five countries based on World Health Organization, 1968.

17.2 Mortality Structure by Cause of Death

Mortality decrease began later in the USSR than in the majority of developed countries. Thus, the first stage of demographic transition in mortality was over at the end of the 19th century – not at the beginning of the 19th century as in Western Europe. Even in 1938–1939 the USSR urban population mortality structure corresponded to a very early stage of demographic transition. (Lack of mortality data prevents us from calculating life table by cause of death for previous periods.) As shown in *Table 17.2*, the deaths of more than 50 percent of males and 45 percent of females were caused by exogenous diseases.

Table 17.4. Basic indexes of life tables by cause of death in the USSR and in five developed countries in the mid-1980s.

Cause of death	Probability of death		Mean age at death	
	USSR	Five countries	USSR	Five countries
Males				
All causes	1,000	1,000	65.0	72.2
Diseases of circulatory system	535	434	73.0	75.7
Neoplasms	181	266	64.4	72.2
Accidents, poisonings, and violence	98	62	44.3	52.2
Diseases of respiratory system	91	90	60.8	77.2
Infectious and parasitic diseases and diseases of digestive system	50	55	56.5	69.5
Others and unknown	45	93	32.5	66.0
Females				
All causes	1,000	1,000	73.6	79.1
Diseases of circulatory system	711	495	79.5	82.5
Neoplasms	123	216	66.0	74.8
Accidents, poisonings, and violence	36	41	52.3	66.5
Diseases of respiratory system	63	72	62.6	81.7
Infectious and parasitic diseases and diseases of digestive system	33	52	54.8	72.0
Other and unknown	34	124	40.6	78.6

Sources: USSR calculations based on State Committee of the USSR of Statistics (Goskomstat), 1988; calculations of five countries based on World Health Organization, 1986.

Basic indexes of mortality tables by causes of death are presented in *Tables 17.3* and *17.4*. The first indicator describes mortality structure by cause of death in a hypothetical cohort for a corresponding period. The second indicator describes age at death by cause.

From 1938–1939 until 1966 life expectancy in urban populations increased by 24 years for males and 25 years for females. *Table 17.2* shows that the trends of mortality by cause of death conforms to the theory of demographic transition. Exogenous diseases determined 86 percent of the total increase for this period including infectious diseases, which were 39 percent.

In the mid-1960s the second stage of transition was over in most of the USSR, and the level of life expectancy nearly approached the level of the major Western countries. Basic indexes of mortality tables by causes of death for the USSR in the mid-1960s are presented in *Table 17.3*. The table also contains similar indexes for five developed countries: the Federal Republic of Germany, France, Japan, the United Kingdom (England and Wales), and the USA. One can see considerable similarity between the USSR and the other five countries, though the USSR mortality structure demonstrates both positive and negative peculiarities which mostly compensate each other (for females to a large extent, for males to a small extent). Among positive peculiarities are the larger death probability from circulatory system diseases and higher mean age at death from this cause of death. Among negative peculiarities are the larger death probability from accidents for males, the increase in respiratory system diseases and infections for both sexes, and the lower mean age at death for all these causes.

The death probabilities of three causes that are treated in this chapter as mostly exogenous diseases were higher for the USSR compared with the five developed countries. Nevertheless, the analysis of regional mortality data by causes proves that in most of the USSR the mortality caused by exogenous diseases nearly lost its role as a factor in the growth of life expectancy.

17.3   Life Expectancy and Cause of Death

In the mid-1960s new negative tendencies emerged from the trend of Soviet mortality which first manifested themselves in the growth of mortality rates for males between ages 15 and 59 and then spread to the female population and older ages (Dmitriyeva and Andreev, 1987). The principal decline of life expectancy for both males and females was connected to an increase in the mortality rate in those over age 35 and to a small increase between ages 15 and 34.

Component analysis of mortality data by cause of death allows one to estimate the impact of different causes of death on the changes of life expectancy at birth in the USSR after 1966. The decline in life expectancy after 1966 was connected first of all with the increase in mortality of circulatory system diseases and of accidents. Thus, by 1980 life expectancy for males had declined by 3.75 years (2.14 years due to an increase in circulatory system diseases, 1.24 years due to accidents). Life expectancy for females declined by 1.74 years. The negative influence on mortality due to circulatory system diseases and accidents grew by 2.09 years, but was partially compensated by the mortality trends of other causes of death. These calculations are based on USSR state statistics.

Component analysis of the impact of a cause of death on life expectancy at birth in two populations is based on Korchak-Chepurkovsky (1968) and Andreev (1982), in which the influence of the force of mortality by cause of death i at age interval $x, x + \Delta x$ on life expectancy is measured by

$$\frac{1}{2}[\mu(x, i, k_1) - \mu(x, i, k_2)][l(x, k_1)e(x, k_2) + l(x, k_2)e(x, k_1)]\Delta x \ ,$$

where $\mu(x, i, k)$ is the force of mortality by cause of death i at age x in population k; $l(x, k)$ is the survival function; $e(x, k)$ is life expectancy at age x in population k.

Comparing the data obtained with the general pattern of demographic transition shows that the trends of mortality in the USSR from the 1960s and the early 1980s fully demonstrate the negative aspects typical of the second stage. From 1980 until 1984 life expectancy in the USSR increased somewhat due to the positive influence of a decrease in respiratory diseases (both sexes) and accidents (males) on mortality dynamics.

The period after 1985 is characterized by substantial growth in life expectancy: by 1987 male life expectancy increased by 2.7 years and female life expectancy by 1.2 years. The major part of the male life-expectancy increase is due to the mortality decline of adults (mostly 40 to 59); with regard to females the impact on the age group 60 and older is more significant. Unfortunately the mortality of younger ages remained practically unchanged.

The significant impact on the changes in life expectancy in 1985 and 1986 was produced by considerable reduction of accident mortality. More than 50 percent of the male life-expectancy growth and 24 percent of female life-expectancy growth is attributed to a decrease in this cause of death. Month-by-month analysis of trends in death rates from accident obviously proves that this mortality decline is the immediate result of drastic measures taken in May 1985 to curb alcohol consumption and alcoholism in the USSR.

Decline of mortality by circulatory system diseases accounts for 58 percent of female life-expectancy increase and 28 percent of male life-expectancy increase.

Changes in age-specific death rates reflect the trends in the general mortality structure by cause of death. It is worth noting the permanent increase in death probability from circulatory system diseases. From the mid-1960s to the mid-1980s this probability increased for males from 473 per 1,000 to 535 per 1,000 and for females from 643 per 1,000 to 711 per 1,000 (compare *Tables 17.3* and *17.4*). This phenomenon was accompanied by a decline in mean age at death: by 2.4 years for males and 0.8 years for females. Probability of death from neoplasms increased for males and decreased for females, but the mean age at death by malignant diseases was lower for both. Certain increases in both probability and mean age at death from circulatory system diseases and neoplasms were observed after 1984.

Mortality-level decline by accidents in 1985 and 1986 contributed to the decrease of death probability (for males from 107 per 1,000 in 1966 to 98 per 1,000 in the mid-1980s) and to the increase of mean age at death. In 1980 death probability from accidents for males was 147 per 1,000.

After 1966 a decline in the probability of death from diseases of respiratory and digestive systems and infectious and parasitic diseases was observed, and a decline in the mean age at death by these causes was also observed. After 1984 mean age at death by exogenous causes of death somewhat increased but remained lower than in 1966.

17.4 Growth of Life Expectancy

In this section several features of the mortality structure by cause of death in the USSR in 1986 are compared with features in five major developed countries: the Federal Republic of Germany, France, Japan, the United Kingdom (England and Wales), and the USA. The central feature is the considerably lower mean age at death from all causes for both males and females in the USSR (*Table 17.4*).

In the USSR there is a higher probability of death from circulatory system diseases and a much lower probability of death from neoplasms. Probability of death from accidents for males is 1.5 times higher in the

Table 17.5. Components of differences in life expectancy in the USSR compared with five developed countries.

Cause of death	Males	Females
All causes	7.2	5.5
Diseases of circulatory system	2.6	3.7
Neoplasms	0.4	−0.1
Accidents, poisonings, and violence	1.7	0.5
Diseases of respiratory system	1.4	1.1
Infectious and parasitic diseases and diseases of digestive system	0.9	0.2
Other and unknown	0.2	0.1

Source: see *Table 17.4*; Methodology of calculation, see Andreev, 1982.

USSR than in the five developed countries. In the five developed countries the category "other and unknown" is more than double for males and about three times larger for females. Results of the analysis are presented in *Table 17.5*.

What causes of death are responsible for the slow decline of mortality rates in the USSR? Exogenous diseases account for 2.3 years in life-expectancy differences (or 32 percent of the total difference) for males and 1.3 years (or 24 percent) for females. The probability of death from this group of diseases in the USSR is practically equal to the average in the five developed countries. These differences are due to lower mean age at death (by 16.3 years for males and 17.8 years for females).

The higher probability of death from accidents accounts for a 24 percent life-expectancy difference for males and for a 9 percent difference for females. Thus direct exogenous factors account for 56 percent of the life-expectancy lag for males and 33 percent for females.

The rest of the life-expectancy lag in the USSR is due to mortality from quasi-endogenous and endogenous diseases. Thus the mean age at death from these causes in the USSR is lower by 4.4 years for males and 4.0 years for females compared with the five developed countries. We consider this evidence of more intensive influence produced by cumulative exogenous factors.

Despite certain efforts to lower mortality levels in 1985–1987, the USSR has only entered into the third stage which is characterized by the restriction of quasi-endogenous mortality; certain regions are still in the second stage of the transition.

The results of our study suggest that the principal cause of the discrepancy between the USSR and Western countries is connected with considerably lower living standards. One can suppose that the USSR mortality-level increases in 1987–1989 are connected with the economic crisis in the USSR. The second principal cause, also an important factor of negative tendencies in mortality, is the environmental situation.

The mortality level in the USSR is visibly influenced by shortcomings in the health-care system. Soviet health-care services are not prepared for solving the problems typical of the third stage of demographic transition in mortality. The situation in mortality is considerably aggravated by the prevalence of hygienically irrational (or even harmful) habits of a considerable part of the population. Positive radical changes in the mortality patterns cannot be achieved without solving a whole set of social problems. Certain positive changes can be achieved provided that the demographic consequences of the decisions taken are considered.

Three concrete tasks concerning a further increase of life expectancy in the USSR must be developed: (1) to continue lowering mortality (especially of children) from exogenous diseases, namely, infections and respiratory diseases (this process began in the 19th century, but it is not over yet); (2) to promote maximum reduction of mortality from accidents (especially for males); (3) to restrict the quasi-endogenous causes of death (mortality from circulatory system diseases and other chronic diseases) in the younger ages. The struggle with premature mortality from chronic diseases will increase along with the decline of exogenous mortality.

References

Andreev, E.M., 1982. Metod komponent v analize prodolzhitel'nosti zhizni (Component Method Applied to Life-Expectancy Analysis). *Vestnik statistiki* (Herald of Statistics) 9:42–47.

Andreev, E.M., 1987. Analiz dozhitiya s ispol'zovaniyem dannykh o prichinakh smerti (The Analysis of Survival Based on Data by Causes of Death). *Populyatsionnaya gerontologiya* (Population Gerontology) T. 6:190–219.

Andreev, E.M., 1990. Prodolzhitel'nosti zhizni i prichiny smerti v SSSR (Life Expectancy and Causes of Death in the USSR). In A.G. Volkov, ed. *Demograficheskiye protsessy v SSSR* (Demographic Processes in the USSR). Nauka, Moscow.

Andreev, E.M., and V.M. Dobrovol'skaya, 1979. Ob odnom metode izucheniya krivykh dozhitiya (One Method of Analysis of Survival Curves). In E.M. Andreev and A.G. Vishnevsky, eds. *Prodolzhitel'nosti zhizni: analiz i modelirovaniye* (Life Expectancy: Analysis and Modeling). Statistika, Moscow.

Bourgeois-Pichat, J., 1952. Essai sur la mortalité biologique de l'homme. *Population* 3:381–394.

Bourgeois-Pichat, J., 1978. Future Outlook for Mortality Decline in the World. *Population Bulletin of the United Nations* 11:12–41.

Dmitriyeva, R.M., and E.M. Andreev, 1987. O sredney prodolzhitel'nosti zhizni naseleniya SSSR (On the Life Expectancy of the USSR Population). *Vestnik Statistiki* (Herald of Statistics) 12:31–39.

Korchak-Chepurkovsky, Y.A., 1968. Vliyaniye smertnosti v raznykh vozrastakh na uvelicheniye sredney prodolzhitel'nosti zhizni (The Influence of Mortality in Different Ages on Increasing of Life Expectancy). In T.V. Ryabushkin, ed. *Sovetskaya demografiya za 70 let* (Soviet Demography over the Past 70 Years). Nauka, Moscow.

Preston, S.H., N. Keyfitz, and R. Shoen, 1972. *Causes of Death: Life Table for National Populations*. Seminar Press, New York and London.

State Committee of the USSR on Statistics (Goskomstat), 1988. *Naseleniye SSSR, 1987: Statisticheskiy sbornik* (Population of the USSR, 1987: Statistical Collection). Finansy i statistika, Moscow.

United Nations, 1982. *Level of Mortality Since 1950*. New York, NY.

Urlanis, B.Ts., 1978. *Evolutsiya prodolzhitel'nosti zhizni* (The Evolution of Life Span). Statistika, Moscow.

Vishnevsky, A.G., 1976. *Demograficheskaya revolutsiya* (Demographic Revolution). Statistika, Moscow.

Vishnevsky, A.G., and A.G. Volkov, eds., 1983. *Vosproizvodstvo naseleniya SSSR* (Reproduction of the Population of the USSR). Finansy i statistika, Moscow.

World Health Organization, 1968. *World Health Statistics Annual 1967*. Geneva.

World Health Organization, 1986. *World Health Statistics Annual 1986*. Geneva.

Chapter 18

A Comparison of Soviet Mortality in the Working Ages: 1959–1988

Barbara A. Anderson and Brian D. Silver

After the mortality rates of men began to increase in the Soviet Union in 1964, demographers both in the Soviet Union and abroad became aware that mortality was not inevitably bound to improve and that the health of the Soviet population was worse than that of many other industrialized countries.[1] When the infant mortality rate for the USSR also rose in the early 1970s (from 22.9 in 1970 to 31.4 in 1976)[2], more people became concerned about the state of Soviet public health, and some published alarmist declarations about poor health conditions in the USSR (Davis and Feshbach, 1980; Eberstadt, 1981). In fact, men also experienced increasing mortality in several Western and Northern European countries in the 1960s and early 1970s (Coale, 1981; Anderson and Silver, 1986a), but the increase was not as large in those countries as in the USSR or in Eastern Europe, especially in Hungary and Poland.

In the 1980s, especially after 1985, mortality rates – as reflected in figures for life expectancy at birth – appear to have fallen in the Soviet Union, especially for males. The infant mortality rate has also fallen, from 31.4 in 1976 to 22.7 in 1989, though the rate remains high in comparison with the rates in most of the industrialized world.[3]

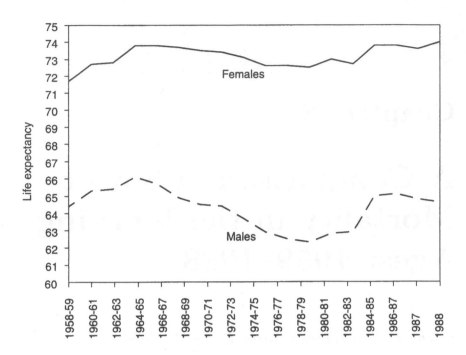

Figure 18.1. Life expectancy at birth for males and females in the Soviet Union, 1958–1959 through 1988.

The changes in overall mortality in the Soviet Union since 1958 are depicted in *Figure 18.1*.[4] For Soviet males, the reported life expectancy at birth (e_0) increased from 64.4 years in 1958–1959 to a post–World War II peak of 66.1 years in 1964. It then fell to the low 60s in 1980–1981, after which it began to rise, with a particularly sharp increase after 1984–1985.[5] For females, life expectancy at birth also reached a peak, at 73.8 years, in the mid-1960s, declined slightly through 1980–1981, and then regained its mid-1960s level in the late 1980s.

Change in summary mortality statistics such as e_0 is a product of very different patterns of change in mortality across age groups. *Figure 18.2* depicts the age-specific death rates (ASDRs) for the Soviet population for the years after 1958–1959 relative to the age-specific death rate in 1958–1959. The ASDRs for Soviet males below age 20 decreased more or less continuously between 1958–1959 and 1988. This generalization applies also to the 0 to 4 age group, even though infant mortality rose between 1971 and 1976. Above age 20, however, especially above age 25, the ASDRs followed an up-and-down path: between 1958–1959 and 1964–1965, the mortality rates decreased; between 1964–1965 and 1980–1981, mortality increased; since 1980–1981, mortality decreased, except for a rise in 1988. For males above age 40, however, the ASDRs in 1988 were higher than the ASDRs in 1958–1959. For Soviet females, the trends in the ASDRs have been very different from those of males. Below age 50, the ASDRs either did not change much or decreased between 1958–1959 and 1988; above age 50, the ASDRs either remained stable or increased slightly over the same years.

The trends in ASDRs show inflection points at 1964 and 1980. An inflection point in 1964 appears for almost all age groups for both males and females, though the change is sharper for males. A previous publication (Anderson and Silver, 1989b) notes that the explanation for this turning point in Soviet mortality trends is not obvious. The turning point is not associated with a known natural disaster or epidemic. Nor can it be a "cohort effect," since it happened simultaneously to people in many different birth cohorts.

Similarly, why mortality rates began to fall in the early 1980s for both males and females is not certain. This turning point occurred during the "period of stagnation" in the last years of Leonid Brezhnev's term as General Secretary of the Communist party. Mortality took an especially sharp downturn for most age groups after 1985. While it is possible that this phenomenon is a result of the anti-alcohol campaign initiated after Mikhail Gorbachev became General Secretary in 1985 (cf. Dmitriyeva and Andreev, 1987), it occurs over a wide range of ages for both males and females and it occurs for subpopulations that are not known for widespread abuse of alcohol. Hence, alternative explanations must be considered. One hypothesis is that the changes in reported mortality are a statistical artifact: perhaps the methods used for constructing the mortality rates changed, or perhaps the use of corrected population figures (such as from the 1979 census results) affected the calculated mortality rates.[6]

18.1   The Goal of the Study

The main goal of this paper is to examine trends in mortality in the working ages in the Soviet Union. We examine the trends between 1958 and 1988 for the USSR, and we compare mortality levels and trends

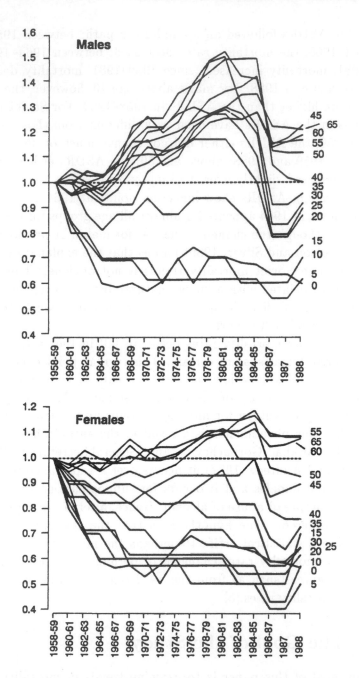

Figure 18.2. Age-specific death rates of the Soviet population, indexed to rates in 1958–1959, by sex (ASDR at given date divided by ASDR in 1958–1959).

in the USSR with the experience of 33 other developed countries. This paper extends an earlier comparative study (Anderson and Silver, 1986a) by adding recently published data for all the countries. Particularly valuable are statistics from the USSR that have been published in the last few years (see State Committee of the USSR, 1988, 1989a, 1989b). These new data refer not only to recent years but also to earlier years for which, until recently, data had not been published – specifically between 1974 and 1985.

We focus on the working ages to assess the effects of mortality change on the labor supply. Of course, mortality in working ages is only one of several factors that affect labor supply, including fertility, migration, mortality before the working ages, time spent in school, health and disability, and the actual work status of people through the life course, including after people reach official retirement age. Nonetheless, a focus on mortality in the working ages answers an important question: Out of the maximum productive life that might be obtained from a given population in a given age range, how much time is lost as a result of mortality? With the loss due to mortality taken as a baseline, the role of other factors or changes of state (such as disability, leaves of absence from work, school, and so on) can be assessed.

We focus on the working ages also for a technical reason – namely, that data in this age range are more accurate than data for older and younger ages. The reported infant mortality rates and mortality rates at old age in the USSR are substantially lower than the actual rates. We have estimated that if one were to adjust Soviet infant mortality rates to take into account differences between the World Health Organization's and the Soviet government's definitions of live birth and infant death, the reported Soviet infant mortality rate would need to be increased by from 22 percent to 25 percent.[7] In addition, mortality above the working ages is substantially underestimated in official Soviet statistics, due to some combination of age overstatement in the censuses, age overstatement in death registration, and underreporting of deaths (Garson, 1986; Bennett and Garson, 1983; Anderson and Silver, 1989a).

We have estimated that if one applies a conservative adjustment for the understatement of infant mortality as well as of mortality above age 60 for Soviet males in 1958–1959, the value of e_0 would have been 62.6 years rather than 64.2 years as reported in official statistics (Anderson and Silver, 1989a). Comparisons of mortality between the USSR and other countries that rely on measures of life expectancy at birth (e_0) or at age 1 (e_1) can lead to erroneous conclusions about the relative levels of mortality in different countries. This is true especially because a majority of the difference between developed countries in life expectancy at birth or expectation of remaining life at age 1 is due to mortality above the working ages.

Ages. In conventional Soviet statistics, the able-bodied (*trudosposobnye*) ages are defined as 16 to 59 for men and 16 to 54 for women. This is not a convenient age range for the study of mortality trends in the USSR because mortality data are almost never reported separately for the 16 to 19 age group, and published values of life-table functions by sex are sometimes given only in 10-year age groups.

Therefore, we refer to the age range 20 to 59 as the able-bodied ages, working ages, or productive ages. This range permits comparison between regions of the USSR and between countries, as well as provides a uniform 40-year interval for comparing the mortality experience of males and females. Furthermore, even though in Soviet statistical usage the able-bodied ages start at 16, the average age of entrants to the labor force is higher: while in 1959, 59.6 percent of the Soviet population age 16 to 19 were employed, in 1970 the figure was only 39.8 percent (Breeva, 1984). This percentage undoubtedly declined further in subsequent years.[8] Moreover, many people continue to work past the formal retirement age of 55 for women and 60 for men.

Mortality indicators. A measure of mortality in an age range that is useful for comparisons is the number of person-years lived between specific ages, either from birth or by a cohort that is initiated at a later age, such as age 20. Using standard life-table statistics, the temporary life expectancy can be calculated on segments of the life span.[9] The value for each age segment would be

$$\frac{T_{a1} - T_{a2}}{l_{a3}},$$

where T_{a1} is person-years lived above age $a1$, T_{a2} is person-years lived above age $a2$, and l_{a3} is the radix of the life table for purposes of the calculation ($a3$ can be any age less than or equal to $a1$). If $a3$ is 0, then the sum of the temporary life expectancies for all age segments over the life span will equal the expectation of life at birth (e_0).

Table 18.1. Number of person-years lived in 20-year age segments for the population of the USSR, by sex.

Age	Males	Females	Males	Females	Males	Females
	1938–1939		*1958–1959*		*1969–1970*	
0–19	14.22	14.74	18.73	18.92	19.19	19.37
20–39	12.79	13.41	17.93	18.45	18.38	19.05
40–59	10.48	11.82	15.94	17.42	15.97	18.07
60–79	5.65	8.08	9.98	13.39	9.45	13.97
80+	0.85	1.64	1.85	3.49	1.38	3.08
Total (e_0)	43.99	49.69	64.42	71.67	64.38	73.53
	1978–1979		*1983–1984*		*1986–1987*	
0–19	19.09	19.28	19.15	19.33	19.19	19.36
20–39	18.21	18.97	18.33	19.04	18.58	19.10
40–59	15.45	17.94	15.54	18.01	16.44	18.25
60–79	8.62	13.54	8.49	13.51	9.50	13.91
80+	1.12	2.90	1.09	2.88	1.33	3.16
Total (e_0)	62.48	72.64	62.60	72.76	65.04	73.78

Source: The figures for 1938–1939 are from Vishnevsky and Volkov, 1983; the figures for later years are from State Committee of the USSR, 1989b.

Table 18.1 shows the results of a calculation of the temporary life expectancies in 20-year age segments for the Soviet population at various dates. In recent years, the temporary life expectancy between birth and the 20th birthday has approached the theoretical maximum of 20 years. This is not contradicted by the fact that infant mortality, which is responsible for more than half of the deaths of persons under age 20, remains high.[10] The temporary life expectancy in the working ages (20 to 59) remains well short of the theoretical maximum of 40 years, primarily because of mortality between ages 40 and 60. Over the nearly 30-year interval between 1958–1959 and 1986–1987, men have gained only half a year in the temporary life expectancy between ages 40 and 60, and women only eight-tenths of a year. Consistent with *Figure 18.2*, during much of that period, the mortality of men in the second half of the working ages was higher than in 1958–1959.

For most of the analysis of mortality in the working ages, we calculate the number of person-years lived using a different radix for the temporary life expectancy. Namely, we calculate it

$$\frac{T_{20} - T_{60}}{l_{20}}.$$

For some purposes, we break working life into 10- and 20-year age segments, where, for example, the person-years lived between birthdays 40 and 60 among those alive at birthday 20 is calculated as

$$\frac{T_{40} - T_{60}}{l_{20}} \quad .$$

An advantage of this measure over measures of overall mortality is that it directly reflects the size of the loss caused by mortality in a theoretically interesting age group, the population of working age. It shows how closely a given population is approaching (or moving away from) the maximum possible number of years that could be lived in the age range given. Finally, comparison of absolute differences between lengths of working life of the sexes is straightforward.

However, the number of person-years lived between, say, the 40th and 60th birthdays among people who survive to their 20th birthday is influenced by the proportion surviving from 20 to 40 (since l_{20} is in the denominator). A previous study (Anderson and Silver, 1986a) examined both the number of person-years lived between ages 40 and 60 and the probability of surviving from age 40 to age 60 (i.e., $_{20}p_{40}$); the substantive results using the alternative measures were similar. We prefer the use of the person-years measure because it is more interpretable in relation to the costs and benefits of population policies. Some Soviet demographers have also found this approach to the study of mortality in the working ages useful (Kruminš, 1976; Pervushin, 1987).

18.2   The USSR and the Developed World

18.2.1   Trends in working-age mortality
            since 1958–1959

Sex differences. Table 18.2a shows the trends between 1958–1959 and 1988 in the average number of person-years lived in the working ages, by sex, for the total, urban, and rural populations of the USSR. Figures are shown for two 20-year age segments as well as for the entire 40-year working-age interval.

For both men and women in the USSR there has been remarkably little change in the number of person-years lived in the first half of the productive ages. The size of the difference between the sexes has also remained quite stable. Among persons who survived to their 20th

Table 18.2a. Number of person-years lived in the working ages in the USSR in 20-year age segments, 1958–1959 to 1988, by sex for total population.[a]

Year	Males			Females			Females − Males		
	20–39	40–59	20–59	20–39	40–59	20–59	20–39	40–59	20–59
Total population of USSR									
1958–1959	19.44	17.25	36.69	19.71	18.60	38.31	0.27	1.35	1.62
1960–1961	19.43	17.29	36.71	19.74	18.70	38.44	0.31	1.41	1.72
1962–1963	19.44	17.29	36.73	19.75	18.71	38.46	0.31	1.42	1.73
1964–1965	19.46	17.30	36.76	19.77	18.77	38.54	0.31	1.47	1.78
1966–1967	19.43	17.15	36.59	19.78	18.77	38.55	0.35	1.62	1.96
1968–1969	19.40	16.93	36.33	19.79	18.78	38.58	0.39	1.86	2.25
1970–1971	19.36	16.78	36.13	19.80	18.79	38.59	0.44	2.02	2.46
1972–1973	19.37	16.81	36.18	19.80	18.80	38.60	0.43	1.99	2.42
1974–1975	19.35	16.67	36.01	19.80	18.77	38.57	0.45	2.10	2.55
1976–1977	19.31	16.49	35.80	19.79	18.73	38.52	0.48	2.24	2.72
1978–1979	19.30	16.34	35.64	19.80	18.71	38.50	0.50	2.36	2.86
1980–1981	19.29	16.28	35.57	19.79	18.69	38.48	0.50	2.41	2.91
1982–1983	19.33	16.43	35.76	19.81	18.76	38.57	0.48	2.33	2.81
1984–1985	19.38	16.51	35.90	19.81	18.73	38.55	0.43	2.22	2.65
1986–1987	19.54	17.27	36.82	19.83	18.93	38.76	0.29	1.65	1.94
1987	19.54	17.28	36.83	19.83	18.94	38.78	0.29	1.66	1.95
1988	19.50	17.19	36.69	19.81	18.89	38.71	0.31	1.70	2.01

[a]Based on $l_{20} = 100,000$.

birthday, women live between one-third and one-half of a year more on average than men between ages 20 and 40.

The situation in the second half of the working ages (40 to 59) is very different. Over the 30-year period from 1958–1959 to 1986–1987, there has been little change in the average number of person-years that women have survived between their 40th and 60th birthdays: the figure has ranged between 18.6 and 18.9 years. While there was very little improvement over time, there was little deterioration during any year in the period since 1958. Men, on the other hand, have gone through several cycles of change. The temporary life expectancy between ages 40 and 60 remained at about 17.3 years between 1958–1959 and 1964–1965, declined by a whole year to 16.3 by 1980–1981, and then rose to 17.3 years by the late 1980s. As a result of the changes for both men and women between 1958–1959 and 1986–1987, the difference in the temporary life expectancy through the productive ages (20 to 59) rose from 1.6 years in 1958–1959 to 2.9 years in 1980–1981, and then declined to about 2.0 years by the late 1980s.

Table 18.2b. Number of person-years lived in the working ages in the USSR in 20-year age segments, 1958–1959 to 1988, by sex for urban and rural populations.

Year	Males			Females			Females − Males		
	20–39	40–59	20–59	20–39	40–59	20–59	20–39	40–59	20–59
Urban population									
1958–1959	19.44	17.22	36.66	19.75	18.65	38.41	0.31	1.43	1.74
1960–1961	19.44	17.29	36.73	19.77	18.73	38.49	0.32	1.44	1.76
1962–1963	19.48	17.36	36.84	19.79	18.78	38.57	0.31	1.42	1.73
1964–1965	19.50	17.41	36.92	19.81	18.84	38.65	0.30	1.43	1.73
1966–1967	19.49	17.30	36.79	19.81	18.84	38.65	0.32	1.53	1.85
1968–1969	19.47	17.11	36.58	19.82	18.85	38.67	0.35	1.73	2.09
1970–1971	19.42	16.91	36.33	19.82	18.84	38.65	0.40	1.92	2.32
1972–1973	19.44	16.97	36.41	19.83	18.87	38.70	0.39	1.90	2.29
1974–1975	19.43	16.86	36.29	19.83	18.85	38.68	0.41	1.99	2.40
1976–1977	19.40	16.70	36.09	19.82	18.82	38.65	0.43	2.13	2.55
1978–1979	19.38	16.55	35.92	19.83	18.80	38.63	0.45	2.25	2.71
1980–1981	19.37	16.47	35.84	19.83	18.81	38.64	0.47	2.34	2.81
1982–1983	19.43	16.68	36.11	19.84	18.85	38.68	0.41	2.17	2.58
1984–1985	19.47	16.76	36.23	19.84	18.84	38.68	0.37	2.08	2.45
1986–1987	19.60	17.44	37.04	19.87	19.03	38.90	0.27	1.58	1.85
1987	19.61	17.49	37.10	19.87	19.03	38.90	0.26	1.54	1.80
1988	19.56	17.39	36.95	19.84	18.98	38.82	0.28	1.59	1.87
Rural population									
1958–1959	19.44	17.27	36.71	19.67	18.54	38.20	0.23	1.26	1.49
1960–1961	19.40	17.27	36.67	19.69	18.62	38.31	0.29	1.35	1.64
1962–1963	19.40	17.20	36.60	19.70	18.63	38.33	0.30	1.43	1.73
1964–1965	19.36	17.11	36.47	19.72	18.68	38.40	0.36	1.57	1.93
1966–1967	19.31	16.86	36.16	19.73	18.66	38.39	0.42	1.81	2.22
1968–1969	19.28	16.63	35.91	19.74	18.67	38.41	0.46	2.04	2.50
1970–1971	19.25	16.57	35.82	19.74	18.66	38.40	0.49	2.10	2.58
1972–1973	19.24	16.53	35.77	19.73	18.64	38.37	0.50	2.10	2.60
1974–1975	19.17	16.29	35.46	19.73	18.59	38.32	0.56	2.30	2.85
1976–1977	19.14	16.11	35.25	19.72	18.54	38.26	0.58	2.43	3.01
1978–1979	19.13	15.97	35.10	19.73	18.51	38.24	0.60	2.54	3.14
1980–1981	19.10	15.83	34.93	19.72	18.47	38.18	0.61	2.64	3.25
1982–1983	19.11	15.89	35.00	19.72	18.50	38.22	0.61	2.61	3.22
1984–1985	19.16	15.93	35.09	19.72	18.46	38.18	0.57	2.53	3.10
1986–1987	19.37	16.79	36.16	19.76	18.69	38.45	0.39	1.90	2.29
1987	19.37	16.78	36.15	19.77	18.71	38.48	0.39	1.93	2.33
1988	19.32	16.66	35.98	19.76	18.69	38.44	0.43	2.03	2.46

Table 18.2b shows data for the urban and rural populations. The basic trends found in the population as a whole are mirrored in each subpopulation. Rural men and rural women have higher mortality than their urban counterparts in the working ages for the Soviet Union as a whole. The sex differential in mortality is also generally higher among rural than among urban residents (since the late 1960s).

Figure 18.3 illustrates the trends over time in the person-years lived in the working ages, by breaking the data into 10-year age intervals. The top panel shows virtually no change in the average number of person-years lived by men between ages 20 and 30. All of the variability occurs at older ages, and the variability increases with age. The 1980–1981 low point in person-years lived (high point in mortality) and the sharp reduction in mortality that occurred in the mid-1980s are noticeable in all three 10-year age intervals between ages 30 and 60.

The trends in the bottom panel are very different from those in the top panel. The scale on the vertical axes of the two panels is the same to make clear the much lower mortality levels of women in each age segment. Moreover, within each age interval one finds virtually no variability in women's mortality over the 30-year period studied. The fall of mortality after 1985 found among Soviet men aged 30 to 59 also appears among Soviet women aged 40 to 59, but in a much more moderate form.

Figure 18.4 shows that the pattern, though not the magnitude, of the changes in mortality in the 1980s is similar for men and women aged 40 to 59. Note the pattern of change for total population (middle line) in each panel. Also note the difference in the scale on the vertical axis of the two panels in *Figure 18.4*. The situation was different in the 1960s and 1970s. During that period, mortality of women aged 40 to 59 fell between 1958–1959 and 1972–1973 and then rose until 1980–1981, while the mortality of men aged 40 to 49 stayed level between 1958–1959 and 1964–1965, then increased until 1980–1981.

The contrast between the existence of a different pattern of change for men and women in 1960s and 1970s and the existence of a similar pattern of change for men and women in the 1980s suggests that the underlying causes of change in men's and women's mortality in recent years may be similar or that common factors are producing a convergence in mortality risks for males and females. This could be reflected in increasing similarity of the patterns of causes of death for men and women. It is possible, however, that the underlying commonality is technical – a result of how the mortality data were generated.

Urban–Rural differences: Mortality crossovers. In principle, it is of great interest to examine differences in mortality between urban and rural areas to understand how differences in styles of life and work, exposure to risk, and other factors associated with the place of residence affect

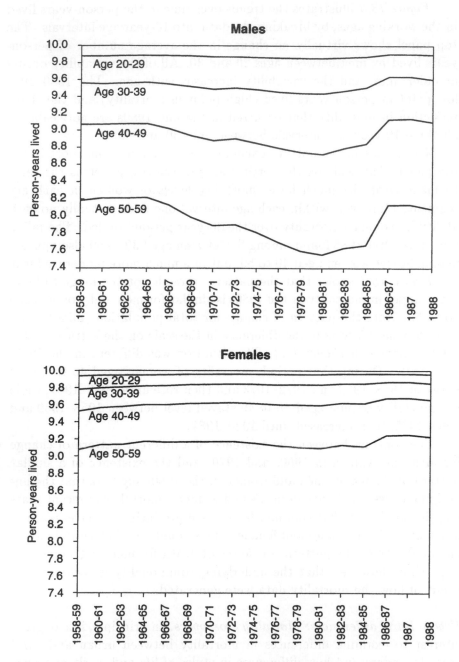

Figure 18.3.   Person-years lived in working ages in the Soviet Union
at 10-year age intervals, 1958–1959 through 1988, by sex.

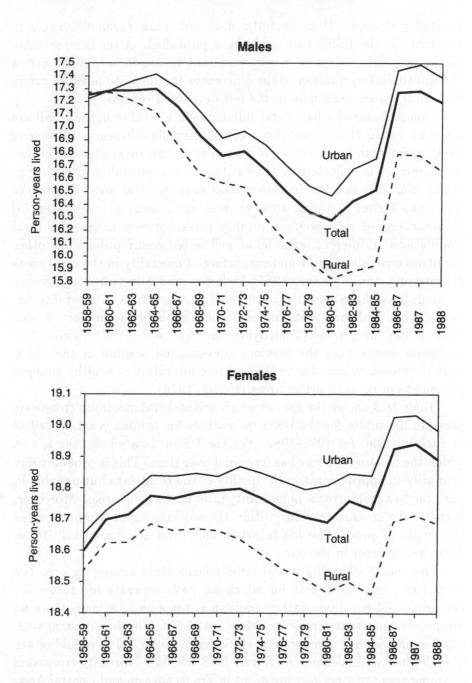

Figure 18.4. Person-years lived age 40 to 59 in the Soviet Union, 1958–1959 through 1988, by sex.

mortality change. Until recently, data on urban–rural differences in mortality in the USSR had rarely been published. After close scrutiny of the new data, however, we are reluctant to use these data to give a substantive interpretation to the differences in mortality between urban and rural places, especially in the less developed regions.

Comparisons of urban–rural differences in mortality in the USSR are more uncertain than comparisons of male–female differences because of likely greater understatement of mortality in the rural areas. The understatement is reflected in the existence of a mortality crossover in which mortality is lower in urban areas than in rural areas at younger ages, and higher in urban areas than in rural areas at older ages.[11] Comparisons of age-specific mortality rates between urban and rural populations in different republics as well as between republics and other countries strongly suggest underreporting of mortality in the rural areas (Dmitriyeva and Andreev, 1987; Anderson and Silver, 1989a). Soviet demographers have recently stated that the crossover in mortality between urban and rural areas in the USSR is probably an artifact of poor data in the rural areas (Dmitriyeva and Andreev, 1987). This interpretation differs from the previous conventional wisdom in the USSR that the crossover was due to the selective migration of healthy younger persons from rural to urban areas (Bedny, 1976).

Table 18.3 shows the ages at which urban–rural mortality crossovers occur in life tables for the USSR as a whole for various years as well as in each republic for 1986–1987. For the USSR as a whole, the age at which the crossovers occur has increased over time. This is probably due primarily to improvement in the quality of the rural data but possibly in part due to a real increase in mortality rates in the rural areas. Moreover, because the crossovers occur within the working ages, comparisons of the length of productive life in urban and rural areas are likely to be distorted by error in the data.

The quality of the data also varies considerably among regions. For 1986–1987, the latest year for which we have separate life tables for the urban and rural populations, only in Estonia and Latvia is there no urban–rural mortality crossover below age 85.[12] This is consistent with other evidence that demographic statistics in these Baltic republics are of high quality (Anderson and Silver, 1988, 1989a). Mortality crossovers at young ages are especially prevalent in Transcaucasia and Central Asia,

Table 18.3. Age interval of urban–rural mortality crossover.

Republic	Year	Sex	Age interval
USSR	1926–1927	Male	25–29
		Female	40–44
	1938–1939	Male	20–24
		Female	45–49
	1958–1959	Male	45–49
		Female	45–49
	1965–1966	Male	50–54
		Female	50–54
	1969–1970	Male	50–54
		Female	50–54
	1978–1979	Male	50–54
		Female	50–54
	1983–1984	Male	55–50
		Female	55–59
	1985–1986	Male	55–59
		Female	60–64
	1986–1987	Male	55–59
		Female	60–64
Russia		Male	60–64
		Female	65–69
West			
Byelorussia	1986–1987	Male	60–64
		Female	65–69
Moldavia	1986–1987	Male	70–74
		Female	75–79
Ukraine	1986–1987	Male	55–59
		Female	55–59
Baltic			
Estonia		Male	>85
		Female	>85
Latvia	1986–1987	Male	>85
		Female	>85
Lithuania	1986–1987	Male	65–69
		Female	65–69
Transcaucasia			
Armenia	1986–1987	Male	45–49
		Female	45–49
Azerbaijan	1986–1987	Male	35–39
		Female	40–44
Georgia	1986–1987	Male	40–44
		Female	40–44
Kazakhstan/Central Asia			
Kazakhstan	1986–1987	Male	50–54
		Female	50–54
Kirghizia	1986–1987	Male	45–49
		Female	75–79
Tajikistan	1986–1987	Male	15–19
		Female	60–64
Turkmenistan	1986–1987	Male	30–34
		Female	70–74
Uzbekistan	1986–1987	Male	35–39
		Female	65–69

but the patterns in the two regions differ from one another: in Transcaucasia, mortality crossovers occur at about the same age for both women and men; in the Central Asian republics, mortality crossovers occur at much younger ages for men than for women. Indeed, the differences in the ages of the mortality crossovers between males and females in Central Asia are so enormous that they suggest a behavioral interpretation rather than one based only on the completeness or accuracy of registration.

We hypothesize that a tendency to favor males over females among traditionally Muslim populations leads to favoritism in access to medical treatment. A male who becomes ill is more likely to receive medical treatment than a female who becomes ill. In general, this would lead to a lower sex differential (or even a reversal in the typical sex differential) in mortality compared with societies that do not have as strong a tradition of favoring males. We shall show later that the size of the sex differential in mortality in Soviet Central Asia is indeed much smaller than the size of the differential in the remaining population of the USSR. We do not have any direct evidence for Central Asia to support this hypothesis. But there is evidence of patterns of differential treatment of illnesses by sex in some less developed societies, especially those with Muslim populations (see, for example, Aziz, 1977; Basu, 1989; Chen and D'Souza, 1981, D'Souza and Chen, 1980; Singh *et al.*, 1962).

Why should favoritism in access to medical treatment lead to a sex difference in the age of the urban–rural crossover in mortality? Rural males who are sick are probably more likely than rural females who are sick to be taken to urban medical centers for treatment. If they die in an urban hospital, or perhaps in the care of relatives who live in an urban area, their death might then be registered as occurring in an urban area rather than a rural one. Although in most republics deaths are supposed to be attributed in official statistics to the place of permanent residence of the decedent or (in the case of infants) of the decedent's parents, we know from other evidence that statistical practice does not always follow legal prescriptions (Anderson and Silver, 1985, 1988).

Our proposed explanation of the urban–rural crossover in Central Asia makes sense in the light of the historical and comparative experience in other countries. Instead of being simply a statistical artifact or a result of the migration of healthy young people from rural to urban places, the crossover could be a result of short-term migration of sick males from rural to urban areas. For example, Ward (1987) has suggested that such a phenomenon may account for higher mortality rates in Mexico City than in the surrounding countryside. Until the alternative explanations are sorted out – and contributions of different factors are determined – we would advise researchers to be cautious about giving substantive interpretations of the urban–rural differences. Similar caution is warranted in the use of cause-of-death data for urban and rural places.[13]

18.2.2   Mortality trends in the working ages in the USSR and in other countries

The discussion of the urban–rural mortality crossovers reminds us that it is often difficult to generalize about Soviet mortality trends as a whole, because the trends and levels of mortality vary greatly. However, to place the trends in mortality in the working ages in the USSR at large in the context of trends in other countries, we have selected a few countries that span the range of mortality experience of the developed world.

Figure 18.5 depicts the number of person-years lived from age 40 to age 59 in six countries at four dates.[14] We restrict our attention to the latter half of the working ages because only in this age range does any substantial difference in mortality appear across the developed countries. The decline in the number of person-years lived from age 40 to age 59 over the period 1958–1959 to 1970 for Soviet men was similar to the experience of men in Finland, Hungary, and the United States, and contrasted with the improving situation for men in Japan. However, unlike American and Japanese men, but like men in Hungary, Soviet men continued to experience increasing mortality in the older working ages between 1970 and 1980. In fact, on average, in 1980 Soviet men lived fewer person-years from age 40 to age 59 than the men in any other developed country; they ranked last out of 34 countries. Only because of a reduction in mortality in the 1980s, and the continuing worsening of the situation for Hungarian men, did the number of person-years lived by Soviet men from age 40 to age 59 climb from last place to second from last place in the developed world by the mid-1980s.

The bottom panel of *Figure 18.5* shows analogous data for women. The number of person-years lived by Soviet women from age 40 to age 59 has changed very little since 1958–1959. Although Soviet women in this age range fell into the bottom ranks of women in the developed world in

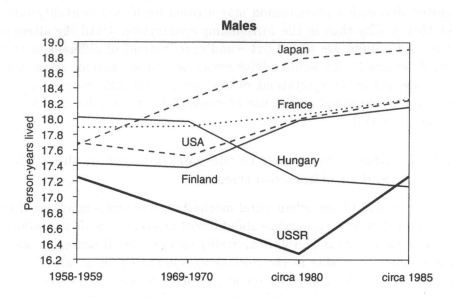

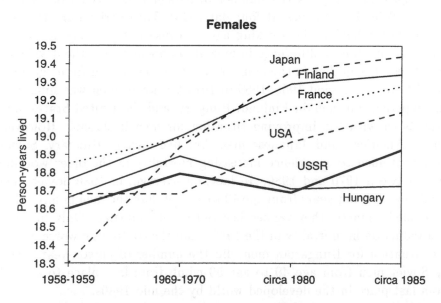

Figure 18.5. Person-years lived, from age 40 to age 59, in the Soviet Union and selected countries, 1958–1959, 1969–1970, 1980, and 1985, by sex.

1958–1959, their mortality was close to the average for these countries. However, between 1958–1959 and 1986–1987, their unchanging mortality level lowered their rank relative to women in other developed countries. This is because of the falling mortality trajectory for women in most developed countries during the same period. As a result, between 1958–1959 and 1986–1987, Soviet women aged 40 to 59 dropped from 29th place to 32nd place among the 34 developed countries on this measure. In the mid-1980s, only women in Hungary and Romania ranked lower than Soviet women on this measure.

18.2.3 Summary

Mortality patterns and trends in the Soviet Union differ greatly between the two sexes. The recent increased availability of Soviet mortality data makes it easier to investigate levels and trends in mortality in the Soviet Union than has been true in the past. We hope that even more data on this topic will be published in the future.

However, publication of the data does not by itself make substantive examination of Soviet mortality an easy task. The more Soviet mortality data that are published, the more clear it is that these data suffer from problems common to such data from other parts of the world. The sex difference in the urban–rural mortality crossover is only one of a number of indications of problems with data quality. At the same time, there are signs of improvement in data quality – which we see, for example, in the progressive increase in the age at which the urban–rural crossover occurs for the USSR, and as we have seen in our previous investigations of trends in infant mortality, the timeliness of the registration of births, and the accuracy of old-age mortality rates. And it is important to note that, on balance, the most likely outcome of improving the completeness and accuracy of the registration of births and deaths is that reported fertility rates and mortality rates go up.

Thus, although the data still contain errors, they are improving over time. And the examination of more detailed Soviet mortality data makes it possible not only to identify areas where improvement is needed but also to address substantive hypotheses – to reject some hypotheses about mortality differences while providing support for alternative hypotheses.

The extreme diversity of the Soviet population means that examination of mortality data for the Soviet Union as a whole, even if such data were completely accurate, is of limited use.

18.3 Comparison of the Soviet Regions: 1970–1988

The population of the Soviet Union is extremely large and diverse. The ancestral homelands of over 90 ethnic groups are located in the territory of the USSR. In 1989, only 50.8 percent of Soviet citizens identified themselves as ethnic Russians; 19.5 percent belonged to a Slavic ethnic group; 19.4 percent belonged to a Muslim group; and 10.3 percent belonged to other ethnic groups.[15] The population of many of the republics, each of which is the traditional homeland of a different ethnic group (titular nationality), exceeds that of many developed countries.

Demographic patterns for the "Soviet" population are a weighted average or an amalgam of the patterns found in very diverse regions and groups. The patterns for the USSR as a whole may not represent the actual experience of any particular significant subpopulation – even of the largest republic.

According to the 1989 Soviet census, 51.6 percent of the population of the USSR lived in the RSFSR; 23 percent in the Soviet "west" (Byelorussia, Moldavia, the Ukraine); 19 percent in the Muslim republics (Azerbaijan, Kazakhstan, Kirghizia, Tajikistan, Turkmenistan, Uzbekistan); 3 percent in the Baltic republics (Estonia, Latvia, Lithuania); and 3 percent in Armenia and Georgia. But the regional shares of births and deaths differ greatly from this distribution. For example, in 1987, the RSFSR contributed 45 percent of births, 55 percent of total deaths, 34 percent of infant deaths, and 35 percent of the natural increase of the population of the USSR. In the same year, the Muslim republics contributed 32 percent of births, 14 percent of total deaths, 51 percent of infant deaths, and 50 percent of the natural increase of the population of the USSR. As a result of differences in mortality, fertility, and age structure, between 1979 and 1987 the RSFSR contributed 15 percent of the net increment to the population in working ages, while the Muslim republics contributed 76 percent (Anderson and Silver, 1990a).

This section examines differences in mortality in the working ages (20 to 59) between Soviet republics and then compares mortality in the Soviet republics with 33 developed countries (as defined by the United Nations). We would prefer to study differences in mortality by ethnic group as well as by region, because republics are not ethnically homogeneous and many of the differences in demographic behavior among

Soviet regions are related more strongly to ethnic differences than to differences in economic structure or development. Unfortunately, however, the Soviet government has published very little information on mortality and fertility by ethnic group. Almost all published statistics on mortality and fertility refer to regions. Therefore, our regional comparisons of mortality in the working ages must rely on regional data rather than ethnic data.[16]

18.3.1   Mortality in the older working ages in the Soviet republics

Table 18.4 shows the trends in the number of person-years lived in the older working ages (40 to 59) in the USSR as a whole as well as in the republics for the years 1970 to 1988.[17] At every date except 1988, men in Russia lived the fewest average number of person-years from age 40 to age 59. In 1988, the figure was smallest in Armenia; this is undoubtedly a result of the earthquake in that year, since in all previous years Armenian men lived more person-years from age 40 to age 59 than men in any other republic.

For Soviet men, there is no correlation between the level of economic development of the republic and the level of mortality in the older working ages. Men in the Baltic republics, which are the most developed regions, generally have mortality that is close to the (unweighted) average for all of the republics, though they have had lower mortality than men in the RSFSR. If we had confidence in the data that imply that mortality of males in the older working ages is lower in the least developed republics in the USSR (in Central Asia) than in the more developed republics, we might propose an explanation based on the comparative risks of urban and rural life. In Section 18.2, however, based on evidence of mortality crossovers, we questioned the validity of the reported differences in mortality between urban and rural areas. We found analogous crossovers in mortality between the less developed republics of the USSR and the more developed ones in both 1958–1959 and 1969–1970 (Anderson and Silver, 1989a). Similar crossovers appear in data for recent years.

Table 18.4 shows that there is generally less regional variability in mortality for Soviet women aged 40 to 59 than for Soviet men. However, these data for women show a more logical relationship between the relative levels of mortality and the levels of economic development of the

Table 18.4a. Number of person-years lived of men, age 40 to 59, in the republics, 1970 to 1988.[a]

	1970–1971	1972–1973	1975–1976	1978–1979	1980–1981	1982–1983	1984–1985	1986–1987	1987	1988
USSR	16.78	16.81	16.58	16.34	16.28	16.43	16.51	17.27	17.28	17.19
Russia	16.39	16.45	16.15	15.91	15.83	16.06	16.11	17.06	17.09	16.99
Byelorussia	17.49	17.43	17.16	16.96	16.82	16.87	16.85	17.42	17.42	17.37
Moldavia	17.43	17.40	17.37	17.17	17.10	17.22	17.39	17.85	17.90	17.76
Ukraine	17.29	17.33	17.09	16.81	16.68	16.76	16.81	17.37	17.37	17.36
Estonia	17.18	17.10	16.87	16.64	16.63	16.73	16.79	17.40	17.40	17.43
Latvia	16.96	16.78	16.49	16.45	16.28	16.42	16.65	17.29	17.32	17.27
Lithuania	17.20	17.11	16.89	16.80	16.59	16.66	16.70	17.39	17.37	17.29
Armenia	18.27	18.30	18.22	18.23	18.25	18.30	18.34	18.53	18.55	16.85
Azerbaijan	17.67	17.80	17.77	17.69	17.65	17.82	17.83	17.93	17.95	17.86
Georgia	17.80	17.94	17.88	17.86	17.67	17.75	17.74	17.89	17.89	17.78
Kazakhstan	16.54	16.57	16.40	16.29	16.23	16.38	16.50	17.24	17.25	17.22
Kirghizia	16.78	16.84	16.68	16.55	16.67	16.88	16.94	17.57	17.54	17.43
Tajikistan	17.81	17.81	17.78	17.79	17.67	17.77	17.93	18.21	18.19	18.11
Turkmenistan	17.32	17.29	17.13	17.11	16.89	17.00	17.15	17.48	17.53	17.51
Uzbekistan	17.51	17.52	17.40	17.32	17.28	17.34	17.47	17.88	17.93	17.86
Mean	17.31	17.31	17.15	17.04	16.95	17.06	17.15	17.63	17.65	17.47
CV[b]	0.028	0.029	0.033	0.037	0.036	0.035	0.022	0.022	0.022	0.019

[a]Based on $l_{20} = 100,000$.
[b]CV is the coefficient of variation, which is calculated as the standard deviation divided by the mean.

Table 18.4b. Number of person-years lived of women, age 40 to 59, in the republics, 1970 to 1988.[a]

	1970–1971	1972–1973	1975–1976	1978–1979	1980–1981	1982–1983	1984–1985	1986–1987	1987	1988
USSR	18.79	18.80	18.73	18.71	18.69	18.76	18.73	18.93	18.94	18.89
Russia	18.77	18.78	18.70	18.67	18.67	18.74	18.73	18.96	18.98	18.96
Byelorussia	19.02	19.01	19.00	18.97	18.94	18.94	18.91	19.07	19.06	19.08
Moldavia	18.58	18.59	18.37	18.22	18.23	18.11	18.10	18.49	18.52	18.55
Ukraine	18.92	18.93	18.91	18.87	18.84	18.88	18.87	18.99	19.00	19.00
Estonia	18.99	19.00	18.95	18.92	18.89	18.96	18.88	19.07	19.12	19.07
Latvia	18.86	18.86	18.90	18.81	18.78	18.85	18.81	18.98	18.95	18.96
Lithuania	18.95	18.97	18.96	18.91	18.88	18.87	18.89	19.05	19.05	19.03
Armenia	19.08	19.11	19.09	19.16	19.15	19.19	19.19	19.20	19.20	16.28
Azerbaijan	18.71	18.71	18.77	18.80	18.82	18.91	18.90	18.92	18.97	18.94
Georgia	18.95	19.06	19.06	19.06	19.03	19.14	19.10	19.13	19.15	19.14
Kazakhstan	18.61	18.60	18.50	18.49	18.46	18.60	18.55	18.77	18.82	18.81
Kirghizia	18.45	18.41	18.38	18.38	18.50	18.52	18.47	18.67	18.65	18.62
Tajikistan	18.26	18.16	18.17	18.33	18.25	18.37	18.41	18.55	18.53	18.64
Turkmenistan	18.30	18.17	18.32	18.24	18.30	18.42	18.45	18.43	18.49	18.45
Uzbekistan	18.48	18.42	18.39	18.44	18.46	18.48	18.51	18.62	18.63	18.66
Mean	18.73	18.72	18.70	18.68	18.68	18.73	18.72	18.86	18.88	18.68
CV[b]	0.014	0.016	0.016	0.016	0.015	0.016	0.015	0.013	0.013	0.036

[a]Based on l_{20} = 100,000.
[b]CV is the coefficient of variation, which is calculated as the standard deviation divided by the mean.

republic than is found among men. The best conditions are found in Armenia (except for 1988); the worst, in Central Asia. The Baltic republics are also among the best, with the RSFSR and the west following.

Figures 18.6 through *18.9* show the average number of person-years lived from age 40 to age 59 by sex within each Soviet region. Each graph also shows the values for the USSR as a whole. The graphs do not have a common vertical scale, so that differences within regions can clearly be seen.

Figure 18.6 shows the values for the RSFSR and the western republics. Women in the more developed republics in this group generally have lower mortality than women in the less developed republics. In Moldavia, the least developed republic in the region, the mortality of women is higher than elsewhere in the west. On average, women in Byelorussia and the Ukraine have lower mortality rates than women in the RSFSR. The situation for men in this region is very different. While men in Byelorussia and the Ukraine have lower mortality than men in the RSFSR, it is paradoxical that men in Moldavia have lower mortality than men in Byelorussia or the Ukraine. Although this difference in the reported mortality levels might be real, its causal basis deserves further exploration.

Figure 18.7 shows the values for the Baltic republics, which are considered the most developed region of the Soviet Union. The comparatively low mortality in Lithuania is somewhat surprising, since Latvia and Estonia have a higher standard of living than Lithuania by most indicators.

The situation of mortality in the Baltic republics could become much more clear if data by nationality were available. Some 80 percent of the population of Lithuania is made up of ethnic Lithuanians, while only 52 percent of the population of Latvia is made up of Latvians, and 61 percent of the population of Estonia is made up of Estonians. When we examine mortality data for Estonia and Latvia as a whole, we confront in microcosm the same situation that we confront when studying the Soviet Union as a whole: the overall values are a weighted average of very different patterns for different ethnic subpopulations. It would be relevant to know how the mortality situation of Latvians in Latvia compares with that of the 48 percent of the citizens of Latvia who are not ethnic Latvians. How does the mortality of ethnic Lithuanians, Latvians, and Estonians within their home republics differ? How does the

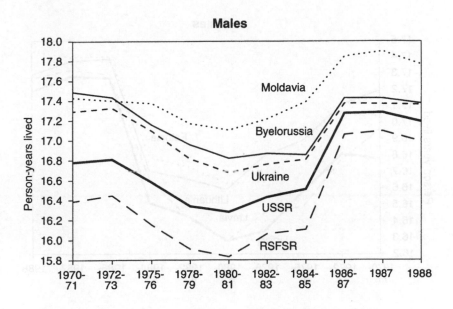

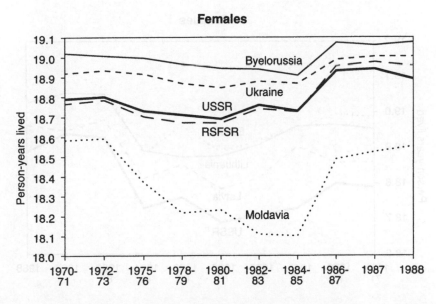

Figure 18.6. Person-years lived, age 40 to 59, in the USSR, the RSFSR, Byelorussia, Moldavia, and the Ukraine, 1970 through 1988, by sex.

Barbara A. Anderson and Brian D. Silver

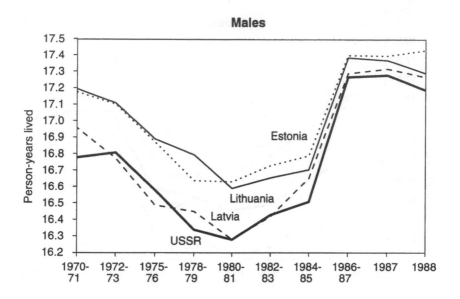

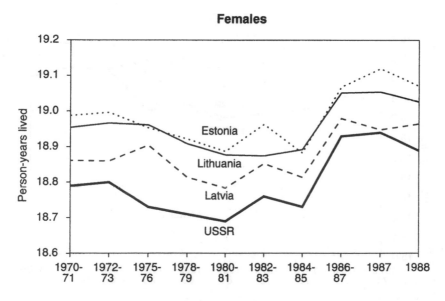

Figure 18.7. Person-years lived, age 40 to 59, in the USSR, Lithuania, Estonia, and Latvia, 1970 through 1988, by sex.

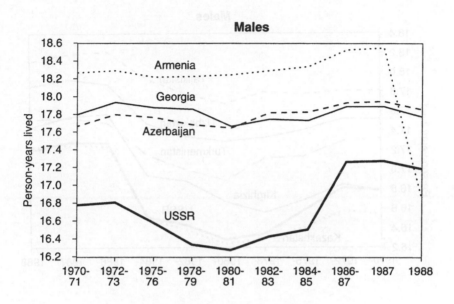

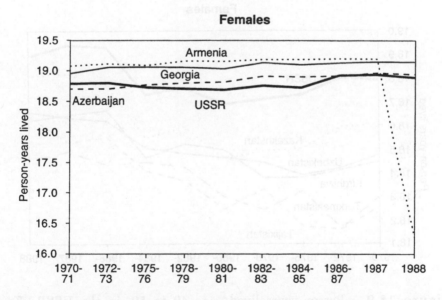

Figure 18.8. Person-years lived, age 40 to 59, in the USSR, Armenia, Georgia, and Azerbaijan, 1970 through 1988, by sex.

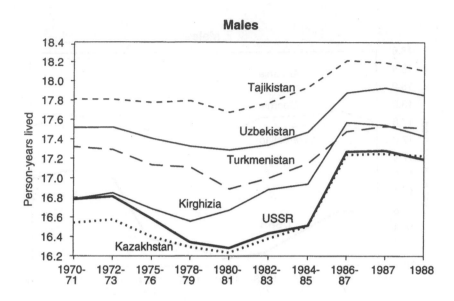

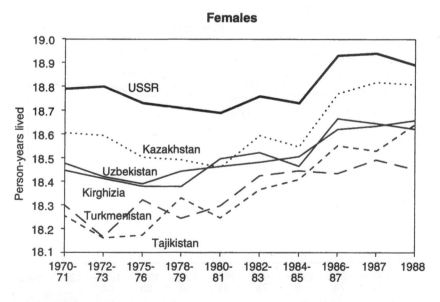

Figure 18.9. Person-years lived, age 40 to 59, in the USSR, Kazakhstan, Uzbekistan, Kirghizia, Turkmenistan, and Tajikistan, 1970 through 1988, by sex.

mortality of Russians in the Baltic republics compare with the mortality of Russians in the RSFSR? Data relevant to such questions could aid in understanding the causal mechanisms behind mortality differentials in the Soviet Union. Some of these questions could be answered from data from the Baltic republics because of the high quality of data from that region.

Figure 18.8 shows the values for Transcaucasia. In Armenia, the sharp drop in the number of person-years of life for both sexes between 1987 and 1988 is a clear indicator of the devastating effect of the earthquake. Until 1988, Armenia had the lowest reported mortality for people in the working ages for both sexes of any Soviet republic. It will be interesting to see how quickly mortality conditions in Armenia return to their former good level.

The data shown in *Figure 18.9* for Kazakhstan and Central Asia are especially interesting. First let us consider the data for women. Mortality rates for Kazakhstan and each Central Asian republic are higher than those reported for the Soviet Union as a whole. This finding is consistent with the generally low level of economic development of that region. Also, it is sensible that mortality of women is lower in Kazakhstan than in the Central Asian republics, given Kazakhstan's relatively high level of development among the republics examined in *Figure 18.9*. It is also sensible that mortality is relatively high among women in Turkmenistan and Tajikistan, since they are the least developed republics in the Soviet Union.[18]

The data for males in *Figure 18.9* portray a completely different picture. The picture is similar to the anomalous position of Moldavia among the Western republics, but in a more extreme form. For men in *Figure 18.9*, the number of person-years lived from age 40 to age 59 is generally inversely related to the level of development of the republics. In addition, the reported average number of person-years lived from age 40 to age 59 by men in every Central Asian republic is higher than in the Soviet Union as a whole. Although we think the actual mortality rates of adult males in Central Asia are probably higher than these values [especially given other evidence of the poor quality of demographic data in the Central Asian republics (see Coale *et al.*, 1979; Anderson and Silver, 1986b, 1988, 1989a; Anderson *et al.*, 1989)], it is not impossible that adult men in Central Asia actually have lower mortality rates than men in the European parts of the Soviet Union.

It is clear from the study of other developed countries that aspects of life-style, such as alcohol consumption and smoking, are extremely important for mortality of adult males. In the Central Asian republics, abuse of alcohol has not been nearly as much of a problem as in Russia, the Ukraine, and other parts of the European USSR. To the extent that mortality of men in the Soviet Union is dominated by these kinds of life-style factors rather than by overall economic development, men in Central Asia could have relatively low mortality. The ranking of the Central Asian republics and Kazakhstan in terms of development is similar to their ranking in the proportion of the population that is made up of members of European nationalities. To the extent that European men in Central Asia and Kazakhstan pursue life-styles similar to men in the European republics, and to the extent that the habits of members of indigenous nationalities are affected by that of Europeans with whom they have contact, the pattern for men in *Figure 18.9* could also reflect reality.

Unfortunately, it is not clear how this hypothesis about life-style and mortality can be tested from available data. Although the rural parts of the Central Asian republics are primarily made up of members of indigenous nationalities, Russians and other Europeans are common in Central Asian cities. In Kazakhstan, however, over half of the rural population is made up of ethnic Russians and Ukrainians. If the data on mortality within republics by urban–rural residence were reliable, then the rural data, at least, could be investigated to gain an indication of the mortality conditions of the indigenous population. However, because of poor data quality, we are skeptical that such a study would be successful in Central Asia.

The sex differential in mortality in the working ages is much smaller in Central Asia than in other parts of the Soviet Union. It is clear from *Figure 18.9* that this is due to a combination of unusually low mortality for men and unusually high mortality for women. Scholarly discussion of sex differentials in mortality in developed countries has usually been on excess mortality among men due to harmful habits that are more frequently engaged in by men than by women, such as smoking and alcohol consumption. In less developed countries, the concern with a sex differential in mortality has often been on excess mortality of females. Most often the concern has been with infant and child mortality. However, as suggested by the earlier discussion of rural–urban mortality crossovers by sex, it would be worthwhile to study the extent to which differential use of medical care or other behaviors that are not favorable to women contribute to higher female mortality in Soviet Central Asia than would occur otherwise.

The data in *Figure 18.4* show a sharp reduction in mortality in the working ages in the Soviet Union as a whole for each sex, in both the urban and the rural populations. A similar sharp reduction between 1984–1985 and 1986–1987 is apparent in the data for each sex in every republic. Although Mikhail Gorbachev's anti-alcohol campaign might be responsible for some mortality decline in that period, it does not seem plausible that this campaign could have had such a similar and virtually simultaneous effect on both males and females in every Soviet republic. We think that further study of the methods of data adjustment is needed before such changes in mortality are accepted as real.

18.3.2 Mortality in the Soviet republics

Table 18.5 places mortality in the working ages for the Soviet population into comparative perspective. For four dates, the table shows the average number of person-years lived from age 40 to age 59 for the USSR as a whole, for individual Soviet republics, and for 33 other developed countries, as defined by the United Nations.[19] For each year for which data are available for all of the Soviet republics, it is possible to compare the mortality levels found in all countries with the range of mortality levels found in the Soviet republics. Values for the other countries that fall within the range of mortality found in the Soviet republics are italicized.

For men in 1969–1970, in 10 other developed countries the number of person-years lived from age 40 to age 59 was greater than the number lived by men in any of the Soviet republics (while in 23 countries the figure fell within the range found among the Soviet republics). By 1980, men in 19 other developed countries aged 40 to 59 had lower mortality than men in any of the Soviet republics, and by 1985 this was true of men in 17 other developed countries.

The situation is similar for women. In 1969–1970, in five other developed countries women lived more person-years from age 40 to age 59 than women in any Soviet republic. By 1980, this increased to 13; and by 1985, it increased to 22. Thus, the lack of decline in mortality of Soviet women in recent decades has left them behind women in most of the rest of the developed world. Soviet women's closest *competitors* in

Table 18.5. Comparison of person-years lived, age 40 to 59, for the USSR and 33 developed countries, 1958 to 1959, 1969 to 1970, circa 1980, and circa 1985.[a] *Italic* values fall within the range of values for Soviet regions for a given year. See Appendix *Table 18A.1* for exact years from which life tables are used.

Republic/Country	Males				Females				Females minus Males			
	1958– 1959	1969– 1970	circa 1980	circa 1985	1958– 1959	1969– 1970	circa 1980	circa 1985	1958 1959	1969– 1970	circa 1980	circa 1985
	(1)	(2)	(3)	(4)	(5)	(6)	(7)	(8)	(9)	(10)	(11)	(12)
USSR	17.25	16.78	16.28	17.27	18.60	18.79	18.69	18.93	1.35	2.02	2.41	1.65
Russia		16.39	15.83	17.06		18.77	18.67	18.96		2.38	2.84	1.90
Byelorussia		17.49	16.82	17.42		19.02	18.94	19.07		1.53	2.12	1.65
Moldavia		17.43	17.10	17.85		18.58	18.23	18.49		1.15	1.13	0.64
Ukraine		17.29	16.68	17.37		18.92	18.84	18.99		1.63	2.17	1.62
Estonia		17.18	16.63	17.40		18.99	18.89	19.07		1.81	2.26	1.67
Latvia		16.96	16.28	17.29		18.86	18.78	18.98		1.90	2.50	1.69
Lithuania		17.20	16.59	17.39		18.95	18.88	19.05		1.76	2.29	1.66
Armenia		18.27	18.25	18.53		19.08	19.15	19.20		0.81	0.90	0.67
Azerbaijan		17.67	17.65	17.93		18.71	18.82	18.92		1.04	1.17	0.99
Georgia		17.80	17.67	17.89		18.95	19.03	19.13		1.15	1.36	1.24
Kazakhstan		16.54	16.23	17.24		18.61	18.46	18.77		2.07	2.23	1.53
Kirghizia		16.78	16.67	17.57		18.45	18.50	18.67		1.67	1.83	1.09
Tajikistan		17.81	17.67	18.21		18.26	18.25	18.55		0.45	0.57	0.34
Turkmenistan		17.32	16.89	17.48		18.30	18.30	18.43		0.98	1.41	0.96
Uzbekistan		17.51	17.28	17.88		18.30	18.46	18.62		0.96	1.18	0.74
Eastern Europe												
Bulgaria	18.23	18.32	*18.11*	*17.97*	18.69	*18.99*	*19.07*	*19.12*	0.46	*0.67*	*0.96*	*1.14*
GDR	18.14	*18.15*	*18.12*	*18.28*	18.72	*18.88*	*19.02*	*19.14*	0.58	*0.73*	*0.90*	*0.86*
CSFR	18.04	*17.74*	*17.79*	*17.85*	18.86	*18.94*	*19.05*	*19.13*	0.82	*1.20*	*1.26*	*1.28*
Hungary	18.02	*17.97*	*17.24*	*17.14*	18.66	*18.89*	*18.71*	*18.73*	0.64	*0.92*	*1.47*	*1.58*
Yugoslavia	17.97	*17.91*	*18.01*	*18.11*	18.34	*18.79*	*19.03*	*19.11*	0.37	*0.88*	*1.02*	*0.99*
Poland	17.81	*17.85*	*17.60*	*17.54*	18.67	*18.94*	*19.03*	*19.05*	0.86	*1.09*	*1.43*	*1.51*
Romania	17.80	*18.04*	*17.74*	*17.65*	18.29	*18.76*	*18.84*	*18.86*	0.49	*0.72*	*1.10*	*1.20*

Northern Europe												
Sweden	18.63	*18.58*	18.61	*18.81*	19.07	*19.15*	19.27	*19.37*	0.44	*0.57*	0.66	*0.56*
Denmark	18.53	*18.49*	18.38	*18.49*	18.94	*18.98*	19.02	*19.09*	0.41	*0.49*	0.64	*0.60*
Norway	18.53	*18.43*	18.64	*18.70*	19.16	*19.23*	19.37	*19.37*	0.63	*0.80*	0.73	*0.67*
Iceland	18.19	*18.23*	18.64	*18.79*	18.94	*19.03*	19.29	*19.41*	0.75	*0.80*	0.65	*0.62*
Finland	17.43	*17.98*	*17.99*	*18.16*	18.76	*19.00*	19.29	*19.34*	1.33	*1.62*	1.30	*1.18*
Anglo-American countries												
New Zealand	18.31	*18.10*	18.36	*18.56*	18.92	*18.84*	*18.99*	*19.10*	0.61	*0.74*	0.63	*0.54*
Engl./Wales	18.30	*18.39*	18.65	*18.82*	18.94	*19.00*	19.17	*19.27*	0.64	*0.61*	0.52	*0.46*
No. Ireland	18.20	*18.17*	18.27	*18.55*	18.81	*18.88*	*19.03*	*19.18*	0.61	*0.71*	0.76	*0.63*
Ireland	18.15	*18.26*	18.34	*18.67*	18.53	*18.84*	*19.04*	*19.28*	0.38	*0.58*	0.70	*0.61*
Canada	18.08	*18.13*	18.38	*18.64*	18.88	*18.99*	19.17	*19.31*	0.80	*0.86*	0.79	*0.67*
Australia	18.03	*17.99*	18.42	*18.64*	18.84	*18.89*	19.19	*19.28*	0.81	*0.90*	0.77	*0.64*
Scotland	17.96	*18.04*	*18.22*	*18.49*	18.65	*18.74*	*18.99*	*19.13*	0.69	*0.70*	0.77	*0.64*
USA	17.69	*17.53*	*18.01*	*18.25*	18.68	*18.68*	*18.97*	*19.13*	0.99	*1.15*	0.96	*0.88*
Western/Southern Europe												
Netherlands	18.66	*18.54*	18.77	*18.90*	19.10	*19.15*	19.30	*19.35*	0.44	*0.61*	0.53	*0.45*
Greece	18.54	*18.71*	18.80	*18.84*	19.01	*19.20*	19.37	*19.43*	0.47	*0.49*	0.57	*0.59*
Italy	18.21	*18.29*	18.45	*18.68*	18.87	*19.06*	19.26	*19.37*	0.66	*0.77*	0.81	*0.68*
Switzerland	18.18	*18.39*	18.55	*18.75*	18.98	*19.13*	19.26	*19.39*	0.80	*0.74*	0.71	*0.63*
Spain	18.18	*18.26*	18.47	*18.62*	18.80	*19.03*	19.26	*19.37*	0.62	*0.77*	0.79	*0.75*
Malta	18.18	*18.41*	18.57	*18.94*	18.64	*19.00*	*19.14*	*19.42*	0.46	*0.59*	0.57	*0.48*
Belgium	18.13	*18.12*	18.30	*18.46*	18.88	*18.95*	*19.06*	*19.21*	0.75	*0.83*	0.76	*0.75*
FRG	18.05	*18.04*	18.30	*18.59*	18.81	*18.98*	*19.14*	*19.30*	0.76	*0.86*	0.84	*0.70*
Luxembourg	18.01	*17.71*	*18.03*	*18.35*	18.82	*18.78*	*19.03*	*19.10*	0.81	*1.07*	1.00	*0.75*
France	17.89	*17.91*	*18.06*	*18.26*	18.85	*18.99*	*19.15*	*19.28*	0.96	*1.08*	1.09	*1.01*
Austria	17.80	*17.86*	*18.03*	*18.26*	18.75	*18.92*	*19.11*	*19.23*	0.95	*1.06*	1.08	*0.96*
Portugal	17.76	*17.89*	*17.94*	*18.19*	18.67	*18.89*	*19.10*	*19.21*	0.91	*1.00*	1.16	*1.03*
Japan	17.67	*18.25*	18.79	*18.91*	18.30	*18.94*	19.36	*19.44*	0.63	*0.69*	0.57	*0.53*
Mean[b] (N=34)	18.08	*18.09*	18.20	*18.39*	18.78	*18.95*	19.11	*19.22*				
CV[b]	0.017	*0.020*	0.027	*0.025*	0.011	*0.007*	0.009	*0.009*				

[a]Based on $l_{20} = 100{,}000$.
[b]The mean and coefficient of variation are calculated based on 34 developed countries, not the separate republics of the USSR.

mortality are women in the countries of Eastern Europe. It is striking that men in the RSFSR had higher mortality than men in any developed country at all three dates. Naturally, the high mortality in the RSFSR exerts a strong influence on mortality for the Soviet population as a whole, since in 1989, 51 percent of all males in the Soviet Union resided in the RSFSR.

It is relevant to determine how many of the other Soviet republics had extremely poor male mortality in comparison with other developed countries. In 1969–1970, men in Finland had the highest mortality in the working ages among all developed countries other than the Soviet Union. At the same time, in addition to the RSFSR, men in Kazakhstan, Kirghizia, the Ukraine, and all the Baltic republics had higher mortality than men in Finland. In about 1980, men in Hungary had the highest mortality in the working ages of any developed country other than the Soviet Union. At that date, men in all Soviet republics except Uzbekistan, Tajikistan, and the Transcaucasian republics had higher mortality rates than men in Hungary. In about 1985, men in Hungary still had the highest mortality rates in the working ages among all developed countries. But by 1986–1987, only men in the RSFSR had higher mortality than men in Hungary.

The mortality experience of women in the RSFSR generally has not looked as poor in comparison to other countries as has the mortality of men in the RSFSR. At all three dates, however, women in Kazakhstan and Central Asia have had higher mortality in the working ages than women in any developed country. In 1969–1970, women in the United States in the working ages had the highest mortality among developed countries. At that date, working-age women in all European republics and in Transcaucasia had lower mortality than women in the United States. By around 1980, women in Romania and Hungary in the working ages had the highest mortality among developed countries other than the Soviet Union. At that date, in addition to Central Asia and Kazakhstan, women in the RSFSR, Moldavia, and the Ukraine had higher working-age mortality than women in Romania. Around 1985, working-age women in Romania still had the highest mortality among developed countries. At that time, besides Kazakhstan and Central Asia, only working-age women in Moldavia had higher mortality than women in Romania.

In four of the seven Eastern European countries, mortality rates for males increased between 1969–1970 and the mid-1980s. This has been described at times as an Eastern European mortality syndrome (Compton, 1985). A combination of poor diet, heavy smoking, alcohol abuse, and industrial pollution has been suggested as creating a situation that leads to extremely high adult male mortality. However, reported mortality of working-age men fell in every Soviet republic except Byelorussia between 1969–1960 and 1986–1987. Thus, if there is an Eastern European male mortality syndrome, this syndrome has not led to increasing mortality among working-age males in most Soviet republics in recent years. It is interesting that even though the level of mortality of working-age men in the RSFSR remains very high even in 1986–1987, the proportionate lowering of male mortality rates in the older working ages between 1969–1970 and the mid-1980s is greater in the RSFSR than in any other European republic of the Soviet Union as well as greater than in any Eastern European country.

It is important to note that, like the relationship between mortality levels and levels of development in Soviet republics, there is not a high correlation between economic development and mortality for the developed countries as a whole. For example, for both sexes, in about 1985, Greece has lower reported mortality in the working ages than Sweden. It seems certain that the quality of Swedish mortality data is better than the quality of Greek mortality data. Although part of the explanation for the lack of a high correlation between development and mortality for developed countries generally may be errors in the mortality data, undoubtedly other factors must be taken into account, including diet and other behavioral patterns. As the economies of several developed countries continued to grow during the 1960s and 1970s, mortality of men in the older working ages rose. *Table 18.5* shows that this occurred between 1958–1959 and 1969–1970 in the United States, Australia, New Zealand, Sweden, Denmark, Norway, Finland, Northern Ireland, the Netherlands, Belgium, the Federal Republic of Germany, and Luxembourg, as well as in Yugoslavia.

18.4   Conclusion

Overall, trends and differentials in mortality in the working ages may appear to be small, but they are related to patterns of mortality over the life span. Most of the differences in life expectation among developed countries are a result of mortality above the working ages. But we also know that the quality of data deteriorates at older ages in virtually all countries, including the Soviet Union (Anderson and Silver, 1989a). Hence, direct comparisons of reported mortality conditions across countries above the working ages are even more risky than comparisons in the working ages.

Mortality in the working ages in much of the Soviet Union, especially for men in the RSFSR, is sufficiently high that concerns about the number of years of life lost which in most other countries only become important above the working ages, are important for the USSR in the working ages. For example, the World Health Organization Project on Healthy Life Expectancy focuses on the role of mortality and disability in determining the number of years that a person can expect to live with a high level of independence. This issue is, of course, also important in the Soviet Union, especially because its population is aging. But it is a much more important issue for people in the working ages in the Soviet Union than it is in most other developed countries. Thus, the study of loss of working life or independent life due to mortality should be supplemented by the study of other conditions, such as disability or chronic health problems, that impair the ability to work or to live a healthy and independent life at all stages in the life span.

Acknowledgment

The authors are grateful to Douglas Johnson, Eduard Ponarin, and Thyne Sieber for research assistance; to Mike Coble for help with the graphs; and to Victoria Velkoff for helpful advice. The research was supported by NICHD Grant Nos. RO1 HD-19915 and P30 HD-10003. An earlier version of this paper was presented at the International Symposium on Demographic Processes in the USSR in the 20th Century in the Context of the European Experience, Tbilisi, 8–12 October 1990.

Notes

[1] The most systematic study of Soviet mortality rates in comparison with those elsewhere in Europe in the 1970s was that of Vallin and Chesnais (1974). Other early studies included that by Dutton (1979). Soviet demographers were among the first to call attention to the reversal of declining mortality that occurred in the USSR in the mid-1960s (see, for example, Bedny, 1972; Zvidriņš and Krumiņš, 1974; Kirdyakin and Fayzulin, 1974;

Kurman, 1974; Kruminš, 1976; Zvidriņš and Shukaylo, 1975; Dmitriyeva and Andreev, 1977).

[2] After 1974, most detailed mortality data for the USSR and individual republics were forbidden to be published. This figure for 1976 was revealed only in 1988.

[3] These changes in infant mortality rates are in the *reported* values. We think that most, if not all, of the rise in infant mortality in the early 1970s was an artifact of more complete registration and improved record keeping (see Anderson and Silver, 1986b; Blum and Monnier, 1989). In a recent publication, Baranov *et al.* (1990) provide new estimates of the actual infant mortality rates for the USSR and for various regions for 1970 and 1986. The estimated rates are much higher than the officially reported rates, especially for 1970 and in the regions of higher infant mortality. Although we think some adjustment of this kind is appropriate, the methods used to develop the new estimates are not described, and it is therefore impossible to judge how reasonable the new estimates are (for further commentary, see Anderson and Silver, 1990b).

[4] The data in *Figure 18.1* and most of the other figures are reported for two-year periods, such as 1980–1981, except 1987 and 1988, for which the reported data refer to a single year. At the time the figure was prepared, the data for 1989 had not yet been obtained. This chapter may occasionally refer to data that have become available recently.

[5] That the figures for 1988 are somewhat lower than those for 1987 is partly due to the natural calamity of the Armenian earthquake, which occurred in December 1988.

[6] For a discussion of some changes in the methods used to construct Soviet life tables, see Andreev *et al.*, 1975; Kingkade, 1987; Anderson and Silver, 1989a. For illustrations of how the availability of new data from the censuses has affected published mortality rates, see Sinel'nikov, 1988.

[7] This adjustment only takes definitional differences into account. A further adjustment would be required to consider the undercounting of infant deaths due to underregistration. There is evidence of substantial underregistration of infant deaths (as well as of births) in some regions of the Soviet Union, particularly in Central Asia (see Coale *et al.*, 1979, Anderson and Silver, 1986b; Anderson *et al.*, 1989).

[8] A survey of the population of the USSR undertaken in 1982–1983 by the Central Statistical Board showed that, whereas in the cohort that entered the labor force in the 1960s, between 37 percent and 40 percent were younger than 18 years old at the time, by the late 1970s less than 30 percent were under 18 years old at age of entry (Filippov, 1989).

[9] Our use of temporary life expectancies is motivated by many of the concerns expressed by Arriaga (1984).

[10] In 1969–1970, for the USSR as a whole, 56 percent of all reported deaths that occurred before age 20 occurred in infancy. In 1986–1987, the corresponding figure was 58 percent.

[11] For a more general discussion of the interpretation of the mortality crossovers as statistical artifacts, see Coale and Kisker (1986). The alternative interpretation of mortality crossovers, the heterogeneity or frailty argument, would imply that crossovers reflect selective mortality of the less healthy at younger ages in a subpopulation, which leads to relatively lower mortality rates of the survivors of those cohorts at older ages (see Vaupel et al., 1979; Manton and Stallard, 1984).

[12] Since the open interval of the published life tables begins at age 85, we do not know whether a mortality crossover occurs above age 85 in Estonia and Latvia.

[13] An additional factor to consider in accounting for the urban–rural crossover in Central Asia is that the ethnic composition of the urban population is much more heterogeneous than the rural population; in fact, a majority of the urban population of Central Asia is not indigenous to Central Asia. Thus, the urban mortality crossovers in Central Asia may be partially caused by a heterogeneity phenomenon (cf. Manton and Stallard, 1984). To test this hypothesis, we would need reliable mortality data by nationality (as well as by gender and residence in urban and rural areas). But we think that differences in the ages of the mortality crossover by sex are not plausibly accounted for by a heterogeneity argument, because differences by nationality in the mortality of females are much smaller than differences by nationality in the mortality of males.

[14] The data on which *Figure 18.5* is based are derived from the *UN Demographic Yearbook*. The years for which we have data differ slightly from one country to the next. This may affect some of the comparisons, especially if mortality levels are changing or if differences in mortality between countries are small. *Table 18.A1* in the Appendix shows the actual years for which we obtained data for each country.

[15] For information on the changing ethnic composition of the Soviet population, see Anderson and Silver, 1989c, 1990a. For a discussion of differences in demographic behavior of Soviet regions and ethnic groups, see also Blum, 1987; Bondarskaya and Darsky, 1988; Vishnevsky, 1988. For a discussion on Soviet labeling of ethnic groups see Silver, 1986.

[16] Bondarskaya and Darsky (1988) have published data on ethnic differences in fertility. We have also derived some estimates of fertility and infant mortality for ethnic groups using regional data and a residual method (Anderson et al., 1989). Dobrovol'skaya (1990) analyzed ethnic differences in mortality in the USSR, based on data that were not previously published.

[17] This number is based on persons who survived to age 20. We begin these comparisons with 1970–1971 because we have life tables for every republic only since 1970, whereas for the USSR at large we have data since 1958–1959.

[18] It is important to bear in mind that unlike the other republics in *Figure 18.9*, a majority of the population of Kazakhstan is not of the titular nationality. Only 40 percent of the population of Kazakhstan is made up of Kazakhs; 38 percent, Russians; 18 percent, Ukrainians; 5 percent, Uzbeks, Tajiks, and members of other nationalities that are indigenous to the region.

[19] The data for the Soviet Union and the republics for circa 1985 refer to 1986–1987. Since mortality in all republics for each sex, except for females in Turkmenistan, improved between 1984–1985 and 1986–1987, the situation of the Soviet Union and individual republics would look worse in comparative perspective if data for 1984–1985 had been used instead of data for 1986–1987. The data for 1984–1985 appear in *Table 18.4*.

Appendix

Table 18A.1. Exact years from which life tables are used in *Table 18.5.*

USSR	1958–1959	1969–1970	1980–1981	1986–1987
Europe and Japan				
Bulgaria	1958–1959	1969–1970	1980–1981	1985
CSFR	1958–1959	1969–1970	1978, 1981	1985
GDR	1958–1959	1969–1970	1980–1981	1985
Hungary	1958–1959	1969–1970	1981–1982	1986
Poland	1958, 1960	1969–1970	1981–1982	1986
Romania	1956	1969–1970	1981–1982	1985
Yugoslavia	1958–1959	1969–1970	1979–1980	1985
Denmark	1958–1959	1969–1970	1981–1982	1986
Finland	1958–1959	1969–1970	1980–1981	1986
Iceland	1955–1959	1965–1969	1980, 1982	1984
Norway	1958–1959	1969–1970	1981–1982	1985
Sweden	1958–1959	1969–1970	1980–1981	1986
Austria	1958–1959	1969–1970	1980–1981	1985
Belgium	1958–1959	1969–1970	1977–1978	1984
France	1958–1959	1969–1970	1979–1980	1984
FRG	1958–1959	1969–1970	1981–1982	1986
Greece	1958–1959	1969, 1971	1980–1981	1984
Italy	1958–1959	1969, 1971	1977, 1979	1982
Luxembourg	1958–1959	1969–1970	1978–1979	1985
Netherlands	1958–1959	1969–1970	1979, 1981	1986
Portugal	1958–1959	1969–1970	1979	1986
Spain	1960	1969–1970	1978	1981
Switzerland	1958–1959	1969–1970	1979, 1981	1986
Japan	1958–1959	1969–1970	1981–1982	1986
Anglo-American countries				
Australia	1958–1959	1969–1970	1979, 1981	1983
Canada	1958–1959	1969–1970	1980–1981	1985
England and Wales	1958–1959	1969–1970	1981–1982	1985
Ireland	1956	1969–1970	1979	1985
New Zealand	1958–1959	1969–1970	1980–1981	1985
Northern Ireland	1958–1959	1969–1970	1980, 1982	1985
Scotland	1958–1959	1969–1970	1981–1982	1985
USA	1958–1959	1969–1970	1981–1982	1985

References

Anderson, B.A., and B.D. Silver, 1985. 'Permanent' and 'Present' Populations in Soviet Statistics. *Soviet Studies* **37**(July):386–402.

Anderson, B.A., and B.D. Silver, 1986a. Sex Differentials in Mortality in the Soviet Union: Regional Differences in Length of Working Life in Comparative Perspective. *Population Studies* **40**(July):191–214.

Anderson, B.A., and B.D. Silver, 1986b. Infant Mortality in the Soviet Union: Regional Differences and Measurement Issues. *Population and Development Review* **12**(December):705–738.

Anderson, B.A., and B.D. Silver, 1988. The Effects of the Registration System on the Seasonality of Births: The Case of the Soviet Union. *Population Studies* **42**(July):303–320.

Anderson, B.A., and B.D. Silver, 1989a. The Changing Shape of Soviet Mortality, 1958–1985: An Evaluation of Old and New Evidence. *Population Studies* **43**(July):243–265.

Anderson, B.A., and B.D. Silver, 1989b. Patterns of Cohort Mortality in the Soviet Population. *Population and Development Review* **15**(September): 471–501.

Anderson, B.A., and B.D. Silver, 1989c. Demographic Sources of the Changing Ethnic Composition of the Soviet Union. *Population and Development Review* **15**(December):609–656.

Anderson, B.A., and B.D. Silver, 1990a. Growth and Diversity of the Population of the Soviet Union. *Annals of the American Academy of Political and Social Science* **510**(July):155–177.

Anderson, B.A., and B.D. Silver, 1990b. Trends in Mortality of the Soviet Population. *Soviet Economy* **6**(July–September):191–251.

Anderson, B.A., B.D. Silver, and J. Liu, 1989. *Mortality of Ethnic Groups in Soviet Central Asia and Northern China.* Research Reports, No. 89–158. University of Michigan Population Studies Center, Ann Arbor, MI.

Andreev, E., A. Kardash, K. Shaburov, and G. Pavlov, 1975. Algoritm rascheta pokazateley tablits smertnosti i sredney prodolzhitel'nosti predstoyashchey zhizni (Algorithm for Calculating Indicators of Mortality Tables and the Average Length of Remaining Life). *Vestnik statistiki* **3**(March):28–35.

Arriaga, E., 1984. Measuring and Explaining Changes in Life Expectancies. *Demography* **21**(February):83–96.

Aziz, K.M.A., 1977. Present Trends in Medical Consultation Prior to Death in Rural Bangladesh. *Bangladesh Medical Journal* **6**.

Baranov, A. A., V. Yu. Al'bitsky, and Yu. M. Komarov, 1990. Tendentsii mladencheskoy smertnosti v SSSR v 70-80-e gody (Tendencies in Infant Mortality in the USSR in the 1970s and 1980s). *Sovetskoye zdravookhraneniye* **3**:3–10.

Basu, A.M., 1989. Is Discrimination in Food Really Necessary for Explaining Sex Differentials in Childhood Mortality? *Population Studies* **43**(July): 193–210.

Bedny, M.S., 1972. *Demograficheskiye protsessy i prognozy zdorov'ya naseleniya* (Demographic Processes and Prognoses of the Health of the Population). Statistika, Moscow.

Bedny, M.S., 1976. *Prodolzhitel'nost' zhizni v gorodakh i selakh* (Expectation of Life in Cities and Villages). Statistika, Moscow.

Bennett, N.G., and L.K. Garson, 1983. The Centenarian Question and Old-Age Mortality in the Soviet Union, 1959–1970. *Demography* **20**(November): 587–606.

Blum, A., 1987. La transition démographique dan les Républiques orientales d'URSS. *Population* 42(March-April):337-358.

Blum, A., and A. Monnier, 1989. Recent Mortality Trends in the USSR: New Evidence. *Population Studies* 43(July):243-266.

Blum, A., and R. Pressat, 1987. Une nouvelle table de mortalité pour l'URSS (1984-1985). *Population* 42(November-December):843-862.

Bondarskaya, G., and L. Darsky, 1988. Etnicheskaya differentsiatsiya rozhdaemosti v SSSR (Ethnic Differentiation of Fertility in the USSR). *Vestnik statistiki* 12(December):16-21.

Breeva, Ie. V., 1984. *Naseleniye i zanyatost'* (Population and Employment). Finansy i statistika, Moscow.

Chen, L., E. Huq, and S. D'Souza, 1981. Sex Bias in the Family: Allocation of Food and Health Care in Rural Bangladesh. *Population and Development Review* 7(March):55-70.

Coale, A.J., 1981. A Reassessment of World Population Trends. *Proceedings, International Population Conference.* International Union for the Scientific Study of Population, Manilla, Liège.

Coale, A.J., and E. Kisker, 1986. Mortality Crossovers: Reality or Bad Data? *Population Studies* 40(November):389-402.

Coale, A.J., B.A. Anderson, and E. Härm, 1979. *Human Fertility in Russia Since the Nineteenth Century.* Princeton University Press, Princeton, NJ.

Compton, P.A., 1985. Rising Mortality in Hungary. *Population Studies* 39 (March):71-86.

Davis, C., and M. Feshbach, 1980. *Rising Infant Mortality in the USSR in the 1970s.* International Population Report Series P-95, No. 74. US Bureau of the Census, Washington, DC.

Dmitriyeva, R.M., and E.M. Andreev, 1977. Snizhenie smertnosti v SSSR za gody Sovetskoy vlasti (Lowering of Mortality in the USSR during the Years of Soviet Power). In A. G. Vishnevsky, ed. *Brachnost', rozhdaemost', smertnost' v Rossii i v SSSR: Sbornik statey* (Nuptiality, Fertility, and Mortality in Russia and the USSR). Statistika, Moscow.

Dmitriyeva, R.M., and E.M. Andreev, 1987. O sredney prodolzhitel'nosti zhizni naseleniya SSSR (On the Life Expectancy of the USSR Population). *Vestnik statistiki* 12(December):31-39.

Dobrovol'skaya, V.M., 1990. Etnicheskaya differentsiya smertnosti (Ethnic differentiation in mortality). In A.G.Volkov, ed. *Demograficheskiye protsessy v SSSR* (Demographic Processes in the USSR). Nauka, Moscow.

D'Souza, S., and L. Chen, 1980. Sex Differentials in Mortality in Rural Bangladesh. *Population and Development Review* 6(June):257-270.

Dutton, J., Jr., 1979. Changes in Soviet Mortality Patterns, 1959-77, *Population and Development Review* 5(June):267-291.

Eberstadt, N., 1981. The Health Crisis in the Soviet Union. *New York Review of Books* (February 19):23-31.

Filippov, F.P., 1989. *Ot pokoleniya k pokoleniyu: Sotsial'naya podvizhnost'* (From Generation to Generation: Social Mobility). Mysl', Moscow.

Garson, L.K., 1986. The Centenarian Question: Old Age Mortality in the Soviet Union, 1897–1970. Ph.D. Dissertation. Princeton University, Princeton, NJ.

Kingkade, W.W., 1987. *Changes in the Treatment of Old-Age Mortality in Soviet Official Life Tables.* Center for International Research, US Bureau of the Census, Washington, DC.

Kirdyakin, V.D., and G.G. Fayzulin, 1974. *Rost narodonaseleniya i formirovaniye trudovykh resursov Moldavskoy SSR* (Osnovnye tendentsii i perspektivy) [Growth of the Human Population and Formation of Labor Resources of the Moldavian SSR (Basic Tendencies and Prospects)]. Karta Moldoveniaske, Kishinev.

Kruminš, Y.K., 1976. Tendentsii rosta sredney prodolzhitel'nosti zhizni muzhskogo i zhenskogo naseleniya Latvii (Tendencies in the Growth of the Average Length of Life of the Male and Female Population of Latvia). In *Uchenye zapiski Latviyskogo gosudarstvennogo universiteta*, Vol. 240, *Voprosy statistiki i problemy demografii i urovnya zhizni naroda*. Latviyskiy gosudarstvennyi universitet, Riga.

Kurman, M.V., 1974. *Aktual'nye voprosy demografii: Demograficheskiye protsessy v SSSR v poslevoennyi period* (Urgent Problems of Demography: Demographic Processes in the Postwar Period). Statistika, Moscow.

Manton, K.G., and E. Stallard, 1984. *Recent Trends in Mortality Analysis.* Academic Press, New York, NY.

Panchenko, N.F., ed., 1978. *Vosproizvodstvo naseleniya v usloviyakh razvitogo sotsializma (Na primere Ukrainskoy SSR)* [Reproduction of Population in the Conditions of Developed Socialism (On the Example of the Ukrainian SSR)]. Naukova dumka, Kiev.

Pervushin, A.S., 1987. Prodolzhitel'nost' ekonomicheski aktivnoy zhizni naseleniya SSSR (Length of the Economically Active Life of the Population of the USSR). Trety Vsesoyuznyi nauchnyi seminar "Metodologiya razrabotki dolgosrochnykh regional'nykh programm razvitiya narodonaseleniya." Mariyskiy nauchno-issledovatel'skiy institut.

Silver, B.D., 1986. The Ethnic and Language Dimensions in Russian and Soviet Censuses. In R.S. Clem, ed. *Research Guide to the Russian and Soviet Censuses.* Cornell University Press, Ithaca, NY.

Sinel'nikov, A.B., 1988. Dinamika urovnya smertnosti v SSSR (Dynamics of the Level of Mortality in the USSR). In L.L. Rybakovskiy, ed. *Naseleniye SSSR za 70 let* (Soviet Population of the USSR over the Past 70 Years). Nauka, Moscow.

Singh, S., J.E. Gordon, and J.B. Wyon, 1962. Medical Care in Fatal Illnesses of a Rural Punjab Population: Some Social, Biological, and Cultural Factors and their Ecological Implications. *Indian Journal of Medical Research* 50.

State Committee of the USSR on Statistics (Goskomstat), 1988. *Naseleniye SSSR, 1987: Statisticheskiy sbornik* (Population of the USSR, 1987: Statistical Collection). Finansy i statistika, Moscow.

State Committee of the USSR on Statistics (Goskomstat), 1989a. *Naseleniye SSSR, 1988: Statisticheskiy ezhegodnik* (Population of the USSR, 1988: Statistical Yearbook). Finansy i statistika, Moscow.

State Committee of the USSR on Statistics (Goskomstat), 1989b. *Tablitsy smertnosti i ozhidaemoy prodolzhitel'nosti zhizni naseleniya* (Tables of Mortality and Life Expectation of the Population). Finansy i statistika, Moscow.

Vallin, J., and J.-C. Chesnais, 1974. Évolution récente de la mortalité en Europe, dans les pays Anglo-Saxons et en Union soviétique, 1960–1970. *Population* 29(July-October):861–898.

Vaupel, J.A., 1986. How Change in Age-Specific Mortality Affects Life Expectancy. *Population Studies* 40(March):147–158.

Vaupel, J.W., K.G. Manton, and E. Stallard, 1979. The Impact of Heterogeneity in Individual Frailty on the Dynamics of Mortality. *Demography* 16:439–454.

Vishnevsky, A., 1988. Révolution démographique et fécondité en URSS du XIXe siècle à la période contemporaine. *Population* 43(July-October): 799–814.

Vishnevsky, A.G., and A.G. Volkov, eds., 1983. *Vosproizvodstvo naseleniya SSSR* (Reproduction of the Population of the USSR). Finansy i statistika, Moscow.

Ward, P., 1987. The Reproduction of Social Inequality: Access to Health Services in Mexico City. *Health Policy and Planning* 2(1):44–57.

Zvidriņš, P.P., and Y.K. Kruminš, 1974. Uroven', dinamika i prichiny smertnosti naseleniya Latviyskoy SSR (Level, Dynamics, and Causes of Death in the Latvian SSR). In *Uchenye zapiski Latviyskogo gosudarstvennogo universiteta*, Vol. 205, *Voprosy statistiki i problemy urovnya zhizni naroda* Latviyskiy gosudarstvennyi universitet, Riga.

Zvidriņš, P.P., and V.F. Shukaylo, 1975. Izmeneniye intensivnosti smertnosti naseleniya Latviyskoi SSR z a 1958–1970 goda (Change in the Intensity of Mortality of the Population of the Latvian SSR, 1958–1970). In *Uchenye zapiski Latviyskogo gosudarstvennogo universiteta*, Vol. 240, *Voprosy statistiki i problemy demografii i urovnya zhizni naroda*. Latviyskiy gosudarstvennyi universitet, Riga.

Chapter 19

Soviet Mortality by Cause of Death: An Analysis of Years of Potential Life Lost

W. Ward Kingkade

19.1 Background

Mortality decline in the developed countries has been associated with a specific course of evolution in the structure of mortality by cause of death. The preindustrial mortality pattern was characterized by high mortality from infectious disease. The historical decline in mortality has been brought about largely by increased control over infectious disease, due as much to improvements in public sanitation and rising living standards as to vaccines (Goldscheider, 1971; United Nations, 1973). Throughout the industrial world degenerative diseases are now the leading causes of death. These differ from infectious disease in that they involve processes which are typically irreversible and which are difficult to detect until they have progressed into rather advanced stages. It is generally agreed that prevention of these diseases requires subtle adjustments in life-style that'fall outside the scope of traditional medical approaches to health care.

While increases in mortality from degenerative disease and accidents are a familiar element of the epidemiological transition in the West, a

339

distinctive aspect of the recent history of Eastern Europe is that in several countries in this region, including the Soviet Union (Dmitriyeva and Andreev 1987), these trends have resulted in an overall rise in total mortality from all causes combined. East European demographers have coined the term "civilization diseases" to cover the types of mortality which become more prevalent with societal modernization (Tolokontsev, 1987). They are the dominant feature in recent Soviet mortality experience.

Among the countries of the world, the Soviet Union encompasses one of the most heterogeneous populations geographically, socioculturally, and socioeconomically. This diversity is reflected by systematic variations in demographic behavior between different segments of the Soviet population. It is widely held that the indigenous populations of the various regions of the Soviet Union are at different stages of the demographic transition from natural to controlled reproduction, one element of which is the shift in etiological structure of mortality from primarily exogenous to endogenous causes of death (Vishnevsky and Volkov, 1983; Bedny, 1979; Karakhanov, 1983). In particular, respiratory disease mortality figures more prominently in the Central Asian mortality pattern than in those of the European regions of the USSR. Certain Soviet authors have attributed the differential importance of this cause category, which tends to play a greater role in Third World mortality patterns than those of highly industrialized countries, to the Central Asian population's lower level of development (Vishnevsky and Volkov, 1983).

Until lately, analysis of regional differentials in Soviet mortality by cause of death has been hampered by scarcity of available data. Goskomstat's 1988 population handbook (State Committee of the USSR on Statistics, 1989) is certainly a welcome contribution, representing the first publication since the Second World War of comprehensive data on the cause structure of mortality for the various regions of the USSR. The data are provided separately for age and sex for the urban and rural populations of each republic as of 1987 and 1988. In this chapter these data are utilized to assess the respective cause-specific mortality levels of the republics, both relative to one another and in comparison with various industrialized countries.

19.2 Methodology

Several alternative methodologies for comparison of cause-specific mortality levels across populations are available. One of the soundest is the familiar technique of direct standardization, in which age schedules of cause-specific mortality rates are applied to a standard age distribution to control differences in age structure between populations (Lilienfield and Lilienfield, 1980). The standardized cause-specific death rates, which represent the overall death rates for each cause category which would prevail if each population possessed the standard age distribution, can be compared to assess relative mortality levels across populations. However, this methodology is subject to the criticism that by weighting all deaths equally it does not accurately reflect the losses of potential life implied by the given mortality schedules.

A second approach is the methodology of cause elimination life tables, which purport to measure the gains in life expectancy that would result from elimination of various causes of death (Elandt-Johnson and Johnson, 1980). A serious obstacle to modern applications of this methodology is the problem of competing risks of mortality from various causes, which necessitates an assumption about the form of interdependence between the respective cause-specific mortality schedules. Unfortunately, the traditional assumption of independence among causes of death, although suitable with respect to the infectious diseases to which the methodology was originally applied, is untenable in relation to the degenerative conditions fundamental to modern mortality patterns.

The measurement of years of potential life lost (Garcia-Rodriguez and da Motta, 1989; Romeder and McWhinnie, 1977) developed in the literature on preventable mortality provides another method for summarizing the impact of a given mortality regime. In this methodology, age-specific death rates for various cause categories are weighted by the difference in years of the midpoint of the given age interval from some specified maximum age. The weighted rates can then be applied to the age distribution of the observed population or to that of any standard population. This approach to measuring loss of life has been adopted in this analysis.

To implement the years of potential life lost (YPLL) methodology, the maximum age used to calculate potential loss must be chosen. Typically, the observed life expectancy of the observed population or a population at a similar level of development is employed using the argument that the attainable maximum is constrained by the state of the country's health system (Lopez, 1989). While this solution is appropriate for health-planning assessments of the performance of national health systems, the sacrifice of global perspective is unsatisfactory for present purposes. Analysis of Soviet mortality in the international perspective is better by choosing a maximum in terms of the present experience of low-mortality countries. Currently attainable minimum mortality schedules implying life expectancies of 81.6 years and 87.2 years respectively for men and women have been obtained by selecting for each of 15 cause of death classes the minimum death rates for each age and sex category in a sample of low-mortality countries (see Appendix *Tables 19A.1* and *19A.2*), summing across causes to obtain mortality rates. The life expectancies derived from these schedules, which are higher than any thus far observed, may be considered indicative of the best attainable under the present state of medicine and public health.

19.3   Years of Potential Life Lost

Estimates of annual years of potential life lost for the USSR and the republics are compared with those obtained for the USA and several European countries in *Table 19.1*, where the figures are expressed relative to the appropriate mid-period populations. The countries have been chosen to represent average to poor performance among developed countries, while providing regional variety. The data reveal that injuries, together with neoplasms and diseases of the circulatory system, account for the majority of years of potential life lost for either sex in every country, including the USSR, while respiratory disease generally makes the smallest contribution. Diseases of the circulatory system are especially prominent among males, contributing the greatest share of YPLL in the four industrialized countries, both in the observed populations and after controlling for differences in age composition through standardization. The departure of Soviet males from this pattern in the actual population is a result of their age distribution, which remains relatively young in comparison with those of the other countries and hence engenders a greater frequency of injuries. As the standardized values demonstrate, the circulatory component of Soviet male YPLL would exceed those of the other cause categories if Soviet males followed the WHO European standard age distribution. Among females, neoplasms become more important than diseases of the circulatory system in three of the industrialized countries, while the USSR and Hungary continue to exhibit greater losses of potential life through circulatory system disease.

When compared with the four industrialized countries in *Table 19.1*, the USSR is found, unsurprisingly, to bear the closest resemblance to Hungary. In most instances the YPLLs for the USSR are intermediate between the (typically higher) Hungarian values and those of the remaining countries. What distinguishes the Soviet case is the vastly higher loss of potential life through respiratory system disease. In addition, the USSR exceeds the other countries in YPLL due to injuries, particularly among males. Among women, the Soviet Union also ranks highest in YPLL from diseases of the circulatory system.

Systematic variations between regions of the USSR in loss of potential life by cause of death are evident in the union republic data in *Table 19.1*. Respiratory disease exhibits the most pronounced regional differentials in YPLL, ranging between levels in the Baltic republics comparable with (among females, frequently lower than) those of the four industrialized countries and values many times greater in Central Asia. The European republics fare considerably better than other republics when looking at this indicator. The unusually high YPLLs through injuries in Armenia undoubtedly reflect the effect of the December 1988 earthquake. The annual data reveal an upsurge of accident deaths in this republic in 1988, principally in urban areas. With this exception, the European republics are distinguished by the highest male YPLLs from injuries; no comparable disadvantage is apparent for the female population of the European republics. Loss of potential life from neoplasms tends to be lower in the Transcaucasia and Central Asia republics than in the European republics, as well as the four industrialized countries in many instances. In general, circulatory system disease accounts for greater loss of potential life in European republics than elsewhere in the Soviet Union, although standardization reduces the differential. Curiously enough, when differences in age composition between republics are controlled by standardization, the greatest circulatory disease YPLLs appear in Turkmenistan, whose neighboring republics exhibit some of the lowest YPLLs for this cause category.

The figures in *Table 19.1* highlight the role of age structure in loss of potential life in Central Asia. The republics in this region, whose fertility levels resemble those of many Third World countries, are growing

Table 19.1. Years of potential life lost by cause of death for the USSR and selected industrialized countries, per 100,000 inhabitants, actual and age-standardized.

	Actual						Age-standardized					
	All causes	Circulatory	Neoplasms	Respiratory	Injuries	Residual	All causes	Circulatory	Neoplasms	Respiratory	Injuries	Residual
Male												
Hungary (1988)	22,126	7,454	4,787	869	4,184	4,832	23,351	7,800	5,035	941	4,181	5,395
UK (1987)	12,632	4,736	3,207	770	1,554	2,367	12,643	4,697	3,183	740	1,495	2,529
Spain (1985)	11,531	2,959	2,899	749	2,188	2,736	12,510	3,264	3,202	822	2,144	3,078
USA (1986)	14,363	3,842	2,682	673	3,704	3,462	15,433	4,547	3,157	752	3,392	3,585
USSR (1987–1988)	23,881	5,576	3,612	2,998	6,017	5,678	25,737	7,509	4,565	2,816	5,750	5,097
Russia	23,754	6,281	4,130	1,701	6,837	4,806	25,687	7,891	4,929	1,843	6,403	4,620
Ukraine	21,129	6,097	4,207	1,453	5,329	4,043	22,610	7,093	4,641	1,604	5,182	4,092
Byelorussia	19,362	5,856	3,702	1,427	5,084	3,293	21,645	7,420	4,415	1,636	4,930	3,243
Estonia	20,794	6,706	3,822	780	5,690	3,797	22,433	7,926	4,374	842	5,509	3,782
Latvia	21,189	6,936	3,967	797	5,990	3,499	22,923	8,187	4,558	894	5,772	3,511
Lithuania	19,226	5,217	3,512	939	6,136	3,421	21,160	6,426	4,187	1,090	6,017	3,440
Moldavia	23,181	4,528	3,111	2,903	5,854	6,784	25,367	6,472	4,015	2,780	5,713	6,387
Georgia	19,078	6,200	2,210	2,696	3,356	4,616	20,460	7,669	2,574	2,453	3,274	4,490
Armenia	24,466	3,056	2,150	2,408	11,971	4,881	24,755	4,990	3,103	2,166	10,289	4,207
Azerbaijan	22,527	4,308	2,138	6,575	3,078	6,428	24,195	8,200	3,484	4,450	2,843	5,218
Kazakhstan	24,429	4,184	2,989	4,256	6,138	6,862	26,545	6,947	4,689	3,587	5,675	5,647
Kirghizia	28,680	3,201	1,907	9,557	5,123	8,892	27,119	6,437	3,343	6,212	4,897	6,230
Tajikistan	34,378	2,111	1,321	10,977	3,174	16,795	24,971	5,005	2,530	5,582	2,919	8,935
Turkmenistan	38,053	4,155	1,666	13,615	4,406	14,211	31,906	8,727	3,307	7,295	3,942	8,635
Uzbekistan	30,238	2,772	1,463	10,123	4,204	11,676	25,740	6,509	2,769	5,469	3,663	7,330

Table 19.1. Continued.

	Actual						Age-standardized					
	All causes	Circu-latory	Neo-plasms	Respir-atory	Injur-ies	Resid-ual	All causes	Circu-latory	Neo-plasms	Respir-atory	Injur-ies	Resid-ual
Female												
Hungary (1988)	18,899	7,870	4,748	587	1,841	3,853	16,997	6,188	4,247	547	1,772	4,243
UK (1987)	13,541	4,837	4,530	875	695	2,604	11,526	3,502	4,112	687	693	2,532
Spain (1985)	9,800	3,341	2,676	571	736	2,476	9,239	2,810	2,632	506	741	2,550
USA (1986)	11,521	3,368	3,322	661	1,345	2,825	11,825	3,322	3,574	658	1,324	2,947
USSR (1987–1988)	19,191	7,249	3,303	2,352	2,000	4,287	18,208	6,563	3,382	2,169	2,016	4,078
Russia	18,002	7,911	3,657	1,141	1,874	3,419	16,469	6,499	3,483	1,096	1,915	3,476
Ukraine	18,062	8,811	3,758	1,048	1,544	2,902	15,753	6,691	3,455	936	1,594	3,076
Byelorussia	15,296	7,053	3,179	1,078	1,409	2,577	14,498	6,214	3,167	1,006	1,457	2,653
Estonia	17,158	7,805	3,785	570	1,868	3,131	15,130	5,932	3,494	564	1,914	3,226
Latvia	17,347	8,117	3,839	557	1,926	2,907	15,188	6,241	3,534	516	1,955	2,941
Lithuania	14,457	5,834	3,418	640	1,784	2,779	13,651	4,966	3,361	613	1,827	2,884
Moldavia	21,106	7,238	3,019	2,141	2,591	6,118	22,295	8,267	3,398	1,984	2,550	6,095
Georgia	15,774	6,967	2,464	2,173	1,097	3,073	15,305	6,507	2,528	2,127	1,098	3,045
Armenia	29,191	3,538	2,120	2,514	16,731	4,288	28,902	4,816	2,744	2,140	15,334	3,868
Azerbaijan	19,553	4,787	1,920	6,094	1,487	5,265	19,977	7,274	2,708	4,339	1,306	4,350
Kazakhstan	18,992	4,821	2,866	3,661	2,150	5,495	19,312	6,015	3,622	2,961	2,029	4,686
Kirghizia	24,530	4,298	1,893	8,394	2,300	7,645	22,784	6,711	2,873	5,543	2,001	5,657
Tajikistan	32,354	2,946	1,281	11,001	1,767	15,359	24,763	5,786	2,299	6,140	1,397	9,142
Turkmenistan	33,372	4,896	1,877	12,170	2,090	12,338	28,556	8,667	3,234	6,940	1,688	8,027
Uzbekistan	27,045	3,723	1,613	9,299	2,206	10,203	23,678	6,965	2,750	5,317	1,682	6,965

Table 19.2. Years of potential life lost per death by cause of death for the USSR and selected industrialized countries.

	Actual						Age-standardized					
	All causes	Circu-latory	Neo-plasms	Respir-atory	Injur-ies	Resid-ual	All causes	Circu-latory	Neo-plasms	Respir-atory	Injur-ies	Resid-ual
Male												
Hungary (1988)	15.421	10.711	14.491	11.688	27.834	26.275	14.908	10.093	14.274	11.379	26.430	26.982
UK (1987)	11.043	8.747	10.569	6.344	34.619	17.808	11.722	9.255	11.342	6.348	34.308	19.391
Spain (1985)	13.229	8.646	13.228	8.220	35.807	17.314	12.534	8.133	13.127	7.756	33.757	16.791
USA (1986)	15.269	9.434	12.409	8.331	40.006	24.066	14.326	9.488	12.644	7.986	36.907	22.088
USSR (1987–1988)	23.661	12.265	19.214	32.136	41.011	44.804	16.781	9.242	16.384	20.901	36.031	34.398
Russia	22.400	13.056	19.045	22.485	40.052	41.356	16.173	9.339	16.076	15.751	35.677	32.963
Ukraine	18.373	10.589	18.797	15.004	38.056	35.617	14.852	8.433	17.054	12.109	35.323	31.472
Byelorussia	19.037	11.523	19.275	14.395	39.582	36.940	14.911	9.294	16.973	10.791	35.768	31.423
Estonia	18.117	10.716	17.087	20.555	37.064	35.569	14.709	8.760	15.342	17.460	33.896	30.440
Latvia	17.371	10.415	16.887	15.028	38.122	32.179	14.837	9.156	15.585	13.093	35.129	27.928
Lithuania	17.698	9.580	16.800	13.015	37.761	34.832	15.625	9.011	15.796	11.837	34.930	30.724
Moldavia	23.212	10.371	21.120	30.035	41.175	38.563	14.347	6.453	18.142	18.247	34.497	28.284
Georgia	20.409	11.444	19.330	40.609	40.164	35.835	15.643	9.226	16.920	30.362	36.241	29.278
Armenia	29.864	11.330	20.964	34.394	47.793	44.606	19.032	8.774	16.971	18.989	38.123	28.813
Azerbaijan	31.533	13.541	21.809	61.116	47.859	50.820	16.865	9.632	16.089	36.727	36.999	30.892
Kazakhstan	30.157	14.394	20.359	41.989	45.880	49.916	18.519	10.488	16.143	22.562	37.989	32.793
Kirghizia	36.774	12.801	22.128	52.615	48.070	57.147	18.911	9.480	17.022	24.728	37.534	35.295
Tajikistan	46.725	11.539	22.108	62.661	51.263	65.626	22.480	10.023	17.024	32.783	38.352	41.332
Turkmenistan	45.141	16.065	22.347	67.027	51.803	64.116	21.183	10.939	16.662	38.325	39.078	39.486
Uzbekistan	41.704	11.865	22.628	63.722	52.624	62.075	19.994	9.671	16.581	35.501	38.985	36.737

Table 19.2. Continued.

	Actual						Age-standardized					
	All causes	Circulatory	Neoplasms	Respiratory	Injuries	Residual	All causes	Circulatory	Neoplasms	Respiratory	Injuries	Residual
Female												
Hungary (1988)	15.541	11.115	19.504	14.550	21.092	28.142	18.446	12.184	22.068	17.908	25.091	35.382
UK (1987)	12.158	8.996	17.124	8.237	26.207	14.564	17.070	11.753	21.789	11.691	34.091	23.136
Spain (1985)	13.000	8.711	19.176	9.657	35.101	16.431	15.320	9.591	21.829	11.083	38.054	20.500
USA (1986)	14.236	8.420	18.478	10.749	39.265	21.119	18.351	11.328	21.968	14.101	42.085	26.721
USSR (1987–1988)	19.274	11.099	23.891	35.536	41.719	47.507	20.766	11.836	25.663	37.210	42.606	47.983
Russia	16.997	10.977	23.116	23.136	37.592	42.183	19.213	11.684	25.289	26.965	40.401	46.227
Ukraine	15.499	10.653	24.098	15.689	37.259	39.237	17.766	11.152	26.672	18.783	41.194	44.901
Byelorussia	15.467	10.389	24.316	13.604	39.092	40.298	17.798	11.629	26.272	16.037	41.270	43.095
Estonia	14.326	9.017	21.993	27.396	33.808	37.349	17.189	9.926	24.647	33.063	38.776	43.322
Latvia	14.321	9.390	22.260	18.591	32.435	34.173	17.445	10.735	25.067	23.225	37.737	39.612
Lithuania	15.202	8.973	23.489	17.644	36.714	39.479	17.993	10.213	25.733	20.964	39.536	43.386
Moldavia	22.824	13.103	28.734	37.296	44.516	40.297	19.019	10.962	27.854	29.945	40.191	36.631
Georgia	18.562	11.504	25.990	43.854	43.136	41.249	19.787	12.102	27.276	45.843	43.834	42.491
Armenia	33.487	11.963	28.292	45.456	53.953	49.117	28.497	12.250	27.084	35.757	47.650	40.475
Azerbaijan	30.763	13.430	27.909	67.530	56.448	56.109	24.411	14.225	26.199	55.684	47.039	44.377
Kazakhstan	26.761	12.898	25.434	47.367	51.173	52.846	24.417	14.077	25.520	39.745	46.118	45.393
Kirghizia	35.133	13.773	27.530	55.133	55.794	61.632	26.123	14.645	26.560	38.714	44.928	47.859
Tajikistan	49.786	14.746	28.518	67.824	61.769	71.648	30.728	15.304	26.370	44.932	46.448	52.559
Turkmenistan	46.054	17.620	28.544	72.182	61.591	69.052	27.970	15.458	25.620	50.737	46.671	49.790
Uzbekistan	41.616	13.621	28.985	69.040	63.592	67.385	26.422	13.819	26.430	47.361	48.647	49.337

rapidly through natural increase and have age distributions weighted heavily in favor of infants and children. As a result, these republics register substantial loss of life from causes of death whose incidence is high in infancy, such as diseases of the respiratory system. According to *Table 19.1*, the Central Asian populations would experience major reductions in YPLL if the age distributions were to resemble the WHO European standard. Thus, one benefit of the fertility decline that is now occurring in Central Asia will be the foreseeable reduction in the child-dependency burden, whose health dimension finds expression here in the loss of human potential inherent in infant and child mortality.

The measure of potential life lost employed so far expresses annual loss of life relative to the mid-period population, indicating the intensity of loss of life per average person at risk. An alternative measure with a different interpretation is YPLL per death, which reflects the differing losses entailed by deaths in the various cause categories. When standardized, this measure becomes a sensitive indicator of the youthfulness of the given mortality schedules; in unstandardized form it reflects the youthfulness of the observed age distributions of deaths. Estimates of YPLL per death are presented for the USSR and the republics along with the four industrialized countries in *Table 19.2*.

According to the figures in *Table 19.2*, the average death entails a greater loss of potential life in the USSR than in any of the four industrialized countries. Some of the differential based on observed deaths is due to the USSR's more youthful age structure, as comparison of the standardized with the unstandardized values indicates. Nevertheless, when differences in age composition are eliminated the USSR continues to exhibit a mortality pattern which indicates more loss of potential life. The importance of respiratory disease and injuries in the Soviet mortality profile undoubtedly contributes to the greater overall loss of life in the average Soviet death, since mortality from these two causes tends to have a younger incidence than mortality from the other causes. However, within cause categories the Soviet average losses are often significantly greater than those obtained for other countries. This is especially well illustrated in the case of respiratory disease, suggesting a reserve of preventable infant and child mortality which has been substantially reduced in the four industrialized countries.

The republic estimates in *Table 19.2* reveal major regional variations in the loss of potential life associated with deaths in the respective cause categories, particularly injuries and respiratory disease. Differences in age structure between republics explain much of the inter-republic differentiation in YPLLs per average observed death. Standardization substantially reduces the variability in union republic values. However, sizable differentials persist in several cause categories after population age structures have been equated. The Central Asian/Transcaucasian (Armenia excluded) respiratory disease mortality pattern evidently entails a considerably younger average age at death than that of European republics, since the average age at death from respiratory disease in the former regions would exceed the corresponding average age in the latter republics by some 15 to 25 years if each population followed the standard age distribution. Appreciable differentials remain after standardization in YPLLs per death from circulatory system diseases and injuries as well, particularly among women, with Central Asia and Transcaucasia again registering greater average loss than European republics.

19.4   Conclusion

A country's achievements in mortality reduction are limited by the present state of technology and medical science, among other things. The measure of preventable mortality developed in this chapter has sought to quantify a society's potential for mortality reduction subject to these constraints. The analysis of Soviet mortality employing this device has indicated the possibility of substantial savings of human potential through mortality reduction. Whether these savings are realized will depend on the progress and quality of the efforts in the republics to develop a health-care system, such as an ambitious campaign to provide the population with medical checkups, as well as their general economic development.

Appendix

Table 19A.1. Minimum cause-specific death rates, per 100,000 inhabitants, in low-mortality countries.

Age	Sum over causes (rate)	Infectious and Parasitic diseases Rate	Country	Year	Neoplasms Rate	Country	Year	Endocrine, nutritional, and metabolic diseases Rate	Country	Year
Male										
0–1	357	2	Sweden	1987	0	N. Zealand	1986	0	N. Zealand	1981
1–4	11	0	Sweden	1979	2	Norway	1986	0	N. Zealand	1984
5–9	9	0	N. Zealand	1983	1	Austria	1985	0	N. Zealand	1986
10–14	7	0	N. Zealand	1985	2	Ireland	1987	0	N. Zealand	1982
15–19	36	0	N. Zealand	1985	4	Ireland	1987	0	N. Zealand	1981
20–24	53	0	N. Zealand	1985	4	Denmark	1985	0	N. Zealand	1980
25–29	43	0	N. Zealand	1978	6	Israel	1986	0	Norway	1982
30–34	45	0	N. Zealand	1978	7	Sweden	1983	0	N. Zealand	1974
35–39	73	0	N. Zealand	1984	16	Ireland	1986	0	Norway	1977
40–44	110	0	N. Zealand	1980	29	Israel	1984	0	Israel	1981
45–49	195	0	Norway	1987	69	Norway	1980	1	Israel	1975
50–54	330	0	Denmark	1984	119	Israel	1985	1	Israel	1981
55–59	571	2	Denmark	1982	237	Sweden	1971	3	Norway	1973
60–64	869	2	Denmark	1988	366	Israel	1983	7	Israel	1981
65–69	1,431	4	N. Zealand	1986	575	Israel	1986	9	Finland	1985
70–74	2,500	9	Denmark	1980	867	Portugal	1975	24	Israel	1981
75–79	4,273	14	Denmark	1986	1,202	Portugal	1975	38	Norway	1971
80–84	6,628	15	Denmark	1984	1,198	Greece	1970	27	Israel	1981
85 +	10,188	24	Denmark	1986	1,131	Japan	1970	71	Austria	1976
Female										
0–1	301	0	N. Zealand	1981	0	N. Zealand	1986	0	N. Zealand	1983
1–4	7	0	N. Zealand	1978	1	Denmark	1982	0	N. Zealand	1986
5–9	4	0	N. Zealand	1986	2	Switzerland	1982	0	N. Zealand	1982
10–14	2	0	N. Zealand	1984	0	Finland	1986	0	N. Zealand	1985
15–19	9	0	N. Zealand	1985	1	Sweden	1986	0	N. Zealand	1984
20–24	14	0	N. Zealand	1986	2	Switzerland	1987	0	N. Zealand	1986
25–29	12	0	N. Zealand	1986	4	Sweden	1979	0	N. Zealand	1986
30–34	19	0	N. Zealand	1982	11	Finland	1985	0	N. Zealand	1981
35–39	35	0	N. Zealand	1983	26	Finland	1985	0	Norway	1985
40–44	60	0	N. Zealand	1986	41	Finland	1986	0	Norway	1975
45–49	114	0	N. Zealand	1979	80	Finland	1980	0	Ireland	1987
50–54	193	0	Switzerland	1986	127	Japan	1988	2	Norway	1977
55–59	300	1	Austria	1988	186	Japan	1986	3	Norway	1981
60–64	474	2	Denmark	1976	250	Portugal	1979	7	Finland	1987
65–69	773	3	Austria	1988	345	Greece	1970	12	Norway	1974
70–74	1,356	4	Switzerland	1988	471	Greece	1971	21	Norway	1981
75–79	2,382	3	N. Zealand	1977	562	Greece	1970	44	Norway	1983
80–84	4,338	15	Denmark	1982	628	Greece	1971	56	Israel	1981
85 +	8,920	17	Austria	1987	632	Greece	1973	69	Japan	1971

Note: In cases of rates of the same values (mainly zeros), the last country/most recent date listed in the WHO data is indicated.

Table 19A.1. Continued.

Age	Diseases of the blood and blood-forming organs			Mental, nervous, and sensory organ diseases			Diseases of the circulatory system		
	Rate	Country	Year	Rate	Country	Year	Rate	Country	Year
Male									
0–1	0	N. Zealand	1986	0	Norway	1976	0	N. Zealand	1986
1–4	0	N. Zealand	1986	0	Belgium	1983	0	N. Zealand	1983
5–9	0	N. Zealand	1986	0	N. Zealand	1986	0	N. Zealand	1982
10–14	0	N. Zealand	1985	0	N. Zealand	1973	0	Switzerland	1987
15–19	0	N. Zealand	1985	1	Ireland	1974	0	Denmark	1985
20–24	0	N. Zealand	1986	1	Norway	1970	1	Finland	1987
25–29	0	N. Zealand	1986	0	Norway	1977	2	Israel	1985
30–34	0	N. Zealand	1985	1	N. Zealand	1971	5	Sweden	1973
35–39	0	N. Zealand	1986	1	Ireland	1981	19	Sweden	1973
40–44	0	N. Zealand	1985	1	N. Zealand	1979	43	Switzerland	1986
45–49	0	N. Zealand	1986	2	Ireland	1983	78	Switzerland	1988
50–54	0	N. Zealand	1984	3	Ireland	1982	148	Japan	1987
55–59	0	N. Zealand	1986	8	Ireland	1982	237	Japan	1987
60–64	0	N. Zealand	1983	12	Israel	1980	354	Japan	1987
65–69	0	N. Zealand	1971	21	Japan	1987	610	Japan	1988
70–74	0	Norway	1977	32	Portugal	1979	1,180	Japan	1987
75–79	0	N. Zealand	1974	32	Switzerland	1977	2,341	Japan	1987
80–84	3	Sweden	1978	35	Portugal	1976	4,211	Greece	1970
85 +	0	Finland	1978	40	Switzerland	1976	6,709	Greece	1970
Female									
0–1	0	N. Zealand	1983	5	Belgium	1983	0	Switzerland	1978
1–4	0	N. Zealand	1986	0	Norway	1977	0	N. Zealand	1986
5–9	0	N. Zealand	1985	0	N. Zealand	1986	0	N. Zealand	1986
10–14	0	N. Zealand	1986	0	N. Zealand	1971	0	N. Zealand	1985
15–19	0	N. Zealand	1986	0	N. Zealand	1984	0	Norway	1983
20–24	0	N. Zealand	1985	0	Switzerland	1976	0	Denmark	1987
25–29	0	N. Zealand	1986	0	Norway	1981	0	Israel	1986
30–34	0	N. Zealand	1986	0	Ireland	1984	2	Denmark	1978
35–39	0	N. Zealand	1986	0	N. Zealand	1975	3	Norway	1984
40–44	0	N. Zealand	1985	0	N. Zealand	1971	10	Norway	1983
45–49	0	N. Zealand	1986	0	N. Zealand	1986	18	Switzerland	1987
50–54	0	N. Zealand	1982	3	Japan	1986	39	Switzerland	1986
55–59	0	N. Zealand	1986	5	Japan	1987	78	Switzerland	1987
60–64	0	N. Zealand	1986	8	Denmark	1976	145	France	1987
65–69	0	N. Zealand	1982	11	Israel	1975	297	France	1987
70–74	0	Denmark	1984	17	Portugal	1975	673	France	1987
75–79	0	N. Zealand	1984	29	Portugal	1988	1,442	France	1987
80–84	4	Finland	1983	26	Portugal	1979	3,034	France	1987
85 +	0	Israel	1978	35	Portugal	1975	6,953	Greece	1970

Table 19A.1. Continued.

Age	Diseases of the respiratory system			Diseases of the digestive system			Diseases of the genito-urinary system		
	Rate	Country	Year	Rate	Country	Year	Rate	Country	Year
Male									
0–1	9	Netherlands	1983	0	Sweden	1984	0	N. Zealand	1981
1–4	0	Finland	1985	0	N. Zealand	1986	0	N. Zealand	1986
5–9	0	N. Zealand	1985	0	N. Zealand	1985	0	N. Zealand	1986
10–14	0	Switzerland	1987	0	N. Zealand	1986	0	N. Zealand	1986
15–19	0	Switzerland	1978	0	N. Zealand	1984	0	N. Zealand	1985
20–24	0	Norway	1985	0	N. Zealand	1986	0	N. Zealand	1986
25–29	0	Switzerland	1982	0	N. Zealand	1981	0	N. Zealand	1984
30–34	0	Norway	1979	0	N. Zealand	1973	0	N. Zealand	1984
35–39	0	Norway	1987	1	Israel	1986	0	N. Zealand	1985
40–44	2	Denmark	1987	4	N. Zealand	1971	0	N. Zealand	1984
45–49	4	Denmark	1988	9	N. Zealand	1985	0	N. Zealand	1978
50–54	11	Norway	1987	10	N. Zealand	1983	0	N. Zealand	1983
55–59	22	Norway	1987	18	Norway	1974	1	N. Zealand	1985
60–64	50	Sweden	1972	33	Norway	1972	5	Switzerland	1986
65–69	107	Sweden	1972	38	Israel	1983	14	Switzerland	1987
70–74	205	Sweden	1972	76	Norway	1977	35	Switzerland	1981
75–79	362	Greece	1986	129	Israel	1986	70	Austria	1988
80–84	663	Sweden	1972	155	Israel	1981	145	Switzerland	1987
85 +	1,339	Sweden	1971	284	Ireland	1971	318	Austria	1987
Female									
0–1	0	Finland	1987	0	N. Zealand	1985	0	N. Zealand	1985
1–4	0	Sweden	1977	0	N. Zealand	1986	0	N. Zealand	1986
5–9	0	N. Zealand	1981	0	N. Zealand	1986	0	N. Zealand	1986
10–14	0	Switzerland	1987	0	N. Zealand	1986	0	N. Zealand	1986
15–19	0	Switzerland	1984	0	N. Zealand	1986	0	N. Zealand	1986
20–24	0	Norway	1987	0	N. Zealand	1982	0	N. Zealand	1986
25–29	0	Norway	1986	0	N. Zealand	1984	0	N. Zealand	1985
30–34	0	Switzerland	1981	0	N. Zealand	1985	0	N. Zealand	1981
35–39	0	Norway	1974	0	N. Zealand	1982	0	N. Zealand	1986
40–44	1	Finland	1987	0	Ireland	1987	0	N. Zealand	1980
45–49	2	Switzerland	1987	2	Finland	1978	0	N. Zealand	1981
50–54	5	Greece	1984	4	Finland	1979	0	N. Zealand	1986
55–59	8	Netherlands	1979	8	Norway	1982	1	Sweden	1987
60–64	20	Netherlands	1977	15	Sweden	1986	4	Norway	1984
65–69	36	Switzerland	1982	31	Denmark	1981	9	N. Zealand	1979
70–74	55	Switzerland	1982	45	Norway	1983	18	Norway	1986
75–79	124	Switzerland	1982	78	Switzerland	1988	31	Switzerland	1988
80–84	254	Switzerland	1982	161	Australia	1978	61	Switzerland	1988
85 +	665	Switzerland	1982	225	Portugal	1976	104	Italy	1976

Table 19A.1. Continued.

Age	Complications of pregnancy/childbirth			Diseases of the skin and muscoskeletal tissue			Congenital anomalies		
	Rate	Country	Year	Rate	Country	Year	Rate	Country	Year
Male									
0–1				0	N. Zealand	1985	168	Japan	1988
1–4				0	N. Zealand	1986	3	Belgium	1984
5–9				0	N. Zealand	1986	0	Israel	1985
10–14				0	N. Zealand	1986	0	N. Zealand	1984
15–19				0	N. Zealand	1986	0	N. Zealand	1977
20–24				0	N. Zealand	1986	0	Switzerland	1987
25–29				0	N. Zealand	1986	0	N. Zealand	1984
30–34				0	N. Zealand	1986	0	N. Zealand	1976
35–39				0	N. Zealand	1985	0	N. Zealand	1986
40 44				0	N. Zealand	1985	0	N. Zealand	1985
45–49				0	N. Zealand	1986	0	N. Zealand	1986
50–54				0	N. Zealand	1982	0	N. Zealand	1983
55–59				0	N. Zealand	1975	0	N. Zealand	1982
60–64				0	N. Zealand	1985	0	N. Zealand	1974
65–69				0	Israel	1977	0	N. Zealand	1984
70–74				2	Greece	1986	0	N. Zealand	1983
75–79				0	Israel	1977	0	N. Zealand	1986
80–84				5	Greece	1986	0	N. Zealand	1986
85 +				0	Finland	1970	0	N. Zealand	1985
Female									
0–1	0	N. Zealand	1986	0	N. Zealand	1986	157	Japan	1987
1–4	0	N. Zealand	1986	0	N. Zealand	1986	3	Sweden	1983
5–9	0	N. Zealand	1986	0	N. Zealand	1986	0	Switzerland	1984
10–14	0	N. Zealand	1986	0	N. Zealand	1986	0	N. Zealand	1986
15–19	0	N. Zealand	1985	0	N. Zealand	1985	0	N. Zealand	1979
20–24	0	N. Zealand	1982	0	N. Zealand	1985	0	N. Zealand	1973
25–29	0	Switzerland	1985	0	N. Zealand	1985	0	N. Zealand	1986
30–34	0	N. Zealand	1981	0	N. Zealand	1979	0	N. Zealand	1985
35–39	0	N. Zealand	1986	0	N. Zealand	1984	0	N. Zealand	1977
40–44	0	N. Zealand	1986	0	N. Zealand	1984	0	N. Zealand	1984
45–49	0	N. Zealand	1986	0	N. Zealand	1985	0	N. Zealand	1983
50–54	0	N. Zealand	1986	0	N. Zealand	1980	0	N. Zealand	1981
55–59	0	N. Zealand	1986	0	Ireland	1977	0	N. Zealand	1976
60–64	0	N. Zealand	1986	1	N. Zealand	1984	0	N. Zealand	1979
65–69	0	N. Zealand	1986	1	Israel	1981	0	N. Zealand	1984
70–74	0	N. Zealand	1986	4	Greece	1986	0	N. Zealand	1985
75–79	0	N. Zealand	1986	9	Greece	1986	0	N. Zealand	1985
80–84	0	N. Zealand	1986	0	Israel	1978	0	N. Zealand	1986
85 +	0	N. Zealand	1986	13	Portugal	1976	0	N. Zealand	1986

Table 19A.1. Continued.

Age	Conditions originating in the perinatal period			Signs, symptoms, and ill-defined conditions			Accidents and injuries		
	Rate	Country	Year	Rate	Country	Year	Rate	Country	Year
Male									
0–1	178	Japan	1988	0	Ireland	1978	0	Norway	1981
1–4	0	N. Zealand	1985	0	N. Zealand	1983	6	Sweden	1986
5–9	0	N. Zealand	1986	0	N. Zealand	1986	7	Denmark	1984
10–14	0	N. Zealand	1986	0	N. Zealand	1986	5	Sweden	1982
15–19	0	N. Zealand	1986	0	N. Zealand	1986	31	Netherlands	1986
20–24	0	N. Zealand	1985	0	N. Zealand	1985	47	Netherlands	1987
25–29	0	N. Zealand	1986	0	N. Zealand	1986	34	Netherlands	1987
30–34	0	N. Zealand	1986	0	N. Zealand	1986	31	Israel	1986
35–39	0	N. Zealand	1986	0	N. Zealand	1985	34	Netherlands	1986
40–44	0	N. Zealand	1986	0	N. Zealand	1986	31	Israel	1982
45–49	0	N. Zealand	1986	0	N. Zealand	1986	33	Israel	1977
50–54	0	N. Zealand	1986	0	N. Zealand	1984	39	Israel	1983
55–59	0	N. Zealand	1986	0	N. Zealand	1986	42	Israel	1980
60–64	0	N. Zealand	1986	0	N. Zealand	1986	40	Israel	1982
65–69	0	N. Zealand	1986	0	N. Zealand	1985	53	UK	1988
70–74	0	N. Zealand	1986	0	N. Zealand	1982	70	UK	1987
75–79	0	N. Zealand	1986	0	N. Zealand	1986	84	Israel	1981
80–84	0	N. Zealand	1986	7	Australia	1983	164	UK	1988
85 +	0	N. Zealand	1986	52	Australia	1986	221	Israel	1982
Female									
0–1	139	France	1987	0	Ireland	1978	0	Norway	1985
1–4	0	N. Zealand	1985	0	N. Zealand	1986	3	Finland	1981
5–9	0	N. Zealand	1986	0	N. Zealand	1986	2	Israel	1976
10–14	0	N. Zealand	1986	0	N. Zealand	1986	2	Switzerland	1987
15–19	0	N. Zealand	1986	0	N. Zealand	1986	8	Ireland	1986
20–24	0	N. Zealand	1986	0	N. Zealand	1984	11	Norway	1970
25–29	0	N. Zealand	1986	0	N. Zealand	1986	8	Norway	1974
30–34	0	N. Zealand	1986	0	N. Zealand	1986	7	Norway	1970
35–39	0	N. Zealand	1986	0	N. Zealand	1986	7	Norway	1973
40–44	0	N. Zealand	1986	0	N. Zealand	1986	9	Greece	1973
45–49	0	N. Zealand	1986	0	N. Zealand	1986	11	Norway	1971
50–54	0	N. Zealand	1986	0	N. Zealand	1986	14	Israel	1982
55–59	0	N. Zealand	1986	0	N. Zealand	1986	11	Ireland	1970
60–64	0	N. Zealand	1986	0	N. Zealand	1985	22	Norway	1977
65–69	0	N. Zealand	1986	0	N. Zealand	1984	28	Netherlands	1986
70–74	0	N. Zealand	1986	0	N. Zealand	1986	48	Spain	1970
75–79	0	N. Zealand	1986	0	N. Zealand	1984	60	Spain	1985
80–84	0	N. Zealand	1986	3	Finland	1976	95	Spain	1985
85 +	0	N. Zealand	1986	70	Australia	1987	138	Spain	1985

Table 19A.2. Life table constructed from minimum mortality schedule in *Table 19.1*.

Age	$m(x)$	$q(x)$	$l(x)$	$L(x)$	$T(x)$	$E(x)$	$k(x)$
Male							
0	0.00357	0.00356	100,000	99,663	8,155,549	81.56	0.05
1	0.00011	0.00044	99,644	398,473	8,055,886	80.85	1.64
5	0.00009	0.00045	99,600	497,890	7,657,412	76.88	2.50
10	0.00007	0.00035	99,556	497,691	7,159,522	71.91	2.50
15	0.00036	0.00180	99,521	497,156	6,661,832	66.94	2.50
20	0.00053	0.00265	99,342	496,051	6,164,676	62.06	2.50
25	0.00043	0.00215	99,079	494,862	5,668,624	57.21	2.50
30	0.00045	0.00225	98,866	493,775	5,173,762	52.33	2.50
35	0.00073	0.00364	98,644	492,321	4,679,987	47.44	2.50
40	0.00110	0.00548	98,284	490,075	4,187,666	42.61	2.50
45	0.00195	0.00970	97,745	486,356	3,697,592	37.83	2.50
50	0.00330	0.01636	96,797	480,025	3,211,236	33.17	2.50
55	0.00571	0.02815	95,213	469,364	2,731,211	28.69	2.50
60	0.00869	0.04253	92,533	452,826	2,261,847	24.44	2.50
65	0.01431	0.06908	88,598	427,688	1,809,020	20.42	2.50
70	0.02500	0.11765	82,478	388,130	1,381,332	16.75	2.50
75	0.04273	0.19303	72,774	328,753	993,202	13.65	2.50
80	0.06628	0.28429	58,727	251,895	664,450	11.31	2.50
85	0.10188	1.00000	42,031	412,555	412,555	9.82	2.50
Female							
0	0.00301	0.00300	100,000	99,718	8,721,577	87.22	0.06
1	0.00007	0.00028	99,700	398,730	8,621,859	86.48	1.52
5	0.00004	0.00020	99,672	498,310	8,223,129	82.50	2.50
10	0.00002	0.00010	99,652	498,235	7,724,819	77.52	2.50
15	0.00009	0.00045	99,642	498,098	7,226,584	72.53	2.50
20	0.00014	0.00070	99,597	497,812	6,728,486	67.56	2.50
25	0.00012	0.00060	99,528	497,488	6,230,674	62.60	2.50
30	0.00019	0.00095	99,468	497,103	5,733,186	57.64	2.50
35	0.00035	0.00175	99,373	496,432	5,236,083	52.69	2.50
40	0.00060	0.00300	99,200	495,255	4,739,651	47.78	2.50
45	0.00114	0.00568	98,902	493,107	4,244,395	42.91	2.50
50	0.00193	0.00960	98,340	489,341	3,751,288	38.15	2.50
55	0.00300	0.01489	97,396	483,354	3,261,948	33.49	2.50
60	0.00474	0.02342	95,946	474,111	2,778,593	28.96	2.50
65	0.00773	0.03792	93,699	459,611	2,304,483	24.59	2.50
70	0.01356	0.06558	90,146	435,950	1,844,872	20.47	2.50
75	0.02382	0.11241	84,234	397,500	1,408,922	16.73	2.50
80	0.04338	0.19568	74,766	337,254	1,011,421	13.53	2.50
85	0.08920	1.00000	60,136	674,168	674,168	11.21	2.50

Table 19A.3. Cause of death classification scheme.

Cause category	WHO code 9th revision (ICD9)	8th revision (ICD8)
Infectious and parasitic	B010-B079	A001-A044
Neoplasms	B080-B179	A045-A061
Endocrine/nutritional/metabolic	B180-B199	A062-A066
Blood/blood-forming organs	B200-B209	A067-A068
Mental/nervous/sensory organs	B210-B249	A069-A079
Circulatory system	B250-B309	A080-A088
Respiratory system	B310-B329	A089-A096
Digestive system	B330-B349	A097-A104
Genito-urinary system	B350-B379	A105-A111
Pregnancy/childbirth	B380-B419	A112-A118
Skin/muscoskeletal	B420-B439	A119-A125
Congenital anomalies	B440-B449	A126-A130
Perinatal conditions	B450-B459	A131-A135
Ill-defined	B460-B469	A136-A137
Injuries	B470-B569	A138-A150

Table 19A.4. Countries included in sample.

Country	Number of observations (years)	Country	Number of observations (years)
Canada	18	Ireland	18
USA	18	Italy	17
Israel	12	Netherlands	18
Japan	19	Norway	18
Austria	19	Portugal	15
Belgium	16	Spain	16
Denmark	19	Sweden	18
Finland	18	Switzerland	19
France	18	UK	19
FRG	19	Australia	18
Greece	17	New Zealand	17

References

Bedny, M.S., 1979. *Mediko-demograficheskoye izucheniye narodonaseleniya.* Statistika, Moscow.

Dmitriyeva, R.M., and E.M. Andreev, 1987. O sredney prodolzhitel'nosti zhizni naseleniya SSSR (On the Life Expectancy of USSR Population). *Vestnik statistiki* (Herald of Statistics) 12:31–39.

Elandt-Johnson, R.C., and M.L. Johnson, 1980. *Survival Models and Data Analysis.* John Wiley and Sons, New York, NY.

Garcia-Rodriguez, L.A., and L.C. da Motta, 1989. Years of Potential Life Lost: Application of An Indicator for Assessing Premature Mortality in Spain and Portugal. *World Health Statistics Quarterly* 42(1):50–56.

Goldscheider, C., 1971. *Population, Modernization and Social Structure.* Little, Brown and Company, Boston, MA.

Hungarian Central Statistical Office, 1988. *Demografiai Evkonw.* Kozponti Statisztikai Hivatal, Budapest.

Instituto Nacional de Estodistica, 1989. *Movimiento Natural de la Poplacion Espanola Ano 1985.* INE Arles Graficas, Madrid.

Karakhanov, M.K., 1983. *Nekapitalisticheskiy put' razvitiya i problemy narodonaseleniya* (Noncapitalistic Way of Development and the Population Problems). Izdatel'stvo "FAN" Uzbekskoy SSR, Tashkent.

Lilienfield, A.M., and D.E. Lilienfield, 1980. *Foundations of Epidemiology.* Oxford University Press, New York, NY.

Lopez, A., 1989. Déçès évitables: vue d'ensemble. *World Health Statistics Quarterly* 42(1):2–3.

Manton, K.G., and E. Stallard, 1984. *Recent Trends in Mortality Analysis.* Academic Press, New York, NY.

Romeder, J.M., and J.R. McWhinnie, 1977. Potential Years of Life Lost Between Ages 1 and 70: An Indicator of Premature Mortality for Health Planning. *International Journal of Epidemiology* 6(2):143–151.

State Committee of the USSR on Statistics (Goskomstat), 1989. *Naseleniye SSSR, 1988: Statisticheskiy ezhegodnik* (Population of the USSR, 1988: Statistical Yearbook). Finansy i statistika, Moscow.

Tolokontsev, N.A., 1987. Zdorov'ye cheloveka i abiologicheskiye tendentsii v sovremennykh usloviakh i obraze zhizni (Human Health and the Negative Tendencies in Modern Mode of Life). In I.I. Smirnov, ed., *Obshchestvennyye nauki i zdravookhraneniye* (Social Sciences and Public Health Care). Nauka, Moscow.

United Kingdom Office of Population Censuses and Surveys, 1989, *Mortality Statistics 1987: Cause.* Series DH2, No. 14, Her Majesty's Stationery Office, London.

United Nations, Department of Social and Economic Affairs, 1973. *The Determinants and Consequences of Population Trends.* United Nations, New York, NY.

US National Center for Health Statistics, 1988. *Vital Statistics of the United States, 1985: Vol. II, Mortality, Part A.* DHHS Pub. No. (PHS) 88–1101. Government Printing Office, Washington, DC.

Vishnevsky, A.G., and A.G. Volkov, eds., 1983. *Vosproizvodstvo naseleniya SSSR* (Reproduction of the Population of the USSR). Finansy i statistika, Moscow.

World Health Organization, 1988. *World Health Statistics.* Geneva.

Chapter 20

Trends in Infant Mortality in the USSR

Natalia Ksenofontova

In the past, the view on infant mortality in the USSR was that infant mortality rates (IMRs) were continuously declining. It was understood that the level in the USSR was one of the lowest in the world. The issue of under-registration and reliability of infant mortality statistics was avoided. However, observed trends of IMR corrected this view. The IMR in the USSR is the highest among industrialized countries, and the increase in this index in the 1970s raised the question of the reliability of both previous and projected infant mortality data.

Unfortunately, for several decades Soviet demographers could not discuss this issue freely in scientific publications. Recent publications in the USSR and abroad show that there is great interest in this issue among scholars.

This chapter presents the general trends in infant mortality rates in the USSR in the 1980s; it attempts to estimate the quality and completeness of available statistics.

20.1 Infant Mortality Rate Trends

Figure 20.1 presents the trends of infant mortality rates in the USSR. In 1950 IMR in urban and rural areas was approximately 90 per 1,000

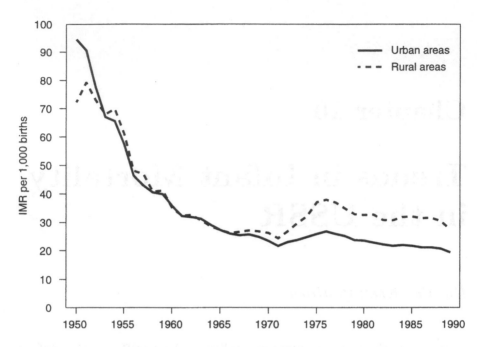

Figure 20.1. Trends in IMRs in the USSR.

live births; in the early 1960s IMR decreased to 35. Infant mortality rates stabilized in the late 1960s at the level of 25 to 30 per 1,000 births. In 1970 IMR decreased again, and in 1971 the lowest levels of the IMR up to that time were registered: 21.6 per 1,000 in urban areas and 24.3 per 1,000 in rural areas. After 1971 IMR increased in urban and rural areas. By 1976 the registered level of IMR increased by 8 percent in urban areas and by 18 percent in rural areas (the indicators returned to the level of the mid-1960s in cities and the late 1950s in rural areas). The gap in IMR between urban and rural areas widened to 11 per 1,000 births.

In 1976 IMR began to decrease again. By 1986 IMR among urban population was lower than the low level reached in 1971. IMR in rural areas declined until the early 1980s and stabilized at the level recorded in the early 1960s. The gap between IMR in urban and rural areas was practically unchanged, and was rather significant. In 1989 the registered IMR in USSR was 22.7 deaths per 1,000 live births; in urban areas, 19.4; and in rural areas, 27.4.

Usually a period of rapid decline of IMR is followed by a period of slower decline with subsequent stabilization of IMR. Such a pattern of infant mortality change was observed in most of the developed countries (Maksudov, 1985). If the stage of rapid mortality decline is explained mainly by a decrease of mortality from infectious diseases, due to increased use of antibiotics and improved medical service, the stage of a reduced decrease of IMR and subsequent stabilization is associated with the fact that the effect of the measures that contributed to the rapid decline of exogenous mortality is to a great extent exhausted (Ksenofontova, 1990).

The trends in IMR in the USSR is beyond the usual scheme of changes. Compared with other developed countries it may be considered extraordinary. No other country experienced such a significant rise in IMR after the decrease began.

20.2   Causes of IMR Increase in the 1970s

Current opinions on the increase in IMR may be divided into two groups. The first and most common one states that the increase of IMR is connected only with registration improvements (Baranov *et al.*, 1990; Borisov, 1988). There are indications that the increase in the registered IMR does not reflect a real mortality increase but is the result of a more complete registration of infant deaths (State Committee of the USSR on Statistics, 1988). This view is shared by some non-Soviet demographers. IMR growth would not have been so great if the figures were not so low in the early 1970s (Maksudov, 1985; Anderson and Silver, 1986).

The second point of view is based on the fact that IMR trends in the USSR as a whole and in the republics reflect both real processes of death probability increase and registration improvements. This view was described by Dmitriyeva and Andreev (1987). Several demographers from outside the USSR also think that registration improvement is not the only reason for rising IMR (Szymansky, 1981; Davis and Feshbach, 1980). We attempted to measure the influence of the variation of these factors on the level of IMR in the republics of the USSR.

The trends in IMR were influenced both by the changes in the distribution of birth numbers in the republics and by the trends of IMR in the republics. To measure the value of each factor, IMR for the USSR was divided into two components (Ksenofontova, 1990). The first component reflects variation of death probabilities under the age of 1 among republics, and the second component measures variation in the birth numbers (*Tables 20.1* and *20.2*). The level of IMR in the USSR as a whole is composed of IMRs for each republic; therefore, the trend of the average rate depends on the variation between regions. If one were to try to figure out which components most influenced the IMR change in the USSR, one would discover substantial differences in the degree of influence of each component upon the IMR change in the USSR and in the groups of republics.

We divided the republics into three groups (*Table 20.2*). The first group includes the republics with a low infant mortality level: the Ukraine, Byelorussia, and the Baltic republics. The second group consists of republics with a high mortality level: Uzbekistan, Kazakhstan, Kirghizia, Tajikistan, and Turkmenistan. The remaining republics (Russia, Georgia, Azerbaijan, Armenia, and Moldavia) make up the third group.

From 1970 to 1975 IMR increased by 5.6 per 1,000 births in the USSR as a whole, by 2.4 per 1,000 in urban areas, and 11 per 1,000 in rural areas. Its increase during this period was mainly caused by an increase of death probability during the first year of life. It was insignificant in the first group of republics (among urban populations, death probability even decreased) and very significant in the second group of republics, especially in rural areas. The impact of changes in the distribution of the number of births by the republics was not significant, because the proportions of births in each group of republics changed differently.

It should be noted that the interrelation of the components of infant mortality change is different in each group. For the first and the third groups of the republics in 1970–1975 the component that reflected an increase in death probabilities in the first year of life raised the IMR for the country as a whole, and the component that reflected a change in the proportion of births lowered it. The exception is the trends of IMRs in urban areas in the first group of republics, where there was no rise in infant mortality during this period. In the second group of republics both components contributed to the rise of the IMR for the whole country. Decisive for this rise was the increase of IMRs in the republics with a higher level.

Table 20.1. Changes in infant mortality rates (additional infant deaths per thousand live births).

	Total population			Urban			Rural		
	1970–1975	1975–1980	1980–1986	1970–1975	1975–1980	1980–1986	1970–1975	1975–1980	1980–1986
Total variation of IMR	5.6	-3.3	-1.9	2.4	-2.4	-1.9	11.0	-4.5	-1.1
Including variations caused by:									
Death probability	5.1	-3.8	-2.6	2.2	-2.4	-2.6	9.1	-5.3	-2.3
Numbers of birth structure	0.2	0.4	0.7	0.1	0.2	0.7	0.8	1.0	1.2
Components interrelations	0.3	0.1	0.0	0.1	-0.2	0.0	1.1	-0.2	0.0

Table 20.2. IMR changes (1970–1975, 1975–1980, and 1980–1986) by groups of republics.

	Total variation of IMR			Including caused by variation of death probability under age 1		
	1970–1975	1975–1980	1980–1986	1970–1975	1975–1980	1980–1986
Total population	5.6	-3.3	-1.9	5.1	-3.8	-2.7
Group of republics:						
1	0.1	-0.8	-0.6	0.4	-0.7	-0.4
2	4.6	-1.2	0.4	3.8	-1.8	-0.7
3	0.9	-1.3	-1.7	0.9	-1.3	-1.6
Urban	2.4	-2.3	-2.4	2.2	-2.4	-1.1
Group of republics:						
1	-1.0	-0.6	-0.5	-0.7	-0.6	-0.4
2	2.7	-0.1	0.2	2.5	-0.3	-0.5
3	0.7	-1.6	-2.1	0.4	-1.5	-0.2
Rural	11.0	-4.5	-1.1	9.1	-5.3	-2.3
Group of republics:						
1	0.0	-1.0	-0.4	0.5	-0.7	-0.1
2	9.4	-1.8	1.1	6.7	-3.0	-0.8
3	1.6	-1.7	-1.8	1.9	-1.6	-1.4

Source: Ksenofontova, 1990.

In the late 1970s the IMR for the USSR began to decrease. This decrease was mainly due to changes in the death probability in the first year of life. Variations in the distribution of the number of births in the republics during this period generally contributed to the increase of the IMR for the USSR. In the republics of the first and third groups the components of infant mortality decreased, thus decreasing IMR for the country as a whole; while in the republics of the second group the components changed in an opposite way (structural component prevented IMR decrease) and thus lessened the influence of the first and third groups of republics on the formation of the IMR for the USSR as a whole.

In the 1980s IMR for the USSR continued to decrease, though slower than in the 1970s. A typical feature of the changes was the growing importance of structural components, which reflect changes in the distribution of the number of births in the republics. The impact of the increase in births of this share of republics with high IMR level was so significant that it exceeded the effect of the death probability decline in this group of republics. As a result, the trends of IMRs in these republics slowed down the decrease of the IMR in the USSR as a whole.

Relative stability of IMR in the USSR from 1981 to 1986 does not reflect, in this case, actual stability of the situation. The situation worsened, on the whole, as compared with 1980. This is indicated in three ways: first, by the instability of IMRs in all republics; second, by some growth of IMRs in certain republics; and third, by the polarization of the republics by the level of IMR.

The decrease of IMRs after their rise in the early 1970s also warrants attention. This lowering could not be compensated so quickly by mortality decline because there were no appreciable improvements in health care.

This point of view is expressed by Borisov (1988), who warns that the subsequent decrease in IMRs could be caused by unreliable statistical data. Such a conclusion calls into question the reliability of the IMR trends of the 1970s. This conclusion allows one to suspect that the problem of the under-registration of infant mortality was not completely solved in the USSR in the 1970s.

20.3   The Problem of Estimating Infant Mortality Levels

The problem of obtaining reliable statistical data on infant mortality has not been solved in the USSR. Both the State Committee of the USSR on Statistics and the USSR Ministry of Health have acknowledged this fact (State Committee of the USSR on Statistics, 1990; Baranov *et al.*, 1990). Unfortunately, several Soviet and non-Soviet demographers have a different point of view: under-registration, one way or another, is connected with the discrepancy in the definitions of *live birth* and *stillbirth* used by the USSR and by the WHO. Under-registration due to this discrepancy is common in all republics, both with adequate registration and low infant mortality level and with inadequate registration and high infant mortality level. But this is only a part of the problem.

Another part is the infant mortality under-registration which is not connected with the divergence of vital registration categories and is the result of inadequate registration, especially in republics with a high level of infant mortality.

The problem of adequate estimation of infant mortality is very important. The problem is not only the absence of uniform criteria that makes IMR comparability impossible, but also the manipulation of statistics to distort the value of IMR.

IMR value depends upon how a death is qualified: whether it was a live birth followed by death or whether it was a stillbirth. Death classification has an influence not only on the computations of rates, which measure the intensity of stillbirths and mortality in the period directly after birth, but also on IMR, as many deaths under age 1 occur in the early period of life. The definitions of live birth and stillbirth used in the USSR have lowered the level of IMR.

Let us consider the correction of absolute numbers of births and deaths of newborns (according to the Soviet criteria of viability) who did not breath but had other signs of life at birth. According to Soviet criteria, this group is counted as stillborn and is excluded from the number of births and the number of deaths before reaching age 1; international regulations count these births as live births and then as deaths under age 1, i.e., it is in both the denominator and the numerator of IMR. The correction of the number of births for this group does not cause any essential changes, but the influence on the number of deaths

Table 20.3. Relative variation of the number of births and number of deaths under age 1 by republic in 1986, in percent.[a]

Republic	Number of births	Number of deaths		
		Under age 1	Under 7 days old	At 1 day old
Russia	0.2	10.1	22.2	202.6
Ukrainia	0.2	13.5	32.8	323.8
Byelorussia	0.2	11.1	24.4	257.8
Uzbekistan	0.2	4.8	27.2	217.5
Kazakhstan	0.2	7.2	24.9	242.0
Georgia	0.2	7.0	22.9	383.9
Azerbaijan	0.2	6.5	38.5	278.4
Lithuania	0.1	12.1	28.7	229.2
Moldavia	0.2	6.6	20.1	160.0
Latvia	0.1	9.5	21.2	355.0
Kirghizia	0.2	4.6	28.0	196.6
Tajikistan	0.3	6.2	42.3	340.7
Armenia	0.2	10.4	36.5	–
Turkmenistan	0.2	4.3	31.9	265.0
Estonia	0.2	11.7	20.3	159.0

[a]Absolute numbers of births and deaths under age 1 are corrected only for infants born not breathing but with other signs of life. Calculated using state statistics.

under age 1 is significant; the number of deaths in the first week of life, especially in the first day, changes significantly in this case (*Table 20.3*).

The question remains, To what extent should IMR be adjusted to eliminate discrepancies in vital registration categories? An answer to this question should be based on the results of a large-scale survey. The available estimates of under-registration are approximate. But if these estimates are similar, they may give an indication of the changes in the IMR levels when the WHO definitions are applied (*Table 20.4*).

However, under-registration connected with the discrepancies in vital registration categories does not cover all under-registration, especially in republics with high infant mortality. That is why methods which estimate the extent of confidence in IMR in general should be used.

Table 20.4. IMR estimates of the USSR adjusted according to the WHO definitions.

Year	Registration data	Estimates 1[a]	2[b]	3[c]
1970	24.7	28.3	30.5	
1971	22.9	26.2	28.3	
1972	24.7	28.3	30.5	
1973	26.4	30.2	32.6	
1974	27.9	31.9	34.5	
1975	30.6	33.6	37.8	
1976	31.4	35.6	38.8	
1977	30.5	34.9	37.7	
1978	29.2	33.4	36.1	
1979	27.4	31.3	33.8	
1980	27.3	31.2	33.7	33.4
1981	26.9	30.8	33.2	33.2
1982	25.7	29.4	31.7	32.0
1983	25.3	28.9	31.2	31.7
1984	25.9	29.6	32.0	32.2
1985	26.0	29.7	32.1	32.2
1986	25.4	29.0	31.4	31.8
1987	25.4	29.0	31.4	31.7
1988	24.7	28.3	30.5	32.6

[a]IMR re-estimated proposed by Davis and Feshbach, 1980; the size of adjustment is 14.4 percent.
[b]IMR re-estimated proposed by Anderson and Silver, 1986; the size of adjustment is 23.5 percent.
[c]Author's estimates.

20.4 IMR Data Reliability

Current approaches of estimating the reliability of IMR levels may be divided into two groups. The first group comprises models which show the interrelation of various demographic characteristics and infant mortality level. Such estimates, as a rule, are based on ad hoc demographic survey data. But these methods cannot be applied fully to Soviet data.

The second group includes estimates based on stable interrelation between certain age-specific rates within the first year of life. This approach uses state statistics and does not require additional surveys. The

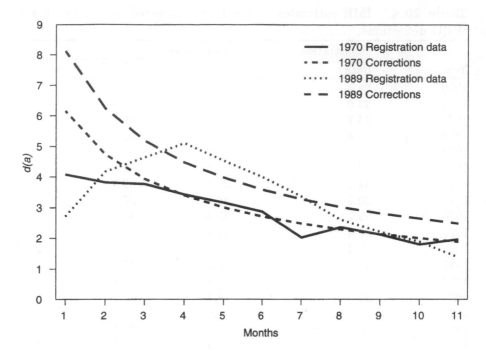

Figure 20.2. Age-specific infant mortality curve in rural areas in Tajikistan.

method is based on well-known statistical regularities, characterizing infant mortality as a demographic process. In this chapter, we use the latter approach.

A peculiarity of infant mortality is a steady decrease of death probability with age. This characteristic of the infant mortality curve was observed in all periods, though the correlation of mortality levels in certain age intervals within the first year of life changes. Death probability within the first year of life decreases with age. Therefore, we assume that divergence from this regularity may be regarded as the result of under-registration. Death probability distribution by months in several republics, especially where IMRs are high, significantly differs from the typical form of the curve; these differences have no explanation. For example, in republics with high infant mortality level, death probability at age 1 to 2 (sometimes 1 to 3) months is lower than at subsequent months and at irregular mortality peaks at 12 months (*Figure 20.2*).

Under-registration of infant deaths varies from age to age. Between the ages of 4 months and 10 months, the change of under-registration is minimal. One reason for under-registration is the fact that the birth had not yet been registered when the child died; the chances of under-registration are greatest during the first 4 months.

The peculiarity of under-registration of deaths at the age of 12 months is the intentional falsification of the age at death by changing the birth or death date so that the age at death is more than 1 year. This decreases the number of deaths under age 1 and increases the number of deaths in the second year.

In such a distribution the most significant under-registration occurs in the first month, especially in the first week. Apparently, the greater part of under-registration consists of cases when the infant died shortly after birth and is classified as stillbirth. That is why, as a rule, this increase is compensated by counting some stillbirths as late fetal deaths, which is allowed by the definitions of live birth and stillbirth existing in the USSR. But it should be emphasized that in republics with a high infant mortality level under-registration in the first month of life is not only due to discrepancies in the vital registration categories.

We tried to correct IMR in two stages: under-registration in neonatal periods and under-registration in postneonatal periods. In the first stage postneonatal mortality rate is estimated and corrected. In the second stage, the estimation and correction of neonatal mortality is made.

As the number of under-registration cases in the postneonatal period is 10 times lower than in the neonatal period we attempted to smooth the infant mortality curve by means of the biometric equation proposed by Bourgeois-Pichat (1951). The analysis shows that the computations done using this equation conform well with the empirical infant mortality curve for ages between 1 and 11 months in regions with an adequate death registration system. In case of supposed under-registration, we calculated the parameters of the formula on the basis of mortality data for the ages between 3 and 10 months, in which the extent of under-registration is minimal. In cases when, according to registration data, death probability at 1 to 2 (sometimes 1 to 3) months is essentially lower than at the subsequent months, rates are computed by extrapolation after the equation parameters were found.

The Bourgeois-Pichat formula allows one to smooth the infant mortality curve and thus to correct an artificial decrease of rates at age 11 months and an increase at age 12 months. This formula makes it possible to consider more accurately both regional peculiarities of under-registration and the form of the infant mortality curve. But it would be wrong to use the Bourgeois-Pichat formula to estimate mortality under the age of 1 month.

It is difficult to estimate the reliability of infant mortality data for the first month of life. Considering the stable interrelation between mortality levels, we estimated the mortality level during the first month on the basis of mortality data for the ages between 1 and 11 months by means of the regression equation system. Two data files formed the basis for this estimate:

- The distribution of deaths under age 1 by the periods of the first year of life for the countries with reliable mortality statistics from 1935 to 1955, the years with relatively high infant mortality levels (80 cases).
- The distribution of deaths at age 1 in countries providing data to the WHO from 1965 to 1991 (496 cases).

Two data files were used for two reasons. First, infant mortality level in the countries with reliable population statistics are, at present, lower than in the majority of the republics. Therefore, we cannot use only recent data to calculate regression equations. Nevertheless, to estimate the rates in republics with lower mortality levels we have to use recent data.

Second, the definitions of live birth and stillbirth used in the USSR are different from the definitions used in most developed countries. The conversion to the new definitions began in the 1950s and took about 30 years. This second type of data allows, to a certain extent, one to measure the influence of the process of the conversion to a new definition of live birth on the infant mortality level.

The regression equations obtained from both files appeared to be rather similar. For the data file for the period from 1935 to 1955, because of its lack of data, we used an ordinary linear regression. Working with the second file of data we considered early and late neonatal mortality separately and analyzed the connection between death rates at ages 0 to 6 days and 7 to 27 days, as well as between 7 to 27 days and over 28 days.

The selection of the logarithmic equation of relationship for ages from 0 to 6 days and from 7 to 27 days is based on graphic analysis.

The final equation of the relationship between the mortality level in the first and subsequent months of the first year is the following:

$$d'(0m.) = 0.348d(1 - 11m.) + \ln[0.247d(1 - 11m.) + 0.947] + 1.075$$

where $d'(0m.)$ is the estimate of the death probability in the first month of life and $d(1 - 11m.)$ is the probability of a newborn dying between 1 and 11 months.

Separately, we analyzed the possibility of using these equations to estimate mortality data for the first month of life in the USSR, where the definitions of vital events are different from those recommended by the WHO. In addition, in the data file for the period from 1965 to 1987, we selected 52 points corresponding to countries or periods with the old definition of live birth. The results differed in the calculations, but insignificantly. Rather unexpectedly, the calculations corresponding to the old definition give higher estimates of infant mortality.

Because of the possibility of random variation and regional peculiarities which could have an impact on the corrections of infant mortality levels during the first year of life, we did not limit ourselves to average correlations from the regression model. The lowest possible values, analogous to confidence intervals in statistics, were also estimated but calculated on the basis of empirical data. Thus we obtained the lowest possible value of each indicator with a certain probability.

If the analysis of the reliability of data on postneonatal mortality does not show any under-registration, then, in further computations, mortality data for ages from 1 to 11 months are used. If under-registration at these ages is observed, then the analysis of reliability of neonatal mortality is made using corrected postneonatal mortality rates obtained in the first stage. If mortality data for the first month of life are reliable, then reliability of late neonatal and then early neonatal mortality must be estimated.

Using these methods, we analyzed the reliability of infant mortality data for all republics during the late 1980s. The analysis shows that it is necessary to correct mortality rates at age 1 to age 11 months only in rural areas of Central Asia and Azerbaijan, in both urban and rural areas. In other regions postneonatal mortality rates may be considered reliable within the assumptions adopted.

The analysis of correlation of mortality levels in the first and last months of the first year of life shows that in urban areas of all the republics (excluding Azerbaijan) and in rural areas of Russia, the Baltic

Table 20.5. Trends in IMR in Central Asian republics and Azerbaijan.

	Urban population		Rural population	
Year	Registered	Corrected	Registered	Corrected
Uzbekistan				
1985	38.50	45.65	48.70	88.99
1986	40.60	51.53	48.80	77.83
1987	39.41	51.96	48.84	78.84
1988	37.63	48.82	40.92	69.40
1989	34.10	40.62	39.36	60.76
Tajikistan				
1985	45.40	61.24	47.30	91.13
1986	43.40	53.44	47.80	87.91
1987	45.67	55.13	50.13	92.45
1988	45.05	52.72	50.27	90.40
1989	39.41	44.77	44.40	78.27
Azerbaijan				
1985	25.60	33.14	33.00	67.55
1986	25.10	34.58	35.60	74.32
1987	23.32	32.32	33.80	69.26
1988	23.60	33.62	28.81	61.95
1989	23.30	31.35	29.00	60.54
Turkmenistan				
1985	49.00	73.39	54.70	105.89
1986	56.50	72.72	59.30	102.28
1987	56.24	65.63	56.26	99.49
1988	52.78	59.21	53.96	96.74
1989	54.17	59.88	54.95	86.95
Kirghizia				
1985	30.00	40.71	47.20	82.49
1986	28.60	37.94	42.40	75.32
1987	31.28	40.74	40.95	74.33
1988	30.39	41.39	39.77	69.56
1989	28.32	30.32	33.94	53.65

Calculated using state statistics.

republics, and Moldavia the registered level of neonatal mortality is reliable. In the other regions this indicator must be corrected. We estimated infant mortality levels only in those republics where the value of correction was the largest (*Table 20.5*).

The IMR estimate for the USSR as a whole was obtained by reweighing the corrected IMRs for the republics by the number of births. The corrected estimate was 30.8 per 1,000 live births in 1989, whereas according to registration data the estimate was only 22.7 per 1,000. The actual infant mortality level, on average, is 25 percent higher than the mortality level according to vital statistics.

In the Central Asian republics and Azerbaijan, where the problem of reliability of infant mortality data is most acute, in rural areas 40 percent to 50 percent of deaths under 1 year of age are not registered (*Table 20.5*). But it should be noted that the corrected IMRs, as well as the percentage of under-registration, tend to decrease.

Let us analyze the estimates of the infant mortality level for the USSR. These estimates refer to different periods of time and are obtained by different methods, making their comparison difficult. But it is possible for us to compare not only levels but also relative under-registration (*Figure 20.3*). The estimate for the 1950s supposed that under-registration of infant deaths in the USSR was not less than 25 percent (Andreev *et al.*, 1990).

The corrected IMRs in the USSR for the period from 1958 to 1974, presented by Anderson and Silver (1986), look quite plausible, and the final estimates of Andreev *et al.* (1990) confirm the initial estimates of Anderson and Silver. Nevertheless, one must be very aware of the increase of relative under-registration, which is presented as deterioration of the quality of vital registration. The registration situation was getting better rather than worse.

More interesting results are given in the comparison between the estimates made by the Soviet Ministry of Health experts (Baranov *et al.*, 1990) for 1970 and 1986 and our estimates, calculated using similar methods. Infant mortality level estimates for the late 1980s are relatively similar; the estimate made by Baranov *et al.* (1990) is 35.8 per 1,000 live births and our estimate for 1989 is 30.8 per 1,000. But the accuracy of the estimate of the infant mortality level in the USSR for 1970 made by Baranov *et al.* seems doubtful.

First, one can hardly suppose that in 1970 half of infant deaths in the USSR were not registered. Such a statement would be correct for republics with high infant mortality level and inadequate registration.

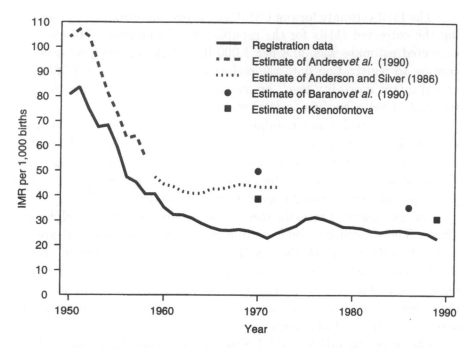

Figure 20.3. Trends in registered and corrected IMRs in the USSR.

But the estimate would mean the same under-registration in the majority of republics in urban and rural areas and in Russia, which has a significant impact on infant mortality level, and that is not plausible.

Second, the discrepancy between the changes in adjusted estimates and vital statistics is evident, especially in the regions with high infant mortality. So, a more detailed analysis of the methods used was made.

In this case the methods proposed by Dellaportas (1965) were used by Baranov *et al.* (1990), and they were based on the hypothesis of linear dependency between the levels of infant mortality up to 6 months and 6 to 11 months. The IMR estimated by these methods consists of two parts: registered rate for ages between 6 and 11 months and the set of estimates for each age interval under 6 months, determined by linear regression equations.

But with a considerable rise of the independent variable (mortality between the ages of 6 and 11 months) in Central Asia, 27 percent for the period 1970–1986, the estimated part of IMR decreased by 40 percent.

That would be possible only if under-registration in 1970 and 1986 was determined by different models.

As the data used to calculate the regression equations were not specified and the equations themselves were not presented, we obtained the values of the parameters of regressions equations using the inverse calculations method (Ksenofontova and Andreev, 1990). Thus, it appeared that the same model was used for calculating the 1986 estimates in all regions because all the points for 1986 lie approximately in a straight line. For 1970 estimates a model was used that was different from the model used for the 1986 estimates, and the calculations for the regions with high infant mortality level were estimated using another model. Such manipulations give the impression that researchers used two different models that slowed the decrease of the rates.

We estimate infant mortality level for the USSR in 1970 to be 38.8 per 1,000 live births; this estimate is essentially lower than the estimate made by Baranov *et al.* (1990) and slightly lower than the estimates of Anderson and Silver (1986). We do not presume that our estimate for 1970 is more correct than the estimates of others. Our methods of estimating IMR in the USSR appear to be less effective, especially in republics with high infant mortality level.

It might be expected that the corrected 1970 IMR for the republics with a high infant mortality level would be higher than or approximately equal to the level of 1989 (*Tables 20.5* and *20.6*), but the estimates obtained were lower. It might be explained as follows. There was spontaneous under-registration in these republics, especially in rural areas and in all periods from the first month to the end of the first year of life; thus, the infant mortality curve in the postneonatal periods was not significantly distorted (*Figure 20.2*). That is why we could not consider the under-registration in the postneonatal period fully, and the neonatal mortality rate was underestimated. In our opinion it decreased the size of the IMR correction rather considerably in the republics with high infant mortality level. But even if the 1970 IMR in these republics were at the 1989 level, then the rate for the USSR would not be more than 40 per 1,000 live births.

A turning point in the improvement of vital events registration system took place in the mid-1970s. In regions with a high infant mortality level and inadequate registration, the spontaneous form of

Table 20.6. Registered and corrected IMR in 1970.

Republic	Urban population		Rural population	
	Registered	Corrected	Registered	Corrected
Russia	21.3	28.3	24.9	38.4
Ukrainia	18.4	29.8	17.0	33.4
Byelorussia	19.0	27.3	18.8	33.5
Uzbekistan	38.8	50.7	28.4	50.8
Kazakhstan	28.0	39.6	24.6	46.9
Georgia	24.7	36.4	28.7	52.4
Azerbaijan	30.7	48.0	38.4	68.0
Lithuania	16.8	26.1	23.4	37.5
Moldavia	23.2	38.3	26.5	45.3
Latvia	16.3	20.7	20.4	27.8
Kirghizia	37.9	55.1	50.0	84.3
Tajikistan	48.8	57.0	40.1	62.5
Armenia	25.7	41.4	23.1	45.3
Turkmenistan	49.6	71.2	46.1	75.7
Estonia	16.9	20.0	18.3	23.2

Calculated using state statistics.

under-registration was overcome by the mid-1970s. We consider this the time of deliberate under-registration.

The existing system of vital events registration is not free of shortcomings but it could provide sufficient full registration of vital events. The infant mortality level is regarded as an indicator of the quality of health-care services; for this reason the health-care system is not interested in the accuracy of IMR. Increases in infant mortality because of the worsening economic, political, and ecological situations in the country and in specific regions with high infant mortality may make medical institutions conceal the under-registration caused by the inadequacy of the registration system.

Our own experience in revising the infant mortality registration shows that the medical staff, as a rule, know about practically all cases of deaths under age 1, even those that are not registered. If the attitude toward registration could be changed and if the definitions of the WHO were introduced, the problem of infant mortality under-registration would be solved rather quickly.

References

Anderson, B., and B. Silver, 1986. Infant Mortality in the Soviet Union. *Population and Development Review* 12:705–738.

Andreev, E.M., L.E. Darsky, and T.L. Kharkova, 1990. Istoriya naseleniya SSSR 1920–1959 gg (The History of the USSR Population, 1920–1959). In *Ekspress-informatsiya, Seriya "Istoriya statistiki."* (Express Information, History of Statistics Series) Issue 3–5 (Part I). Goskomstat SSSR, Moscow.

Baranov, A.A., V.Yu. Al'bitsky, and Yu.M. Komarov, 1990. Tendentsii mladencheskoy smertnosti v SSSR v 70–80e gody (The Tendencies in Infant Mortality in the USSR in the 1970s and 1980s). *Sovetskoye zdravookhraneniye* (Soviet Public Health) 3:3–10.

Borisov, V.A., 1988. Prodolzhitel'nost' zhizni v SSSR. In T.D. Ivanova, ed. *Naseleniye i obschestvennoye razvitiye* (Population and Social Development). Institut Sotsiologii AN SSSR, Moscow.

Bourgeois-Pichat, J., 1951. La mesure de la mortalité infantile. Principes et méthodes. *Population* 2:233–248.

Davis, C., and M. Feshbach, 1980. *Rising Infant Mortality in the USSR in the 1970s.* International Population Report, Series P-95, No. 74. US Bureau of the Census, Washington, DC.

Dellaportas, G.J., 1965. Correlation-based Estimation of Early Infant Mortality. *Public Health Report* 87(1–3):275–281.

Dmitriyeva, R.M., and E.M. Andreev, 1987. O sredney prodolzhitel'nosti zhizni naseleniya SSSR (On the Life Expectancy of the USSR Population). *Vestnik statistiki* (Herald of Statistics) 12:31–39.

Ksenofontova, N.Y., 1990. Nekotorie tendentsii mladencheskoi smertnosti v posledneye desyatiletie (Some Tendencies of Infant Mortality during the Last Decade). In A.G. Volkov, ed. *Demograficheskiye protsessy v SSSR* (Demographic Processes in the USSR). Nauka, Moscow.

Ksenofontova, N.Y., and E.M. Andreev, 1990. *Ob otsenkakh dinamiki mladencheskoy smertnosti v SSSR* (About the Estimates of Infant Mortality Dynamics in the USSR). Vsesoyuznaya nauchnaya konferentsiya "Naseleniye i sotsialnoye razvitiye" (The All Union Scientific Conference "Population and Social Development"), Moscow.

Maksudov, S., 1985. *Vyzov so storony demografii* (The Demographers' Challenge). Chalidze Publication, Benson, VT.

Rybakovskiy, L.L., ed., 1988. *Naseleniye SSSR za 70 let* (Population of the USSR over the Past 70 Years). Nauka, Moscow.

State Committee of the USSR on Statistics (Goskomstat), 1988. *Naseleniye SSSR, 1987: Statisticheskiy sbornik* (Population of the USSR, 1987: Statistical Collection). Finansy i statistika, Moscow.

State Committee of the USSR on Statistics (Goskomstat), 1990. Novaya informatsiya Goskomstata SSSR (New Information of the State Committee of the USSR on Statistics). *Vestnik statistiki* (Herald of Statistics) 4:49–64.

Szymansky, A., 1981. On the Uses of Disinformation to Legitimize the Revival of the Cold War: Health in the USSR. *Science Society* 4:453–474.

Chapter 21

Spatial Differences in Life Expectancy in European Russia in the 1980s

Vladimir M. Shkolnikov and Sergei A. Vassin

In the early 1980s, the trend toward a decrease in life expectancy at birth
(e_0), prevalent in the USSR over a long period of time, was replaced
by a slight increase. From 1985 to 1987 there was a relatively strong
rise. In 1988, life expectancy again registered a slight decrease. Upon
publication of demographic data on the USSR after 1985, the situation
was frequently discussed in research (see Chapter 18; Dmitriyeva and
Andreev, 1987; Sinel'nikov, 1988; Blum and Pressat, 1987; Blum and
Monnier, 1989). But little is known about the situation in individual
republics and their provinces.

However, the observed change of life expectancy of the USSR population is in many respects determined by mortality changes in Russia, the
largest and most populous republic. The republic comprises dozens of
ethnic groups, and the 73 provinces account for about half of all territorial units in the country. Russia is heterogeneous not only geographically
and socioeconomically, but also demographically. It would be logical to
expect that Russia's regions have different levels of life expectancy.

This chapter describes and analyzes the changes in life expectancy
and its spatial variation for urban and rural populations of Russia in

1979 and 1988. It also examines the impact on regional differences in the decrease in mortality in the 1980s.

Taking into account underestimation of the statistics on mortality of children and the elderly in the USSR, one could be skeptical of using e_0 as the basis for analysis as well as the study of urban and rural mortality statistics. Nevertheless this topic should be investigated for three reasons. First, the study of regional mortality differentiation has not yet been discussed in scientific publications, and it seems reasonable to assume that the general idea can be conveyed using an integral measure of mortality such as e_0. Second, in contrast to most developed countries, the differences in the life-style, social status, and mortality of urban and rural populations in Russia are great, and they should not be ignored. Third, vital registration in the rural areas of European Russia was somewhat improved in the 1980s.

21.1   Data and Methods of Calculation

All measures were calculated on the basis of statistical reports from the Russian State Committee on Statistics on the number of deaths by sex, age, and (for Russia as a whole) causes of death, together with data on population by sex and age, from the 1979 and 1989 censuses.

The abridged life tables were constructed by Chiang's method (Chiang, 1984), with the use of Keyfitz's iteration procedure for the evaluation of the average fraction of the last age interval of life, $_na_x$ (Keyfitz, 1977). To calculate age-specific death rates (ASDRs), the age structures were recalculated for mid-1979 and mid-1988.

The age-component analysis of e_0 change was made using the technique developed by Andreev (1982). The principle of other versions of this method is the same (Pollard, 1982; Valkovics, 1984). The age components of e_0 change were estimated by

$$e_0 - e_0' = \sum_0^{80} (_nc_x) = \sum_0^{80} [l_x(e_x - e_x') - l_{x+n}(e_{x+n} - e_{x+n}')] \ , \quad (21.1)$$

where $_nc_x$ are age components of life-expectancy change, l_x is the number of survivors at age x, e_x is life expectancy at age x, and l_x' and e_x' correspond to the reference life table; for ages over 85 the difference of

e_{85} was used. The age components of e_0 change by causes of death were estimated by

$$e_0 - e_0' = \sum_i \sum_{x=0}^{85+} \frac{\left(_n m_x^i -_n m_x^{i'}\right)}{\left(_n m_x -_n m_x'\right)}_n c_x \tag{21.2}$$

where $_n m_x$ and $_n m_x^i$ are the age-specific mortality rates from all causes of death and from i-th cause of death, respectively.

21.2 Changes in Life Expectancy and Its Components in the 1980s

In the 1980s the profiles of changes in life expectancy in Russia and in the USSR were very similar. The difference was minor: e_0 in Russia increased faster and reached the highest level in 1986–1987. As a result the lag of the life expectancy of the Russian male from the national level was reduced from 0.7 years in 1979–1980 to 0 in 1988 and excess of life expectancy for Russian females over the USSR level increased from 0.5 to 0.8 years.

The similarity between the Russian level and the national level could also be seen in the ratio of urban–rural mortality. *Table 21.1* shows that rural life expectancy (especially for males) was lower than urban life expectancy. There was almost no urban–rural difference in the e_0 increase between 1979 and 1988. For urban males, the growth was 3.25 years; for rural males, 3.46. For females it was 1.50 and 1.21, respectively.

Nonetheless, decomposition of the life-expectancy increase by age and by leading causes of death indicates existence of some structural differences between these subpopulations. Compared with the urban decline, the rural mortality decline in ages above 60 was very modest. The decrease of urban mortality from circulatory and respiratory diseases played equally important roles in the life-expectancy increase between 1979 and 1988, while in the rural areas the impact of circulatory diseases was two times smaller for men and seven times smaller for women than the influence of respiratory diseases.

But in general the urban–rural differences in mortality are small as compared with differences by sex. In 1979, female life expectancy was higher than male life expectancy by 10.8 years in urban areas and 13.1 years in rural areas. By 1988, this gap was reduced by 1.7 years and 2.2 years, respectively. The difference in life expectancy between sexes

Table 21.1. The increment of life expectancy in Russia in 1979–1988 and its components by leading causes of death and by age.

| | Urban | | Rural | |
	Male	Female	Male	Female
e_0 in 1979	62.17	73.03	59.40	72.53
e_0 in 1988	65.42	74.53	62.86	73.74
e_0 increment	3.25	1.50	3.46	1.21
Cause of death				
Infections	0.24	0.15	0.22	0.09
Neoplasms	0.01	0.09	−0.29	−0.11
Circulatory	0.74	0.56	0.48	0.10
Respiratory	0.73	0.50	0.97	0.77
Injury/poisoning	1.60	0.37	2.19	0.54
Residual	−0.07	−0.17	−0.11	−0.18
Age				
0–14	0.37	0.33	0.43	0.43
15–29	0.43	0.05	0.69	0.11
30–44	1.07	0.28	1.47	0.47
45–59	0.72	0.36	0.59	0.23
60–74	0.44	0.21	0.22	−0.08
75+	0.22	0.27	0.06	0.05

in the USSR is larger than the difference often observed in developed countries (five to seven years).

The analysis of the changes in expectation of life by leading causes of death (*Table 21.1*) demonstrates that for both men and women a positive contribution was made by the decline of mortality from all causes of death except neoplasms and the group of residual causes of death.

Practically in every age and for each cause of death the increase in e_0 was higher for men than for women. For men, the maximum increase in life expectancy at birth was connected with the drop in mortality from injury and poisoning. This cause of death accounts for about half of the overall increment of the urban male life expectancy and for nearly 63 percent of the rural male life expectancy. The overall contribution in the fall of mortality due to circulatory and respiratory diseases was 1.47 in cities and 1.45 in rural areas. For females, the main component of the life-expectancy increment corresponded to the reduction of mortality from circulatory (in urban areas) and respiratory diseases (in rural areas).

Distinct differences by sex also exist in the age composition of the increase of e_0. Due to the decline of mortality in the working ages (15 to 59), male life expectancy increased by 2.2 years in urban areas and by 2.75 years in rural areas. Actually, the mortality decline in that age range provided almost 70 percent of the reduction of the overall gap in the male and female life expectancy. Thus, it was the decrease of injury and poisoning mortality and mortality in the working ages that was the source of the sharp rise of the male e_0 and the reduction of sex differentiation in mortality in Russia in the 1980s.

A sharp rise of the life expectancy took place in the short time interval between 1984–1985 and 1987. Life expectancy increased by 2.7 years for male and by 1.3 years for female. According to several studies, this sudden fall in mortality was facilitated by the anti-alcohol campaign launched by the Soviet government in May 1985 (see Chapter 18; Dmitriyeva and Andreev, 1987; Blum and Monnier, 1989). The dynamics of the monthly mortality index from January 1979 to December 1988 (*Figure 21.1*) conforms to this argument (index is adjusted to the number of days in each month and the season).

Figure 21.1 depicts clearly a sharp fall in male mortality beginning in May–June 1985, immediately after the launching of the anti-alcohol campaign. Between the second half of 1985 and November 1987, the mortality index was about 10 percent lower than the average 10-year level taken as 100. In late 1987 and 1988, the index of male mortality increased somewhat, but did not reach the previous level. For females the fall of the monthly index in late 1985 and 1986 was not that great, and soon (approximately in the last quarter of 1986) the previous mortality level was re-established.

So, in 1985–1986 not only ASDRs declined but also the monthly number of deaths fell. This does not support the hypothesis that in these years the expectation of life rose due to the changes in the methods used for constructing the mortality rates (see Chapter 18). However, it is not sensible to conclude that the anti-alcohol campaign was the only factor contributing to the increase in life expectancy in this decade. This is confirmed by the earlier emergence of the trend studied, which may be dated as far back as 1981, and the declining infant mortality, which is hardly directly connected with alcohol consumption.

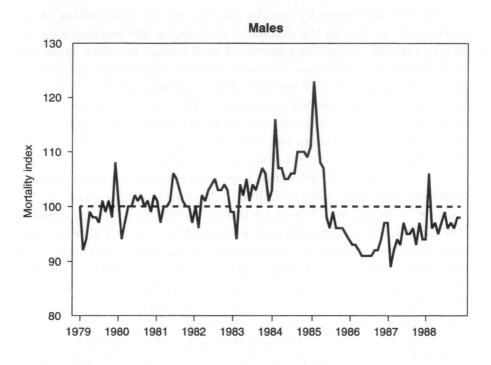

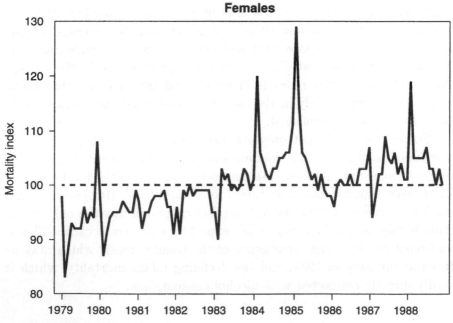

Figure 21.1. Monthly index of mortality in Russia from January 1979 to December 1988 (expected level = 100).

21.3 Regional Variation of Life Expectancy in European Russia in 1979 and 1988

This study considers only the European part of Russia. This region consists of 53 territorial units where about 80 percent of the population lives.[1] A list of the 53 territories is given in *Figure 21.2* with a map showing their boundaries. From north to south, European Russia covers the territories from the Barents Sea to the Black Sea and Caspian Sea. In the west it borders with Estonia (the northwest), Byelorussia, and the Ukraine; in the east it borders with Siberia; in the south it borders with Georgia and Azerbaijan; and in the southeast it borders with Kazakhstan.

21.3.1 Patterns of regional variation of life expectancy

The territorial variation of life expectancy at birth in European Russia in the late 1980s was considerable, despite the reduction in regional differences as compared with 1979. For urban males, the standard deviation of life expectancy was 1.0 year in 1988 and 1.5 years in 1979; for urban females it was 0.5 years and 1.2 years, respectively; for rural males, 2.0 years and 2.8 years, respectively; for rural females, 1.2 years and 2.0 years, respectively. The differences between the territories with the maximum and minimum values of e_0 were especially great: nine and ten years for men and five and seven years for women. Regional variation were greater for the rural population than for the urban population, and they were greater for men than for women.

Which age groups account for the greatest contribution to this variation? The deviations of life expectancy from the average level of e_0 (i.e., e_0 of corresponding subpopulation of the Russia as a whole) in all territories were divided into two groups: in the first group expectation of life is higher than that of the Russia as a whole and in the second group it is lower (*Table 21.2*). The totals of the age components of e_0 for these two groups can be used to find the ages that play the most significant role in classifying the territories into one of the two groups and to assess the age components of the life-expectancy variation.

Table 21.2 shows that for urban males the differences in the e_0 level are determined by mortality in the ages 45 to 74; for their rural counterparts they are determined in the ages 30 to 74, among which the ages 45 to 59 are particularly significant. The distribution of age

Northern Region
1 Arkhangelsk
2 Vologda
3 Murmansk
4 Karelia ASSR
5 Komi ASSR

Northwestern Region
6 Leningrad
7 Novgorod
8 Pskov

Central Region
9 Bryansk
10 Vladimir
11 Ivanovo
12 Tver
13 Kaluga
14 Kostroma
15 Moscow
16 Orel
17 Ryazan
18 Smolensk
19 Tula
20 Yaroslavl

Volga-Yatska Region
21 Gorky
22 Kirov
23 Mari ASSR
24 Mordov ASSR
25 Chuvash ASSR

Central Blackearth
26 Byelgorod
27 Voronezh
28 Kursk
29 Lipeck
30 Tambov

Volga Region
31 Astrachan
32 Volgograd
33 Kuybyshev
34 Penzen
35 Saratov
36 Ulyanovo
37 Kalmyc ASSR
38 Tatar ASSR

North Caucasus Region
39 Krasnodar kray
40 Stavropol kray
41 Rostov
42 Dagestan ASSR
43 Kabardino-Balkar ASSR
44 North Osetin ASSR
45 Checheno-Ingush ASSR

Ural Region
46 Kurgan
47 Orenburg
48 Perm
49 Sverdlovsk
50 Chelyabinsk
51 Bashkir ASSR
52 Udmurt ASSR

Baltic Region
53 Kaliningrad

Figure 21.2. Map of 53 administrative areas of European Russia.

Table 21.2. Age-specific components of e_0 in 53 provinces of European Russia and Russia as a whole in 1988.

Age group	Urban Males Years	%	Females Years	%	Rural Males Years	%	Females Years	%
Areas with life expectancy higher than in Russia								
0–4	0.99	4.3	3.58	13.8	4.22	12.6	5.21	18.0
5–14	0.02	0.1	0.40	1.6	1.09	3.2	1.03	3.6
15–29	3.92	17.1	0.97	3.8	4.02	12.0	0.54	1.9
30–44	3.26	14.1	2.36	9.1	6.12	18.3	2.70	9.3
45–59	6.81	29.5	2.82	10.9	8.44	25.2	5.66	19.6
60–74	6.63	28.7	8.65	33.5	5.97	17.8	5.11	17.7
75+	1.44	6.2	7.06	27.3	3.65	10.9	8.65	29.9
Total	23.07	100.0	25.84	100.0	33.51	100.0	28.90	100.0
Areas with life expectancy lower than in Russia								
0–4	−0.30	1.7	−3.32	35.2	2.21	−5.5	1.05	−4.9
5–14	0.26	−1.4	−0.15	1.6	−1.16	2.9	−1.57	7.4
15–29	−1.57	8.7	−0.42	4.5	−5.40	13.4	−0.69	3.2
30–44	−2.83	15.6	−0.45	4.8	−8.82	21.9	−1.88	8.9
45–49	−5.80	32.0	−1.23	13.0	−15.15	37.5	−4.85	22.8
60–74	−5.46	30.1	−1.88	19.9	−8.14	20.2	−5.67	26.6
75+	−2.41	13.3	−1.98	21.0	−3.87	9.6	−7.67	36.0
Total	−18.11	100.0	−9.43	100.0	−40.33	100.0	−21.28	100.0

components of variation for men is almost identical in the groups with a high and low life expectancy.

A key role in the female mortality differentiation belongs to the ages above 60 in the urban areas and above 45 in the rural areas, especially in the ages above 75.

The situation of child mortality (0 to 4) requires an additional explanation. In the group with a high level of e_0, child mortality accounts for a great part of all deviations (from 12.6 percent for rural males to 18.0 percent for rural females). But as a factor of the differentiation of territories by groups with a relatively high and low life expectancy, child mortality has almost no influence. In the territories with negative deviations from the average, the contribution of ages 0 to 4 is either positive (rural population) or close to zero (urban males). Hence, relatively low child mortality does not prevent a region from having exceptionally low life expectancy whereas high mortality in that age group does not inhibit relatively high life expectancy at birth.

The mortality of urban females is an exception. Their mortality is manifested not only by the fact that child mortality is the most important determinant of life expectancy for urban females, but also by the fact that the sum of negative deviations in all ages except 0 to 4 in the second group of territories is relatively small. The latter circumstance is evidence of the extreme heterogeneity of that group and of the coexistence of different mortality patterns in this group.

21.3.2   Geography and expectation of life

Cartograms of male and female life expectancy and urban and rural life expectancy are given in *Figures 21.3* to *21.6*. To construct the cartograms a uniform scale with five divisions was applied. The width of each division is $2s/5$, where s is the standard deviation of e_0 for 53 provinces. As a result the width of divisions for each map is different.

The standard errors of the life-expectancy estimations were calculated by using Chiang's method (Chiang, 1984). For all 53 territories the deviations in life expectancy from unweighted average, which are shown by the shadings on the maps, are statistically significant ($p < 0.05$).[2]

Figures 21.3 to *21.6* show that the maps for 1979 follow a common pattern. All cartograms depict, as the main feature of the territorial differentiation, the confrontation of the location of relatively low life expectancy in the northeast with the location of relatively high life expectancy in the southwest. The territories to the north and east of the Moscow province (15) were characterized mainly by a lower or a much lower life expectancy. More favorable locations had the shape of a T tilted to the right stretched from the Smolensk (18) and Bryansk (9) provinces in the west to Bashkir (51) and Chelyabinsk province (50) in the east (stem of the T) and from Moscow province (15) in the north to the North Caucasian region in the southwest. This is the overall picture of life expectancy in 1979. But it is altered from one subpopulation to another mainly due to the change of the boundaries of the T shape with a relatively high life expectancy (particularly due to changes of the areas located in the southwest).

An exception to this pattern is evident in the map for rural male life expectancy (*Figure 21.3*). The dark shaping covers not only the entire north and east, but also many provinces in the central region and about all of the Volga-Vyatka region. This map has the most shading in the entire series.

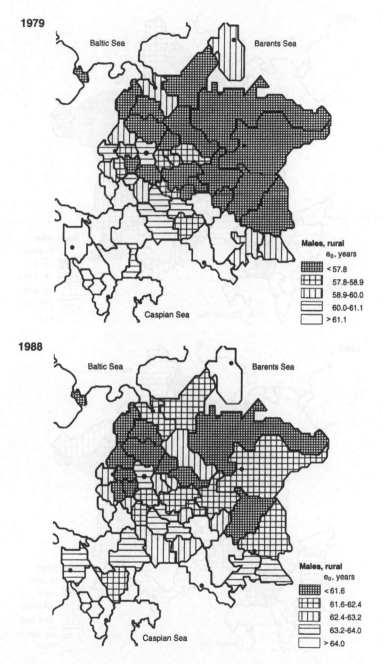

Figure 21.3. Life expectancy at birth of rural males by territories of European Russia in 1979 and 1988.

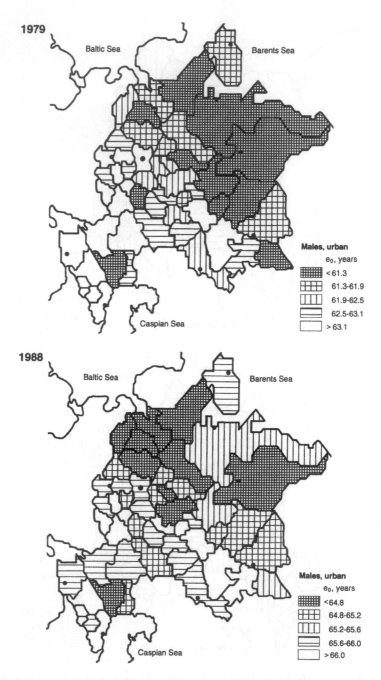

Figure 21.4. Life expectancy at birth of urban males by territories of European Russia in 1979 and 1988.

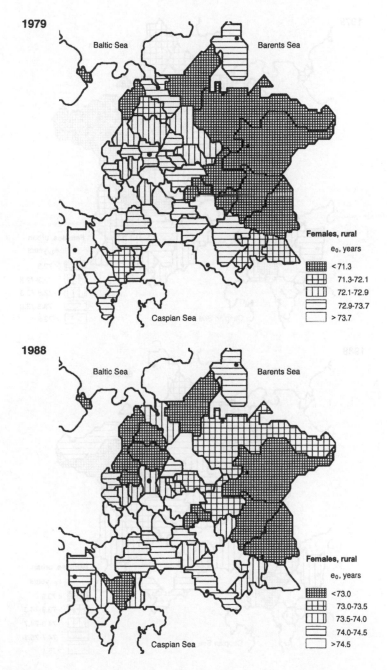

Figure 21.5. Life expectancy at birth of rural females by territories of European Russia in 1979 and 1988.

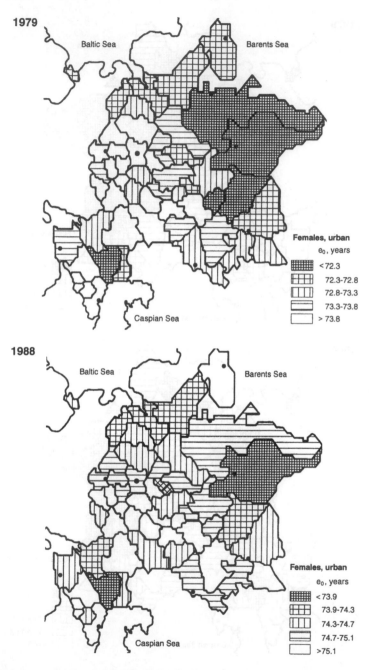

Figure 21.6. Life expectancy at birth of urban females by territories of European Russia in 1979 and 1988.

For urban males (*Figure 21.4*), the boundary of the favorable location is farther north, covering part of the provinces in the central region (Bryansk, Kaluga, Orel, and Moscow provinces).

The shading for rural females (*Figure 21.5*) is similar to the shading for urban males. However, there are several territories with moderately high e_0 level (central region, Volga-Vyatka region, and Vologda). These areas are situated between the location with high and low mortality. As a result, the cluster of territories with the lowest life expectancy lies more to the east than to the north.

The T shape stretches the farthest north in the urban female map (*Figure 21.6*). In fact, the entire western part of European Russia is lightly shaded, which is evidence of the emergence of a new cluster of low mortality comprising eight provinces of the central region. In general, the maps for female life expectancy have much lighter shadings than the maps for male life expectancy.

In 1988, the e_0 increased. The size of the increase was the biggest in the group with the lowest level of life expectancy and the smallest in that with the highest level of e_0. The interval between the divisions narrowed. The geographical picture of life expectancy became more complex, with the main changes observed in the cluster of territories with the highest mortality in 1979.

Thus, in 1988, for the rural males of European Russia (*Figure 21.3*), the cluster of high mortality shrank and moved to the west, while the locations of the high life expectancy remained the same. A similar situation is observed for their urban counterparts. But here a new western cluster of higher mortality is more distinct. It stretches from the northwest to the center.

For rural females two clusters of the lowest life expectancy are concentrated in the northwest and east. The high mortality cluster practically disappeared from the map of their urban counterparts; it now comprises only three territories: Komi ASSR, Ivanovo, and Kalmyc ASSR.

Compared to 1979, the picture of regional differentiation of e_0 in European Russia in 1988 became less distinct and comprehensible. The maps, in general, show a tendency of darkening in the southern provinces and the central region of European Russia.

21.3.3   Factors of the spatial differences in life expectancy

The most powerful factor of the territorial differences in male mortality in 1979 was the worsening of climatic conditions and the general low level of socioeconomic development that moved north and west in Russia (Shkolnikov, 1987). The second significant factor, particularly in rural areas, was the high level of alcohol consumption. The differentiation of accuracy of mortality statistics may be considered a third factor of spatial variation of life expectancy in Russia.

What was the role of these factors in the spatial mortality variation in European Russia in the 1980s? Despite the changes in mortality geography in the 1980s, the relationship between the expectation of life and geographical zones was preserved. So, in 1988, the contingency criterion (X^2) and Cramer's contingency coefficient in the 9-by-5 table contingency[3] were 59.6 and 0.54, respectively, for urban males; 57.9 and 0.53 for urban females; 54.9 and 0.52 for rural males; and 57.9 and 0.53 for rural females. In the urban and rural male and rural female cases, there is a statistically significant relationship $(p < 0.01)$. But the matter is not so much the climatic conditions that either abate or increase the impact of unfavorable conditions such as the socioeconomic development. In this case, the social environment and life-style do not compensate for harsh climate.

The widespread abuse of alcohol may be regarded as an indicator of social discomfort. The anti-alcohol campaign led to an increase in life expectancy in Russia, particularly among men between ages 15 and 59 in rural areas. The effect of this action might be reflected in the regional peculiarities of the life-expectancy changes and the age components in European Russia in the 1980s, especially in working-age males in regions where much alcohol is consumed.

During this period the highest increment of e_0 due to the decrease of mortality in working ages was registered in the Arkhangelsk, Murmansk, Kirov, Kurgan, Perm, and Sverdlovsk provinces and Komi, Mari, Chuvash, and Udmurt ASSR. The increment of e_0 due to the decline of mortality in the ages between 15 and 59 fluctuated between 2.5 and 4.0 years for urban males, 3.4 and 4.5 years for rural males, 1.0 and 1.6 years for urban females, and 1.3 and 2.6 years for rural females. These territories belong to the North, Volga-Vyatka, and the Ural regions, and had the lowest e_0 level in Russia in 1979. This group registered the biggest life-expectancy increment in the 1980s.

This result may be interpreted as a manifestation of the impact of the anti-alcohol campaign on the value of regional variation of life expectancy and the change in certain areas in European Russia in 1988. In particular, the unexpected moderate success of the campaign in the provinces of the central region moved these provinces into the group with a lower level of life expectancy.

Thus, if our assumptions are valid, the analysis allows one to evaluate the approximate size and age range, as well as the spatial localization of the "upper layer" of the alcohol-correlated male mortality in European Russia that seemed to be easily eliminated by administrative actions.

The problem of the uneven accuracy of reporting of the demographic statistics becomes particularly acute in the mortality analysis at the provincial level. This applies to the poor quality of data in the rural areas for infants and the elderly. Although the demographic statistics in European Russia are better than, say, in Central Asia or Azerbaijan, they vary from one province to another and from period to period.

A puzzling issue of Soviet demography is the urban–rural mortality crossover: urban mortality is lower in young ages and in older ages it is higher (Bedny 1976; Dmitriyeva and Andreev, 1977). Recently, Coale and Kisker (1986) have found that often mortality crossovers are an artifact engendered by poor (or relatively poor) data on one of the populations. Therefore, the existence of urban–rural mortality crossovers may be regarded as an indicator of a understatement of rural mortality (Chapter 18; Anderson and Silver 1989a, 1989b), for it is likely that urban data on population are more precise than rural data.

As the first step to test rural data quality in European Russia we used the following measure:

$$w = \sum_{x=60}^{85+} {}_n c_x$$

$${}_n m_x^u > {}_n m_x^r , \quad x \geq 60$$

where ${}_n m_x^u$ and ${}_n m_x^r$ denote urban and rural age-specific death rates, accordingly; ${}_n c_x$ are age components of difference in e_0 between urban and rural population. This "index of unreliability" of the quality of data on the rural population above 60 is the sum of age components of the difference of life expectancy of rural and urban populations in ages above 60 in those age ranges, where the urban mortality is higher than the rural mortality. It is supposed that the bigger that index is, the poorer the quality of rural reporting (Chapter 18; Anderson and Silver, 1989a, 1989b).

Figure 21.7 depicts the distribution of the 53 territories by the level of this index in 1979 and 1988 (the territories are sorted by the index level). Three factors are evident:

- An improvement in the accuracy of mortality statistics in European Russia occurred between these two dates.
- The improvement was especially marked for women.
- The reporting quality is worse for women than for men.

If the underlying hypothesis is correct these findings have several implications. First, the improved accuracy of the mortality level in the 1980s (a positive phenomenon) negatively affected the level of rural life expectancy in this period. Moreover, if the quality of rural data continues to improve it would prolong the negative influence on the changes in the e_0 level of the rural population, especially female life expectancy.

Second, part of the tremendous gap in life expectancy of rural male and female is imaginary, being related to inaccurate female mortality data. The reduction of the gap in 1988 was determined both by the effect of the anti-alcohol actions and by the improvement in the quality of data for women. Nevertheless, in 1988, a part of the sex differentiation remained imaginary because the data on male mortality were still more accurate than the data on female mortality.

Third, eight maps of life expectancy in European Russia are not equally reliable: the maps of urban population are more reliable than those of rural population; the maps of male populations are more reliable than the maps of female populations. The 1988 maps are closer to the real situation in 1988 than the 1979 maps to the real situation in 1979.

In the 1980s the increase of mortality in ages above 60 led to the decline of rural life expectancy in 15 provinces for males and in 30 provinces for females, while for their urban counterparts that was not the case at all. This might be interpreted as indirect evidence of improved statistics in rural European Russia in ages above 60. Among territories, a considerable part is made up of provinces with the highest level of e_0 in 1979, which frequently have the worst quality of mortality statistics. Hence, if in 1979 the e_0 value in these territories were overestimated, then the

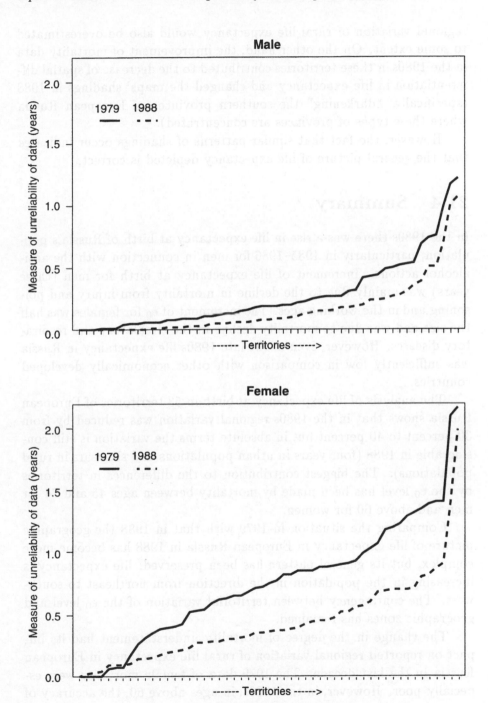

Figure 21.7. The sum of differences in rural and urban life expectancy for ages 60 and above with lower rural death rate in 53 areas of European ⁊9 and 1988 (areas sorted by value

regional variation of rural life expectancy would also be overestimated to some extent. On the other hand, the improvement of mortality data in the 1980s in these territories contributed to the decrease of spatial differentiation in life expectancy and changed the maps' shadings in 1988 (specifically, "darkening" the southern provinces of European Russia where these types of provinces are concentrated).

However, the fact that similar patterns of shadings occur indicates that the general picture of life expectancy depicted is correct.

21.4   Summary

In the 1980s there was a rise in life expectancy at birth of Russia's population, particularly in 1985–1986 for men in connection with the anti-alcohol actions. Increment of life expectancy at birth for men (three years) was mainly due to the decline in mortality from injury and poisoning and in the working ages. The increment of e_0 for females was half the size and was due to mortality decline in the circulatory and respiratory diseases. However, even in the late 1980s life expectancy in Russia was sufficiently low in comparison with other economically developed countries.

The analysis of life expectancy at birth in 53 territories of European Russia shows that in the 1980s regional variation was reduced by from 30 percent to 40 percent but in absolute terms the variation is still considerable in 1988 (four years in urban populations and six years in rural populations). The biggest contribution to the differences in territories by the e_0 level has been made by mortality between ages 45 and 74 for men and above 60 for women.

Comparing the situation in 1979 with that in 1988 the geographic picture of life expectancy in European Russia in 1988 has become more complex, but its general pattern has been preserved: life expectancy is increasing in the population in the direction from northeast to southwest. The contingency between territorial variation of the e_0 level and geographic zones has remained.

The change in the degree of mortality understatement had its impact on reported regional variation of rural life expectancy in European Russia in the last decade. The 1979 data of female mortality were especially poor. However, in the 1980s, in ages above 60, the accuracy of rural data has been improved.

Acknowledgment

The authors are grateful to Natalia Saulskaya for help with the cartograms.

Notes

[1] There are three kind of territorial-administrative units: province (*oblast*), autonomous republic, and *kray*. Province is an ordinary administrative unit; Autonomous Soviet Socialist Republic (ASSR) is an ethnic territorial entity; *kray* and province are the same units but the former is often bigger and is located near boundaries of Russia. In some cases *kray* comprises an autonomous area with ethnic status called autonomous *oblast*.

[2] In 1988, the smallest urban population was in Kalmykia (147,000); the smallest rural population (90,000) was in the Murmansk province. Accordingly, the standard error of life expectancy was 0.198 years for men and 0.131 years for women in urban Kalmykia; 0.231 years and 0.161 years for rural male and female in Murmansk province.

[3] The territories were classified by nine geographical zones of European Russia and by five divisions of e_0 level shown on the maps.

Appendix

Table 21A.1. Life expectancy in 53 provinces of European Russia

	Urban				Rural			
	Male		Female		Male		Female	
Provinces	1979	1988	1979	1988	1979	1988	1979	1988
Northern region								
Arkhangelsk	60.2	65.4	71.8	74.8	55.2	61.3	71.1	73.1
Vologda	61.7	64.4	73.4	74.5	57.5	62.6	72.2	74.5
Murmansk	61.4	65.9	72.4	75.1	59.2	65.6	72.8	74.2
Karelia ASSR	60.9	64.3	72.7	74.0	56.7	61.6	70.0	72.9
Komi ASSR	59.9	64.2	71.2	73.1	56.4	61.6	69.1	71.8
Northwestern region								
Leningrad	61.6	64.3	72.5	74.1	59.5	62.3	73.2	73.8
Novgorod	60.5	64.5	72.9	74.3	54.2	58.4	71.3	72.7
Pskov	61.8	64.4	72.5	74.0	55.7	61.0	70.3	72.5
Central region								
Bryansk	63.5	65.9	74.1	75.0	60.5	62.5	73.7	74.4
Vladimir	61.9	64.9	74.1	75.1	58.4	62.6	73.4	73.5
Ivanovo	61.6	64.0	73.0	73.7	59.0	61.1	73.5	73.2
Tver	61.5	64.6	73.8	74.4	55.9	59.4	72.3	72.6
Kaluga	63.2	65.7	74.0	74.7	57.9	60.7	72.8	72.5
Kostroma	61.0	64.9	72.7	74.7	58.4	61.2	72.3	73.2
Moscow	63.5	65.7	74.0	74.8	60.9	63.2	73.4	73.9
Orel	63.2	66.5	75.0	75.7	58.2	62.0	72.5	73.9
Ryazan	62.3	65.9	73.1	75.3	57.6	62.1	74.0	74.7
Smolensk	62.7	64.9	74.2	75.0	59.4	61.1	72.9	72.9
Tula	62.5	64.9	74.1	74.9	56.4	60.0	72.2	73.5
Yaroslavl	61.3	64.9	73.4	75.0	56.5	60.8	72.9	74.0
Volga-Vyatka region								
Gorky	62.3	64.6	73.7	74.6	56.5	62.2	73.2	74.5
Kirov	61.1	65.5	72.8	74.7	56.3	62.3	70.5	73.4
Mari ASSR	62.7	65.8	72.7	74.9	55.2	61.9	67.3	71.1
Mordov ASSR	62.6	66.1	74.4	76.5	58.3	63.5	74.8	74.7
Chuvash ASSR	61.7	67.3	73.3	75.8	56.7	63.1	68.6	72.8

Table 21A.1. Continued

	Urban				Rural			
	Male		Female		Male		Female	
Provinces	1979	1988	1979	1988	1979	1988	1979	1988
Central Blackearth region								
Byelgorod	63.9	66.5	75.0	74.4	62.7	65.1	76.4	75.3
Voronezh	63.4	66.6	74.4	75.7	61.7	64.1	75.6	74.6
Kursk	63.2	65.9	73.4	75.1	59.0	62.6	73.4	74.3
Lipeck	64.1	66.6	74.5	75.6	60.1	62.7	74.9	75.4
Tambov	60.9	65.0	72.8	75.2	59.3	61.8	72.2	74.5
Volga region								
Astrachan	62.6	64.8	72.8	74.6	61.2	64.4	73.6	73.6
Volgograd	64.1	65.9	74.2	75.1	61.5	63.7	73.4	74.1
Kuybyshev	62.5	65.7	73.4	74.6	59.2	63.0	73.2	73.6
Penzen	62.4	65.9	74.2	75.6	60.7	63.3	74.0	74.9
Saratov	62.6	64.8	73.9	74.4	60.6	62.6	74.6	74.3
Ulyanovo	62.1	65.9	74.5	75.4	60.5	63.9	73.2	74.2
Kalmyc ASSR	57.9	62.1	70.4	70.9	61.3	62.2	71.6	72.7
Tatar ASSR	63.5	66.7	75.0	75.8	61.6	64.6	74.5	75.6
North Caucasus region								
Krasnodar kray	63.7	65.7	73.8	74.3	62.2	63.7	74.0	74.1
Stavropol kray	64.7	66.7	74.5	75.5	62.9	64.0	74.2	73.9
Rostov	63.1	65.6	73.2	74.0	63.0	64.3	74.8	74.9
Dagestan ASSR	64.5	68.4	75.4	77.3	63.5	66.9	74.4	75.9
Kabardino-Balkar ASSR	64.0	66.6	74.9	76.5	64.7	66.0	76.8	76.4
North Osetin ASSR	66.3	66.9	75.4	75.3	65.0	64.4	76.6	76.0
Checheno-Ingush ASSR	65.3	65.8	82.0	74.3	67.1	65.0	76.8	75.7
Ural region								
Kurgan	60.7	65.4	72.2	75.4	59.1	63.8	71.7	75.0
Orenburg	62.0	65.7	72.9	74.9	62.5	65.1	74.3	74.2
Perm	60.5	65.0	71.2	74.2	55.0	60.5	67.4	71.8
Sverdlovsk	61.3	65.1	72.3	74.4	56.5	61.7	70.3	72.7
Chelyabinsk	62.4	66.1	73.1	74.9	59.8	63.7	71.7	73.8
Bashkir ASSR	63.6	66.2	73.6	75.2	61.3	64.4	73.7	75.0
Udmurt ASSR	60.7	65.1	71.8	74.7	54.8	62.4	69.3	73.6
Baltic region								
Kaliningrad	63.0	66.0	72.7	74.6	57.4	60.0	70.6	71.4

References

Anderson, B. A., and B.D. Silver, 1989a. The Changing Shape of Soviet Mortality, 1958–1985: An Evaluation of Old and New Evidence. *Population Studies* 43(2):243–265.

Anderson, B.A., and B.D. Silver, 1989b. Patterns of Cohort Mortality in the Soviet Population. *Population and Development Review* 15(3):471–501.

Andreev, E., 1982. Metod komponent v analize prodolzhitel'nosti zhizni (Components Method Applied to Life Expectancy Analysis). *Vestnik Statistiki* (Herald of Statistics) 9:42–47.

Bedny, M., 1976. *Prodolzhitel'nost' zhizni v gorodakh i selakh* (Expectation of Life in Cities and Villages). Statistika, Moscow.

Blum, A., and A. Monnier, 1989. Recent Mortality Trends in the USSR: New Evidence. *Population Studies* 43(2):243–266.

Blum, A., and R. Pressat, 1987. Une nouvelle table de mortalité pour URSS (1984–1985). *Population* 42(2):843–862.

Chiang, C.L., 1984. *The Life Table and its Applications*. Krieger Publishing, Malabar, FL.

Coale, A.J., and E. Kisker, 1986. Mortality Crossovers: Reality or Bad Data? *Population Studies* 40(2):389–402.

Dmitriyeva, R., and E. Andreev, 1977. Snizhenie smertnosti v SSSR za gody Sovetskoy vlasti (Lowering of Mortality in the USSR during the Years of Soviet Power). In A.G. Vishnevsky, ed., *Brachnost', rozhdaemost', i smertnost' v Rossii i v SSSR: Sbornik statey* (Nuptiality, Fertility, and Mortality in Russia and in the USSR). Statistika, Moscow.

Dmitriyeva, R., and E. Andreev, 1987. O sredney prodolzhitel'nosti zhizni naseleniya SSSR (On the Life Expectancy of the USSR Population). *Vestnik Statistiki* (Herald of Statistics) 12:31–39.

Keyfitz, N., 1977. *Introduction to the Mathematics of Population*. Addison-Wesley, Reading, MA.

Pollard, J.H., 1982. The Expectation of Life and its Relationship to Mortality. *Journal of the Institute of Actuaries* 225–240.

Shkolnikov, V., 1987. Geograficheskye factory prodolzhitel'nosti zhizni (Geographical Factors of Life Expectancy). *Izvestiya AN SSSR* (Proceedings of the USSR Academy of Sciences) 3:35–45.

Sinel'nikov, A., 1988. Dinamika urovnuya smertnosti v SSSR (Dynamics of Mortality Levels in the USSR). In L.L. Rybakovskiy, ed., *Naseleniye SSSR za 70 let* (Population of the USSR over the Past 70 years). Nauka, Moscow.

Valkovics, E., 1984. L'évolution récente de la mortalité dans les pays de l'est: essai d'explication à partir de l'exemple hongrois. *Espace. Population. Sociétés.* III:141–168.

Chapter 22

Changing Mortality Patterns in Latvia, Lithuania, and Estonia

Juris Kruminš

In Eastern Europe and the former USSR there is evidence of an interest in understanding the determinants of mortality on a wide scale, along with comparative analyses of the effectiveness of social policies and health-care systems. Newly accessible data have piqued the interest of many health-policy analysts, epidemiologists, and demographers, who until recently could merely speculate on the trends that were occurring in the USSR in the 1970s and the 1980s. This is especially true for the Baltic republics of Estonia, Latvia, and Lithuania, for which detailed demographic statistics for the postwar period were previously unavailable. Researchers have had to rely on aggregated Union-wide data which could not give accurate insight into the situation in the Baltic republics.

Statisticians in the Baltic republics have been collecting and analyzing mortality and other demographic data for many years, yet during those years little attention was paid to research in this field. Access to demographic and socioeconomic data was severely limited. The available data were restricted to the aggregate level of the Soviet Union, thereby obscuring the diverse patterns of mortality within the individual republics. However, during the late 1980s, under glasnost, statistical data became more readily available to researchers interested in demography and health status, thereby yielding not only renewed interest but more importantly, new research in this field. This chapter examines the changing patterns of mortality in Latvia, Lithuania, and Estonia over the past three decades. Because the information on Latvia was the most exhaustive, analysis about the Latvian population is the most detailed. This analysis is mainly based on information obtained before 1990 and does not reflect the changes that have taken place after the declarations of independence in spring 1990.

Estonia, Latvia, and Lithuania have much in common in their historical, social, and economic developments. The relatively advanced state of public health care and the economic development in the Baltic region have contributed to a higher level of life expectancy at birth (e_0) of its population in comparison with other regions of European Russia at the end of the 19th century. Life expectancy in the Kovno, Kurland, and Livland provinces was the highest among 11 nationalities living in separate administrative districts of former European Russia in 1896–1897 and exceeded the life-expectancy level in some developed nations. The highest indexes were among Latvians: 43.1 years for males and 46.9 years for females (Ptoukha, 1960).

During the 1920s and 1930s, when the Baltic republics were independent, the level of life expectancy steadily rose. The best situation according to many sociodemographic and health indicators was in Latvia, with Estonia and Lithuania following (*Table 22.1*). In 1940 the Baltic states were incorporated into the USSR; nevertheless, they retained their leading position over the other republics of the USSR in the level of development of health services and other socioeconomic indicators.

22.1   Changes in Life Expectancy

Life expectancy at birth was higher in the Baltic republics than in Russia, the Ukraine, and the European part of the USSR in the 1920s (*Table 22.2*). Life expectancy in Estonia for males in the 1930s was 53.1 years, for females it was 59.6 years (State Bureau of Estonia on Statistics, 1937). In Latvia, however, it was 55.4 years and 60.9 years, respectively (State Office of Latvia on Statistics, 1938). It was higher in Latvia than in several Eastern and Southern European countries (such as Hungary,

Table 22.1. Sociodemographic and health-service indicators in the three Baltic republics, 1930s.

Indicator	Year	Estonia	Latvia	Lithuania
Urban population	1939	34%	35%	23%
Infant mortality rate				
per 1,000 births	1930–1934	97	87	151
	1935	89	79	123
	1938	77	68	113
Population per				
physician	1938	1,183	1,247	2,727
Population per				
dentist	1938	5,525	2,366	4,056
Population per				
drugstore	1938	5,333[a]	3,793	7,761
Births per midwife	1938	102	42	47

[a]1937

Source: State Office of Latvia on Statistics 1940a, 1940b.

Table 22.2. Changes in life expectancy at birth (both sexes) in the three Baltic republics, Byelorussia, Russia, the Ukraine, and total USSR, 1926–1927 to 1989.

	1926– 1927	1958– 1959	1978– 1979	1986 1987	1988	1989
Baltic republics	51.9	68.3	69.6	71.6	71.3	70.8
Estonia	51.7[a]	68.4	69.6	71.0	71.0	70.6
Latvia	53.7[b]	69.1	69.1	70.9	71.0	70.4
Lithuania	50.5[c]	68.4	70.6	72.5	72.4	71.8
Byelorussia	52.7	69.8	71.4	72.0	71.7	71.8
Russia	42.9[d]	67.9	67.7	70.1	69.9	69.6
Ukraine	47.1	69.3	69.8	71.1	70.9	70.9
USSR	44.4[d]	68.6	67.9	69.8	69.5	69.5

[a]1922.
[b]1924–1925.
[c]1925–1926.
[d]European part.

Note: If the life expectancies for both sexes were not available, they were computed on the average of men's and women's life expectancy at birth as follows:

$$e_0 = 0.512em + 0.488ef .$$

Sources: Ptoukha, 1960; State Committee of the USSR on Statistics, 1988, 1989b, 1990; National life tables for the 1920s and 1950s.

Poland, Greece, Spain, and Italy) and approximately corresponded to the levels in Austria, Belgium, Finland, France, and Scotland.

No data on the development of mortality are available for the Baltic republics between 1940 and 1957. The first life tables during the Soviet period were calculated on the basis of the 1959 census. Since then life-table calculations have regularly been computed.

During the postwar years the gap between the level of life expectancy in the Baltic republics and the level in total USSR rapidly diminished. In 1978–1979 e_0 was 1.7 years longer in the Baltic republics than the overall level in the USSR. By 1989 this difference again decreased to 1.3 years.

In such comparisons it is necessary to take into consideration possible differences in the completeness of death registration and the calculation of total population (see Dmitriyeva and Andreev, 1987; Anderson and Silver, 1989). According to some calculations e_0 in the USSR was overestimated by at least one year during the 1960s (Sinel'nikov, 1988). Data about the Baltic republics must also be critically evaluated. Statistics seem to have been better organized there than in other republics. For this reason the Baltic republics can partly serve as a model for a more precise survey of the evolution of mortality in the European USSR.

In the Baltic republics, and in the USSR as a whole, three periods can be discerned in the postwar evolution of mortality by studying the trends in life expectancy as well as the change in age patterns of mortality: first, a period of increasing e_0 up to the mid-1960s; second, a period of stagnation and decline of e_0 from the mid-1960s to the late 1970s; third, a period of a moderate rise in e_0 from the late 1970s and a decrease in the late 1980s. Within these periods there are modifications depending on sex and region. These changes are more prominent for males than for females.

The level of life expectancy at birth in Latvia is nevertheless somewhat lower than it was in the mid-1960s. In Lithuania the highest e_0 (77 years for females and 68 years for males) was registered between 1986 and 1988. These were among the highest levels in the USSR.

Table 22.3. Infant mortality rate by cause of death in the three Baltic republics and the USSR in 1989, per 10,000 births.

Cause of death	Estonia	Latvia	Lithuania	USSR
Total	147.2	111.4	106.8	226.8
Infectious, parasitic diseases	6.2	10.7	4.7	32.2
Respiratory system diseases	11.9	6.9	3.2	67.9
Congenital anomalies	34.9	35.6	39.9	31.2
Perinatal complications	73.4	45.0	45.1	70.7
Accidents, poisoning, injuries	9.4	4.8	3.8	8.4
Other causes	11.4	8.4	10.1	16.4

Sources: State Committee of the USSR on Statistics, 1990.

22.2   Trends in Infant Mortality

Infant mortality rates (IMRs) have always been considerably lower in the Baltic republics than in the USSR as a whole. In the USSR infant mortality in 1940 was 182 per 1,000 births (State Committee of the USSR on Statistics, 1988). Infant mortality rates were 1.5 to 2.5 times higher in total USSR than in Estonia, Latvia, and Lithuania. In the early 1950s these differences diminished. Only in Lithuania (disregarding under-registration of infant deaths) were IMRs higher than the average level in the USSR until 1964. A sharp fall in infant mortality occurred in the mid-1960s, after which a certain stabilization set in (*Figure 22.1*). The main causes of this anomalous increase have been discussed (Jones and Grupp, 1983; Anderson and Silver, 1986; Ksenofontova, 1990).

The sharp short-term increase in infant mortality rates in Latvia as well as in Lithuania and Estonia during the late 1970s may be due to an increase in mortality from exogenous causes (Krumiņš and Zvidriņš, in press). The increase in IMR came to an end in the 1970s and in the 1980s a decrease was observed. In the USSR as a whole this trend was very weak. As a result the difference between the IMR in the Baltic republics and the average rate in the USSR grew, especially in the 1980s.

IMRs from infectious and parasitic diseases and respiratory system diseases in the Baltic republics were essentially lower than the average level in the USSR in 1989 (*Table 22.3*). At present, the IMR in the Baltic republics is almost double the level reached in developed European countries.[1]

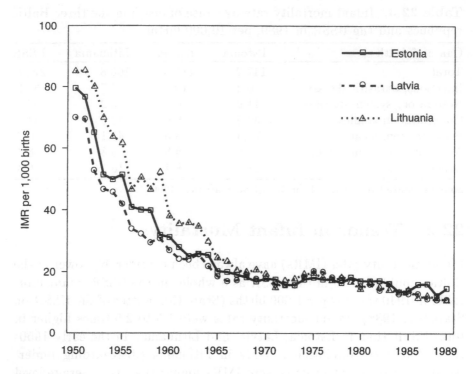

Figure 22.1. Trends in infant mortality rates in the Baltic republics, 1950–1989.

22.3 Changing Mortality Patterns

To study the Baltic republics' mortality patterns a comparison with the United Nations and Coale-Demeny model life tables was carried out. For calculations the *Mort Pak-Lite* package was used (United Nations, 1988).

There was a radical change in the pattern of mortality among males in Estonia, Latvia, and Lithuania in the postwar period. In the 1930s there was a difference in the pattern of mortality among males in the Baltic republics and the USSR. In the 1970s and 1980s the mortality rate of males in the Baltic republics assumed many of the characteristics of the Far East and Chilean patterns in the UN model life tables.

Until the late 1950s the Estonian and Latvian male mortality pattern had the best fit with the Coale-Demeny West model. Later changes

Table 22.4. Contribution of age groups to changes in life expectancies at birth in 1958–1959 to 1978–1979 and 1978–1979 to 1986–1987 in the three Baltic republics, in years.

Age	1958–1959 to 1978–1979			1978–1979 to 1986–1987		
	Estonia	Latvia	Lithuania	Estonia	Latvia	Lithuania
Males						
0	1.25	0.89	2.39	0.20	0.44	0.45
1–14	0.19	0.46	0.39	0.13	0.12	0.18
15–24	0.10	−0.12	−0.10	0.19	0.21	0.27
25–34	−0.12	−0.27	−0.28	0.49	0.66	0.54
35–44	−0.54	−0.66	−0.50	0.49	0.55	0.40
45–54	−0.66	−0.71	−0.52	0.43	0.33	0.15
55–64	−0.30	−0.62	−0.31	0.13	0.15	−0.03
65–74	0.11	−0.26	−0.15	−0.09	0.02	−0.03
75+	−0.12	−0.10	−0.05	0.12	0.04	0.18
Total[a]	−0.08	−1.39	0.87	2.11	2.51	2.11
Females						
0	1.36	0.77	2.13	−0.07	0.43	0.37
1–14	0.22	0.24	0.48	0.11	0.17	0.11
15–24	0.26	0.14	0.11	0.04	0.10	0.04
25–34	0.23	0.18	0.17	0.07	0.07	0.10
35–44	0.18	0.07	0.13	0.08	0.10	0.08
45–54	0.07	−0.11	0.05	0.22	0.18	0.14
55–64	0.01	−0.06	0.25	0.16	−0.06	0.05
65–74	0.31	0.22	0.73	−0.05	0.01	−0.05
75+	0.06	−0.03	−0.14	0.16	0.14	0.45
Total[a]	2.80	1.43	3.93	0.72	1.15	1.29

[a]Because of rounding the figures may not add up to the total.
Sources: State Committee of the USSR on Statistics, 1989b. National life tables for 1958–1959.

were due to the very high mortality at older working ages. It can be observed that up to the middle-age groups the fit of other life-table models is better than that of the UN Far East and Chilean models.

In the case of female mortality the fit of life-table models to empirical mortality patterns is generally better. In the past 30 years changes in mortality patterns for females were not as pronounced as they were for males. In the late 1970s and mid-1980s the Baltic republics' female mortality pattern had the best fit with the Coale-Demeny West and

Table 22.5a. Contribution by cause of death and age group to life expectancies at birth for males in Latvia, 1958–1959 to 1986–1987, in years.

Cause of death	1958–1959 to 1964–1965				1964–1965 to 1978–1979				1978–1979 to 1986–1987			
	Total	0–24	25–64	65+	Total	0–24	25–64	65+	Total	0–24	25–64	65+
All causes	1.73	0.98	0.43	0.32	-3.12	0.20	-2.61	-0.71	2.51	0.77	1.69	0.05
Infectious and parasitic diseases	0.51	0.12	0.33	0.06	0.44	0.01	0.34	0.09	0.22	0.16	0.04	0.02
Malignant neoplasms	-0.25	–	-0.20	-0.05	-0.16	–	-0.22	0.06	-0.29	-0.03	-0.13	-0.13
Circulatory system diseases	-0.10	–	0.21	-0.31	-2.00	–	-1.18	-0.82	0.43	-0.01	0.25	0.19
Respiratory system diseases	0.33	0.24	0.05	0.04	-0.10	0.07	-0.14	-0.03	0.39	0.28	0.11	0.00
Digestive system diseases	0.14	0.09	0.02	0.03	-0.02	0.02	-0.08	0.04	0.01	-0.01	0.06	-0.04
Congenital anomalies	0.38	0.38	–	–	0.11	0.11	–	–	0.01	0.01	–	–
Accidents, poisonings, and injuries	0.32	0.05	0.23	0.04	-1.89	–	-1.78	-0.10	1.61	0.26	1.34	0.01
Other causes	0.40	0.10	-0.20	0.50	0.50	-0.01	0.46	0.05	0.08	0.06	0.02	0.00

Sources: State Office of Latvia on Statistics 1964; life table and cause of death statistics in Latvia, 1958–1987.

Table 22.5b. Contribution by cause of death and age group to life expectancies at birth for females in Latvia, 1958-1959 to 1986-1987, in years.

Cause of death	1958–1959 to 1964–1965				1964–1965 to 1978–1979				1978–1979 to 1986–1987			
	Total	0–24	25–64	65+	Total	0–24	25–64	65+	Total	0–24	25–64	65+
All causes	2.08	0.92	0.81	0.35	−0.64	0.07	−0.55	−0.16	1.15	0.48	0.52	0.15
Infectious and parasitic diseases	0.54	0.12	0.38	0.04	0.14	−0.01	0.12	0.03	0.09	0.08	0.01	0.03
Malignant neoplasms	−0.01	–	−0.16	0.15	0.15	0.01	0.03	0.11	−0.11	–	−0.04	−0.07
Circulatory system diseases	−0.32	–	0.34	−0.66	−0.65	–	−0.28	−0.37	0.36	–	0.08	0.28
Respiratory system diseases	0.39	0.26	0.04	0.09	−0.04	0.03	−0.07	0.00	0.11	0.02	0.06	0.03
Digestive system diseases	0.10	0.08	0.01	0.01	0.07	0.01	0.00	0.06	−0.04	–	0.02	−0.06
Congenital anomalies	0.33	0.33	–	–	0.03	0.03	–	–	0.03	0.03	–	–
Accidents, poisonings, and injuries	0.16	0.10	0.06	0.01	−0.52	0.01	−0.47	−0.06	0.34	0.02	0.33	−0.01
Other causes	0.89	0.03	0.14	0.72	0.18	−0.01	0.12	0.07	0.37	0.33	0.06	−0.02

Sources: State Office of Latvia on Statistics 1964; life table and cause of death statistics in Latvia, 1958–1987.

North models. To identify the effect of changing age patterns in mortality, the method of decomposition of life expectancy at birth differences was used (Arriaga, 1984). Since no published life tables are available on the mid-1960s the changes in e_0 in the Baltic republics from 1958–1959 to 1986–1987 were analyzed.

According to *Table 22.4* the influence of age-specific changes in mortality on the change in e_0 for males and females coincides only in younger age groups from 1959 to 1979. A considerable increase in age-specific death rates is evident in working-age males (specifically between ages 35 and 64). This increase has consequently led to a decline in their life expectancy by almost two years. Negative changes for females were registered beginning with the 45–54 age group (that is, worsening of the situation coincided with preretirement and retirement ages – age of retirement is 60 for men and 55 for women).

The nature of the changing intensity of mortality in working-age males has drastically changed in the 1980s. As a result of the decline of mortality in working-age males in all Baltic republics a considerable growth in e_0 has been attained. For females these changes are not as essential, except in the older age groups which have also contributed to an increase in e_0 in the 1980s. The negative digression in the first year of life in Estonia, as well as in older age groups in Lithuania, is obviously accidental. The 1988–1989 data confirm this.

22.4 Changes in Mortality by Cause of Death

Since no published cause-of-death data are available for the 1960s and 1970s for Estonia and Lithuania, attention will be paid to mortality changes in Latvia (*Table 22.5*). Decrease in e_0 from 1964 to 1979 was mainly due to an increase in mortality from circulatory system diseases and from accidents, poisonings, and injuries. The contribution was more pronounced for males than for females. These two types of causes later were of great importance to the increase in e_0 in the mid-1980s.

The radical change in the evolution of mortality in working ages, especially from accidents, injuries, and poisonings, which took place in the mid-1980s is mainly connected with the 1985 restrictions imposed to reduce alcohol consumption. However, in this period the amount of home-made alcohol increased. It must be noted that even with the favorable tendency of declining mortality from accidents, injuries, and poisonings, the mortality level in 1988–1989 in the Baltic republics was still higher than in 1986. This trend is evidently connected with the new rise in the consumption of alcohol and with subsequent traumatism.

Until the late 1970s mortality from accidents was the main reason for the great gap in male and female life expectancy in Latvia. Circulatory system diseases overtook accidents in the mid-1980s as the leading cause of death only after an essential decrease occurred in the latter cause for male mortality. The share of malignant neoplasms in the male and female e_0 difference gradually increased.

The age-specific excess of male mortality by cause of death in Latvia mainly corresponds with the trends that were widespread in the 1960s and 1970s among developed countries (United Nations, 1982). Yet, the excess of male mortality from neoplasms and diseases of the respiratory system in the Baltic republics is more prominent in older age groups. In the middle-age groups the excess stems from diseases of the circulatory system and from accidents, poisonings, and injuries.

22.5   Urban–Rural Mortality Differentials

In spite of little consistency to the patterns of differential mortality between rural and urban areas in more developed countries in the postwar period, the gap is narrowing. The pattern is shifting toward higher rates in urban environments (United Nations, 1982).

Age-specific probabilities of death are higher in the Baltic urban areas in the postwar period than in rural areas for practically all ages, excluding the oldest age groups (*Figure 22.2*). The excess e_0 in urban areas of the Baltic republics from 1958–1959 to 1978–1979 has changed: in Estonia from 1.4 to 2.8 years, in Latvia from 1.0 to 3.0 years, and in Lithuania from 1.2 to 2.6 years. In the 1980s the life-expectancy gap between urban and rural populations slightly decreased (excluding Lithuania) and in 1989 it expanded to 2.3 years; in Lithuania it expanded to 3.0 years (State Committee of the USSR on Statistics, 1990).

The rural population has a different pattern of mortality. For example, at age 10 and over age 65 age-specific mortality rates in rural areas of Latvia fit the UN Chilean life-table model, but in urban areas the rates fit the UN Far East model for males and Coale-Demeny West model for females. The largest differences between rural and urban mortality rates appear in the categories of respiratory system diseases

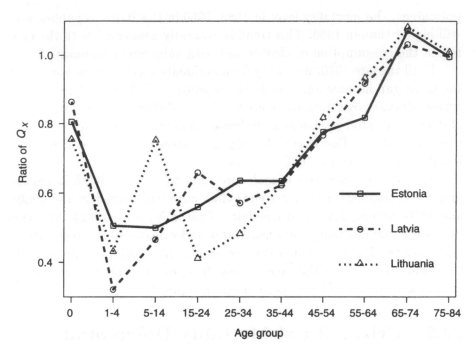

Figure 22.2. Ratio of probabilities of death (Q_x) in urban areas to those in rural areas in the Baltic republics.

and accidents, poisonings, and injuries (*Table 22.6*). Age-standardized death rates (ASDR) from respiratory illness were about 35 to 50 percent lower in the cities of the Baltic republics (excluding females in Estonia) than those in the rural areas in 1988.

An anomaly in the general pattern of higher rural mortality is evident for neoplasms; the rural rates are slightly lower than urban rates. However, in this case significant changes in the 1980s took place in rural areas. The available data do not support any definitive conclusions for this trend, but it is reasonable to suspect that differences in diagnostic and treatment capabilities between rural and urban medical facilities, as well as environmental factors, may play a significant role in this discrepancy (Pulcins and Kruminš, 1990).

The residual difference between urban and rural mortality rates in the Baltic republics can be explained as follows: first, the level of socio-economic development is still lower in rural regions than in urban areas; second, medical care is not readily available in rural regions; third, in

Table 22.6. Ratio of age-standardized death rates (ASDR) by main cause of death in urban areas to those in rural areas in the three Baltic republics, 1980 and 1988, in percent.

	Estonia		Latvia		Lithuania	
Cause of death	1980	1988	1980	1988	1980	1988
Males						
Neoplasms	121	107	134	111	118	106
Circulatory system diseases	84	95	86	85	97	95
Respiratory system diseases	107	65	78	67	61	51
Accidents, poisonings, injuries	73	75	60	65	66	60
Females						
Neoplasms	127	133	153	139	130	138
Circulatory system diseases	78	81	84	85	89	88
Respiratory system diseases	75	128	80	65	53	56
Accidents, poisonings, injuries	91	68	80	81	77	75

Sources: State Committee of the USSR on Statistics, 1988.

rural regions the male population is subject to the effects of an unhealthy life-style (alcoholism, injuries, and so on).

22.6   Mortality Differentials by Ethnic Groups

Differences in mortality can also be observed among ethnic groups. The ethnic differential in mortality is one of the least analyzed topics in the demography of the Soviet period in the Baltic republics.

In 1978–1979 life expectancy at age 5 for titular nationalities in all Soviet republics, except Byelorussia and Moldavia, was higher than for Russians. In the Baltic republics the difference was between 0.9 and 1.8 years for males and between 0.4 and 1.2 years for females (Dobrovol'skaya, 1990). The Estonian population has had a slightly higher life expectancy compared with the non-Estonian population for every postwar census year. In Latvia the situation is similar (*Table 22.7*).

Although, the difference is small it must be taken into consideration when analyzing the urban–rural differences. Non-Estonians make up 90 percent of the urban population in Estonia, but only 60 percent of Estonians live in urban areas. The share of the urban population among Latvians and Russians in Latvia has changed over the years. Some 47 percent of the Latvian population lived in urban centers in 1959 and 60

Table 22.7. Ethnic differentials in life expectancy at birth in Estonia and Latvia, 1958–1989.

	Estonia[a]		Latvia			
	Actual e_0		Actual e_0		Standardized e_0[b]	
Year	Estonians	Non-Estonians	Latvians	Russians	Latvians	Russians
1958–59	68.0	67.8	69.5	69.8	69.5	69.4
1969–70	69.9	69.3	70.3	70.3	70.3	69.8
1978–79	69.4	68.4	69.3	68.6	69.6	67.9
1988–89	70.9	70.6	71.0	70.2	71.4	69.7

[a]Life expectancy for both sexes are computed on the average of men's and women's e_0, see note in *Table 22.2*.
[b]Standard urban–rural distribution of the total population.
Sources: for Estonia, Katus, and Puur, 1990; for Latvia, author's calculations.

percent lived in urban centers in 1989. Some 73 percent of the Russian population lived in urban centers in 1959 and 85 percent lived in urban centers in 1989.

This residual factor accounted for in the difference between life expectancies by nationality seems to be important. Urban–rural standardized life expectancy for both sexes shows that Latvians have had higher life expectancy compared with Russians for the whole postwar period.

There is reason for concluding that mortality is slightly higher for the population of non-titular nationalities than for the population of titular nationalities. The high proportions of immigrants who have not adapted to urban conditions might partly explain this phenomenon.

22.7 Conclusion

Because of initially higher level of socioeconomic development in the Baltic republics until 1940 and because of their leading position in many social and economical developments during the postwar period, the pattern of mortality in this region differs from the general pattern of mortality in the USSR. In the Baltic republics mortality has been considerably lower for children and it is slightly higher for older age groups which is typical for the later stage of demographic transition.

During the postwar period Estonia, Latvia, and Lithuania began to lose their leading position in life expectancy among Eastern European countries. A pattern of increasing male mortality was gradually becoming more prominent.

Against the background of a continuous decline of mortality in younger age groups, a general growth of mortality from the mid-1960s to the late 1970s took place for working-age males, but for females an increase occurred in the period preceding retirement and the first few years after retirement. Such a situation was evidently due to drawbacks in social policy. The causes of such increases in mortality also are to be looked for in individual behavior and life-style. The influences of long-term consequences of the two world wars (especially World War II), mass repressions in 1940–1941 and 1945–1953, and increased migration activity are indisputable.

The rise in life expectancy in the 1980s proved to be selective. It was more pronounced in working-age males. It was more evident in rural populations than in urban populations. The decline of deaths from exogenous causes played a significant role. The campaign against alcohol consumption, which began in the mid-1980s, undoubtedly contributed to a decrease of mortality from accidents, poisonings, and injuries. However, it would be a simplification of the problem to attribute the change in mortality to this cause alone. The exaggerated importance that was ascribed to measures against alcoholism along with the recent deterioration of social structures has again contributed to a rise in mortality and stopped the growth of life expectancy. The situation can be improved only by a radical revision of the measures aimed at providing services in all spheres of activities and improving public health.

Note

[1] The Soviet definition of infant mortality used in the Baltic republics in the postwar period differs from the WHO recommendations. If the IMR in the Baltic republics were adjusted to fit the WHO definition, it would be 10 to 25 percent higher.

References

Anderson, B.A., and B.D. Silver, 1986. Infant Mortality in the Soviet Union: Regional Differences and Measurement Issues. *Population and Development Review* **12**:705–738.

Anderson, B.A., and B.D. Silver, 1989. The Changing Shape of Soviet Mortality, 1958–1985: An Evaluation of Old and New Evidence. *Population Studies* 43:243–265.

Arriaga, E., 1984. Measuring and Explaining Changes in Life Expectancies. *Demography* 21:83–96.

Dmitriyeva, R., and E. Andreev, 1987. O sredney prodolzhitel' nosti zhizni naseleniya SSSR (On the Life Expectancy of the USSR Population). *Vestnik Statistiki* (Herald of Statistics) 12:31–39.

Dobrovol'skaya,V., 1990. Etnicheskaya differentsitsya smertnosti (Ethnical Differentiation in Mortality). In A.G. Volkov, ed. *Demograficheskiye protsessy v SSSR* (Demographic Processes in the USSR). Nauka, Moscow.

Jones, E., and F.W. Grupp, 1983. Infant Mortality Trends in the Soviet Union. *Population and Development Review* 2.

Katus, K., and A. Puur, 1990. General Mortality Trend: Case of Estonia 1897–1989. Paper presented at the International Conference on Health, Morbidity, and Mortality. Vilnius.

Kruminš, J., and P. Zvidriņš, in press. Recent Mortality Trends in the Three Baltic Republics. *Population Studies*.

Ksenofontova, N., 1990. Nekotorie tendentsii mladencheskoi smertnosti v poslednee desyatiletie. In A. Volkov, ed., *Demograficheskiye protsessy v SSSR* (Demographic Processes in the USSR). Nauka, Moscow.

Ptoukha, M.V., 1960. *Otcherki po statistike naseleniya* (Essays on Population Statistics). Gosstatizdat, Moscow.

Pulcins, I., and J. Kruminš, 1990. Death and Dying under Perestroika, Paper presented at the 12th Conference on Baltic Studies. Seattle, WA.

Sinel'nikov, A.B., 1988. Dinamika urovnuya smertnosti v SSSR (Dynamics of Mortality Levels in the USSR). In L.L. Rybakovskiy, ed. *Naseleniye SSSR za 70 let* (Population of the USSR over the Past 70 Years). Nauka, Moscow.

State Bureau of Estonia on Statistics, 1937. Tallinn.

State Committee of the USSR on Statistics (Goskomstat), 1988. *Naseleniye SSSR, 1987: Statisticheskiy sbornik* (Population of the USSR, 1987: Statistical Collection). Finansy i statistika, Moscow.

State Committee of the USSR on Statistics (Goskomstat), 1989a. *Naseleniye SSSR, 1988: Statisticheskiy ezhegodnik* (Population of the USSR, 1988: Statistical Yearbook). Finansy i statistika, Moscow.

State Committee of the USSR on Statistics (Goskomstat), 1989b. *Tablitsy smertnosti i ozhidaemoy prodolzhitel'nosti zhizni naseleniya* (Tables of Mortality and Life Expectation of the Population). Finasy i statistika, Moscow.

State Committee of the USSR on Statistics, (Goskomstat), 1990. *Demograficheskiy ezhegodnik SSSR, 1990* (USSR Demographic Yearbook, 1990). Finansy i statistika, Moscow.

State Office of Estonia on Statistics *Tablitsy smertnosti i srednei prodolzhitel-nosti zhizni naseleniya Estonskoi SSR 1958–1959 gg* (Life Tables of the Estonian SSR in 1958–1959). TsSU ESSR, Tallinn.

State Office of Latvia on Statistics, 1938. *Menesa biletens* (Monthly Bulletin) 2:206–208.

State Office of Latvia on Statistics, 1940a. *Tautas veselibas statistika par 1938. gadu* (Health Statistics in 1938). Valters un Rapa, Riga.

State Office of Latvia on Statistics, 1940b. *Menesa biletens* (Monthly Bulletin) 2:135–136.

State Office of Latvia on Statistics, 1964. *Tablitsi smertnosti i srednei prodol-zhitel'nosti zhizni naseleniya Latvijskoi SSSR 1958–1959 gg.* (Life Tables of the Latvian SSSR in 1958–1959). Riga.

State Office of Lithuania on Statistics, 1963. *Tablitsi smertnosti i srednei prodolzhitel'nosti zhizni naseleniya Litovskoi SSR 1958–1959 gg* (Life Tables of the Lithuanian SSSR in 1958–1959). TsSU, Vilnius.

United Nations, 1982. *Levels and Trends of Mortality Since 1950*. New York, NY.

United Nations, 1988. *Mort Pak-Lite: Population Studies* **104**, New York, NY.

Part IV

Changes in Population Structure

Chapter 23

Population Dynamics: Consequences of Regular and Irregular Changes

Evgeny M. Andreev, Leonid E. Darsky, and
Tatiana L. Kharkova

A number of gloomy periods mark the USSR history in the 20th century, and their impact is quite noticeable on the age–sex structure of the population. The tragic past still has an effect on the demographic situation today. The effects of three demographic catastrophes are evident in the age pyramid based on the 1989 All-Union census data.

- World War I and the civil war following the revolution caused considerable human losses among men born in the late 19th century and a significant fall in the size of cohorts born in 1914–1924.
- Mortality caused by the famine in the early 1930s peaked in the spring of 1933; this period was followed by a considerable decrease in the number of births.
- World War II caused extensive human losses and decreased the fertility level by more than half.

From the late 1920s until the mid-1950s there were no reliable data on population size. Demographic statistics were subject to intentional or unintentional distortions and direct falsification. In the early 1930s vital statistics were inadequate or unavailable. The available fragmentary data on vital events and information on the size of the population contradict one another. More precise definitions and corrections are required, and the 1926, 1937, and 1939 census data need to be re-estimated.

Both, Soviet and foreign scientists have shown interest in Soviet demographic history, but even the most skillful and fundamental works (Biraben, 1976; Kozlov, 1989; Kvasha, 1990; Lorimer, 1946; Maksudov, 1989; Rybakovskiy, 1989) have not addressed all the problem. In the past demographers did not have access to the state archives. Data, available today, allow demographers to estimate human and demographic losses in the USSR caused by the famine in the 1930s and World War II.

This chapter critically evaluates the results of the 1926, 1937, and 1939 population censuses and makes necessary corrections. We use modern methods to provide a picture of possible population trends in each cohort during these years.

23.1   Population Size in Census Years

23.1.1   1926 population census

In the past, the 1926 census data have been used by Novosel'skiy and Payevskiy (*Smertnost' i prodolzhitel' nost' zhizni*, 1930) and Ptoukha (1987) without any corrections. Only Korchak-Chepurkovskiy (1970) wrote about the possibilities of under-registration in this census. Kurman (1937) explained the discrepancies between the 1937 census results and the available estimates as the over-registration of 1 percent or 1.5 million people in the 1926 census. To our knowledge, there is no reason for assuming that double-counting in the 1926 census would have exceeded the usual under-registration or would have been comparable with it in scale. We thought it necessary to correct the registered size of only two population groups.

First, considering that under-registration of children usually takes place in censuses, we increased the number of children under age 3, based on the comparisons between census data for these ages and the data on births in 1926, 1925, and 1924. We estimated under-registration in the 0 to 2 age group to be equal to 10 percent or 1.412 million people. Second, based on the analysis of the age–sex structure of the population in Transcaucasia (mainly, in Azerbaijan) and Central Asia, we increased the number of females between ages 8 and 27 by almost 90,000 or 0.12 percent of the total number of females registered in the USSR (Andreev *et al.*, 1990a).

These adjustments added 1 percent to the population total of the 1926 census: 148.530 million instead of the published result of 147.028 million (Vsesoyuznaya perepis' naseleniya 1926 goda).

23.1.2   1937 population census

The fate of the 1937 population census was tragic. The results were declared to be wrong; the census itself was declared methodologically fallacious; and its organizers were repressed. Recent studies have shown that its methodological blemishes appear to have been greatly exaggerated (Livshits, 1990; Tolts, 1991; Volkov, 1990). Until recently census data, not completely preserved, were unavailable to investigators. The total population size registered in the 1937 census, 162 million people (Iz arkhivov Goskomstata SSSR, 1990) was considerably less than earlier published estimates, which became the reason for the disavowal of the census results.

For the 10 years preceding the 1937 census, registration of vital events was far from complete. Under-registration of deaths exceeded under-registration of births, which led to the overestimation of natural increase and population sizes. Particularly big discrepancies appeared to be in the data from Kazakhstan, the Ukraine, the North Caucasus region, and the Volga region (the regions most affected by the 1933 famine). Data by sex and age from the 1937 census were compared with the 1939 census data (corrected independently) and with the data on births and deaths in 1937 and 1938. The under-registration of the 1937 census was estimated at 0.43 percent or 700,000 people.

23.1.3   1939 population census

Study of the 1939 census, and the documents connected with it, shows that 167.309 million people were registered, which is less than the officially released figures for 1939 [TsGANKh (Central State Archives of National Economics), f. 1562, op. 329, yed. khr. 256, 277; op. 336, yed. khr. 95].

The population size was announced before census results were obtained; Stalin (1952) announced that the size of the population was 170 million people on 10 March 1939 in his report to the 18th Congress of VKP(b).

The USSR State Planning Committee chairman, Voznesenskiy, and the TsUNKhU head, Sautin, on 21 March 1939 (TsGANKh, f. 1562, op. 329, yed. khr. 256, l. 38) in a note to Stalin and Molotov estimated the population size to be 170.13 million people. They arrived at this figure by adding 1.142 million people to the registration results and adjusting the figure by 1 percent (1.7 million people) for possible under-registration. Their calculations increased the figure by almost 2.8 million. Analysis shows that not more than 806,000 people should have been added to the census figure.

With all corrections considered, the population size according to the 1939 census was estimated to be 168.871 million people. The results of all three censuses were shifted to the beginning of 1927, 1937, and 1939, correspondingly, smoothed and linked with each other to get vital statistics and plausible dynamics of certain cohort sizes. For further calculations we considered the USSR population size at the beginning of 1927 to be equal to 148.656 million people; at the beginning of 1937, 162.500 million people; at the beginning of 1939 (within borders before 17 September 1939), 168.524 million people. We used these estimates as a basis for further calculations without any modification. Population size at the beginning of each year, after adjustments for age and sex, has been published in Andreev *et al.* (1990b).

23.2   Vital Statistics

23.2.1   Number of births and deaths: 1927 to 1936

Statistical birth and death under-registrations in the prewar years were explained by two factors: first, registration was not carried out or was incomplete in some regions; second, under-registration occurred in some regions.

Analysis of vital statistics data shows that there were death and birth under-registrations. Our adjustment for the registered number of births was based on the estimated size of certain cohorts in the 1937 and 1939 censuses and on the number of deaths in these cohorts from birth.

To obtain a final estimate we combined the data from three sources: the population size in certain cohorts from the 1937 and 1939 censuses;

the number of births from vital statistics; and an estimate of the number of births based on 1960 fertility survey data (Sifman, 1974).

The sharp decrease in the number of births in 1932, 1933, and 1934 only partly reflects the actual situation. In 1933 famine affected many in rural areas in the Ukraine, Kazakhstan, the Volga region, and the North Caucasus. According to available data, the famine was first evident in the spring and summer of 1933. One can hardly estimate its significant effect on the number of pregnancies in 1932 and the number of births in 1933.

One should also bear in mind that the population in that period, especially in rural areas, had no experience in practicing birth control. Therefore the low number of births registered in 1933 and partly in 1934, to a great extent, is an indication of the high share of unregistered births and infant deaths in the period of famine; this is confirmed by the sizes of these cohorts counted in the 1937 census. Using these corrections, we calculated that the total number of births from 1927 to 1936 was 60.97 million instead of the 47.54 million recorded in vital statistics.

Our next task was to obtain annual numbers of deaths from 1927 to 1936. According to registration data, 28.5 million deaths were recorded in this period. Assuming that the percentage of population covered by civil registration is approximately the same for births and deaths we estimated the number of deaths to be 34.3 million people. Considering the growth of population size from 1927 to 1936 to be 13.8 million people and the number of births in the same period to be 60.9 million people, the sum of deaths and net-migration must be 47.1 million people.

Studies have estimated that during this period emigration was equal to 0.2 million people (Maryański, 1966; Tatimov, 1989). Using this figure we estimated the total number of deaths to be 49.6 million, i.e., average under-registration of deaths in this period appears to be 30.7 percent.

As a result, the estimates of the annual number of deaths were obtained; they did not contradict the data on population size and structure in 1927 and 1937 or the estimates of the number of births.

Data on the number of births and deaths for each month remain in the archives for some regions. Analysis of these data shows that the period of catastrophe-level mortality was relatively short. The famine started in spring 1933 and ended in late 1934. The mortality level in 1933 was essentially higher than in 1932 and 1934. A high increase in the mortality level was first evident among the rural population, though cities also experienced a sharp rise in mortality. The rise in mortality appears to have taken place mainly in 1933. Comparing each cohort-size estimate of early 1927 and early 1937 and the total number of births and deaths for each year of the period from 1927 to 1936, we calculated the number of age-specific deaths for each year. We used life tables for 1926–1927 and 1938–1939 and the data on the total number of deaths. The results are presented in *Table 23.1*.

23.2.2 Vital statistics in 1937–1938

In 1937–1938 registration coverage of territories was incomplete, and some under-registration remained. We compared the number of registered births with the number of children under 2 according to the 1939 census data and the number of infant deaths registered. Thus we obtained reliable estimates of the number of births in these years. A fertility-level rise is confirmed by the 1960 survey results; this rise is due to the effect of the USSR Central Executive Committee resolution prohibiting abortions, adopted on 26 June 1936.

As archive data reveal, deaths of exiles and prisoners were not included in the civil registration system, at least up to 1937. There is no reason for concluding that this policy changed in 1938–1939. According to the data on GULAG's contingent replacement (*Argumenty i Fakty*, 1989), 115,900 people died in the camps of GULAG during 1937–1938. The repressed died not only in camps but also in prisons and isolation. In addition, mass executions occurred with and without court sentences. Thus, the number of deaths not included in the statistics of those years was much higher. According to published KGB material, of the 3.778 million people repressed between 1930 and 1953, 786,000 were shot (USSR State Security Committee, 1990). A considerable number of these executions occurred in 1937–1938. Therefore, a great part of the death underestimation years is connected with mass repressions rather than with inadequate registration.

Further adjustments on age-specific mortality rates of male and female in 1937–1938 were made using these data. We obtained the estimates of age-specific birth rates by proportionally adjusting the rates calculated on the basis of 1960 survey data.

Recorded natural increase of the USSR population in 1939 and 1940 was 5.89 million people. However, birth and death registrations in this period were inadequate.

Table 23.1. Estimates of population size in the USSR and main indexes of its reproduction from 1927 to 1940.

Year	Population (in millions)	Number of births (in millions)	Number of deaths (in millions)	Natural increase (in millions)	Total fertility rate	Life expectancy (years)	Infant mortality rate (per 1,000)
1927	148.656	6.950	3.984	2.965	6.396	37.5	182
1928	151.622	6.944	3.878	3.066	6.155	38.9	182
1929	154.687	6.876	4.132	2.745	5.944	37.4	190
1930	157.432	6.694	4.284	2.410	5.680	36.5	196
1931	159.841	6.510	4.501	2.009	5.365	35.0	210
1932	161.851	5.837	4.786	1.051	4.666	32.8	213
1933	162.902	5.545	11.250	-5.705	4.470	11.6	317
1934	156.797	4.780	3.410	1.369	3.796	38.2	204
1935	158.167	5.249	3.282	1.967	4.042	39.6	198
1936	160.134	5.589	3.223	2.366	4.207	41.1	186
1937	162.500	6.549	3.557	2.992	4.874	39.9	184
1938	165.492	6.516	3.483	3.033	4.736	41.4	174
1939[a]	168.524						
1939	188.794	7.634	3.829	3.805	4.944	43.6	168
1940	192.598	6.999	4.205	2.794	4.531	41.2	184
1941	195.392						

[a]USSR borders changed in 1939.
Figures may not add up because of rounding.

Table 23.2. Population size by age group in the USSR in mid-1941, in millions of inhabitants.

Age	Total	Males	Females
0–4	26.514	13.324	13.189
5–9	18.463	9.228	9.235
10–14	22.325	11.102	11.222
15–19	20.914	10.451	10.462
20–24	14.950	7.231	7.718
25–29	17.436	8.342	9.094
30–34	16.618	8.163	8.456
35–39	13.622	6.625	6.997
40–44	11.043	5.085	5.959
45–49	8.242	3.805	4.437
50–54	6.978	3.106	3.872
55–59	5.850	2.482	3.368
60–64	4.926	1.974	2.952
65+	8.835	3.419	5.416
Total[a]	196.716	94.338	102.378

[a]Totals may not add up because of rounding.

After September 1939 the USSR borders changed, and modern borders were fixed by 1946. According to available data, in the autumn of 1939, the population size in the USSR increased by 20.1 million people, i.e., it was 190.7 million. All further computations were made for the population within modern borders.

In some parts of the country, including the newly joined regions of the Ukraine and Byelorussia, registration was inadequate or nonexistent. Analysis of recorded data allows one to make adjustments for territories not covered by registration: 3.5 percent for 1939 and 10 percent for 1940. Based on this analysis the following adjustment was chosen for the first 10 years after the war: 5 percent for births and 10 percent for deaths. The best estimate of the population size in the USSR in mid-1941 is 196.7 million people (*Table 23.2*). In 1941 the USSR population age structure was already marked with two demographic catastrophes. World War I and the civil war caused a considerable disproportion in the age group over 40 and a decrease in the size of the cohorts born between 1914 and 1924. The 1933 famine, which led to a fertility decline and a sharp mortality increase (especially among children), caused a decrease in the age structure of children between the ages 5 and 9.

23.3 Human Losses and Demographic Deficit in the 1930s

By reconstructing the age–sex structure, as well as vital statistics estimates, we can approximate the population losses in the 1930s. Here we distinguish between human losses caused by mortality increase and losses from net migration and demographic deficit (including the consequences of fertility decline).

Human losses were estimated by population projections by ages from 1927. The estimate is based on annual reconstruction of the USSR population age–sex structure and age-specific fertility and mortality rates in the period from 1927 to 1941. Archive data on population statistics and balanced cohorts size with the 1926, 1937, and 1939 censuses were used. To estimate the size of these losses, reconstructed age-specific fertility and mortality rates during the periods of catastrophes were compared with the outcomes of their linear interpolation between the first and the last years of the period (*Figure 23.1*).

Thus, we estimated total demographic deficit in the period from 1927 to 1941 at 13.5 million people; of this human losses were estimated at about 7 million people. This indicates general collectivization and the repressions connected with it, as well as the 1933 famine, may be responsible for 7 million deaths. *Figure 23.2* presents the adjusted age structure of the USSR population according to the 1939 census data.

23.4 Population Size in 1946

We thought it expedient to estimate losses for the period from the end of June 1941 until the end of December 1945. We shifted the upper limit of the period from the end of the war to the end of the year to account for the number of deaths of the wounded in hospitals, repatriation of war prisoners and displaced civil persons to the USSR, and repatriation of citizens of other countries.

According to our estimation, population size in the USSR at the beginning of 1946 was 170.5 million people (74.3 million males and 96.2 million females). Population size in the USSR at the beginning of 1946 was calculated using the reverse-survival method according to the population-size estimate of the 1959 population census, using vital statistics and data on migration from and to the USSR during the

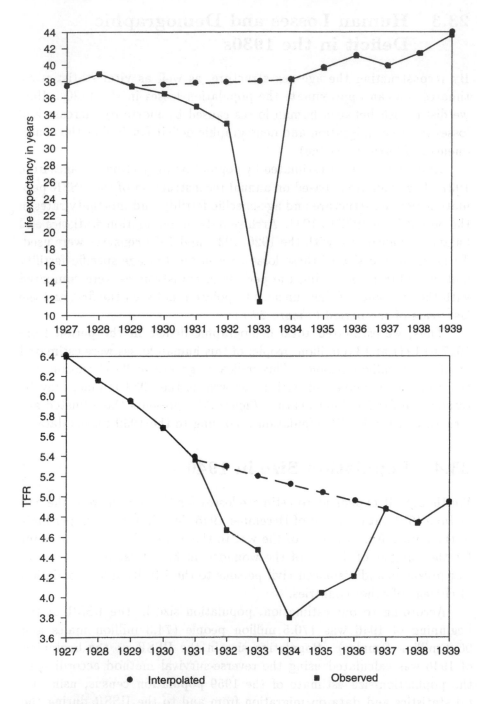

Figure 23.1. Linear interpolations of total fertility rate and life expectancy from 1927 to 1939.

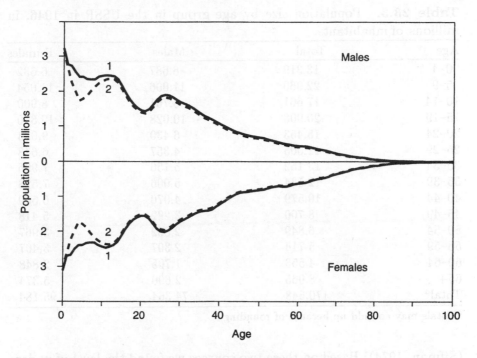

Figure 23.2. Human losses in the USSR populations in the mid-1930s:
(1) hypothetical structure in case of no human losses and (2) actual age
structure in early 1939.

period 1946–1958. Calculations were based on a comparison of cohort
size, according to the 1959 census, adjusted for childhood ages, and vital statistics for 1946–1958, as well as on estimates of birth and death
records made by statistics officials during this period.

The materials show that during this period the share of unregistered
deaths was much higher than that of unregistered births. It follows
from the same data that correlation of the shares of underestimated
deaths among children and adults was relatively stable. We took this
interrelation as a basis for correcting data on the number of births and
deaths in the period 1946–1958.

The mortality curve for the reverse-survival method calculations of
the 1959 census was modified for each year using the Brass (1971) model.
The number of deaths of children and adults had to correspond to the
adjusted registration data. The number of births was controlled using birth rates obtained during the 1960 retrospective fertility survey

Table 23.3. Population size by age group in the USSR in 1946, in millions of inhabitants.

Age	Total	Males	Females
0–4	13.319	6.687	6.632
5–9	22.060	11.006	11.054
10–14	17.661	8.761	8.900
15–19	20.908	10.028	10.880
20–24	15.453	6.430	9.023
25–29	11.005	4.357	6.648
30–34	13.152	5.156	7.996
35–39	12.534	5.006	7.528
40–44	10.579	4.070	6.509
45–49	8.700	3.282	5.418
50–54	6.849	2.882	3.967
55–59	5.714	2.307	3.407
60–64	4.553	1.705	2.848
65+	8.065	2.690	5.374
Total[a]	170.548	74.364	95.184

[a]Totals may not add up because of rounding.

(Sifman, 1974). Based on these two sources we found the level of under-registration for the years 1946–1958 to be 6.9 percent for births and 18.5 percent for deaths.

Population age structure at the beginning of 1946 presented in *Table 23.3* is marked by the demographic catastrophe that the country experienced. The total number of males and that of females in 1946 was 21 percent and 6 percent, respectively, lower than in 1941. Because of the decrease in fertility and increase in child mortality level during the war the number of children under age 5 was one-half of what it was in 1941. Of total population number at the beginning of 1946, 159.456 million people (68.783 million males and 90.673 million females) were born before the war.

23.5 Human Losses during World War II

Our estimates show that the total number of deaths during the period 1941–1945 (including Soviets outside the country) of those born before July 1941 was 37.2 million people. But this number, of course, cannot be considered as giving only losses from World War II; people die in peacetime as well. If age-specific death rates of the USSR population in 1941–1945 were the same as in prewar 1940, then the number of deaths during four and a half years would be 11.9 million people. Thus war-related human losses of these cohorts are 25.3 million people (37.2 minus 11.9).

Naturally a question arises, should the number of human losses include deaths caused by illegal repressions during war years and deaths of prisoners who had been in camps when the war began? To include these deaths in the number of war victims is by no means justified. It should be taken into account that the 1940 death rate, which we took as the standard, includes some deaths of those repressed. From memoirs of those imprisoned, one may conclude that the mortality level in the camps, being very high in peacetime, increased during the war years. One can regard such excess in mortality as a consequence of the war, and it is quite possible to include it in the total number of human losses from the war, as it is in our calculations.

To the 25.3 million war-related human losses of those born before the war, one must add losses due to higher (than before the war) mortality of children born during the war. Two methods to estimate the number of births during the war were used: an analysis of mortality correlations in different age groups in 1941–1945 in comparison with the numbers of those who survived up to the beginning of 1946 and a 1960 retrospective fertility survey. The number of births in the war years according to these data was estimated at 16.5 million people. It is quite close to Biraben's (1976) estimate of 16.25 million births during the period 1941–1945. The excess in the number of deaths in comparison with the standard level is 1.28 million people. This 1.3 million should be added to human losses.

Thus, total human losses in the population of the USSR during World War II (1941–1945), estimated by the method of demographic balance, are approximately 26.6 million people. As the calculation method shows, this estimate is approximate and may be adjusted in future investigations. But it is quite close to the estimates obtained by others using different methods. In any case differences between various estimates cannot be large. For example, in estimating the 1939 census data, we considered an alternative variant, according to which the USSR population size within the borders before 17 September 1939 would be 167.937 million people. In this case population size at the beginning of 1941

Table 23.4. Human losses in the USSR during World War II (1941–1945) by age and sex, in million of inhabitants.

	Males			Females		
Age	Synthetic size	Actual size	Human losses	Synthetic size	Actual size	Human losses
0–4	7.334	6.687	0.647	7.293	6.632	0.661
5–9	11.591	11.006	0.585	11.684	11.054	0.630
10–14	8.954	8.760	0.194	9.007	8.900	0.107
15–19	11.092	10.028	1.064	11.220	10.880	0.340
20–24	9.839	6.430	3.409	9.911	9.023	0.888
25–29	6.871	4.357	2.514	7.437	6.648	0.789
30–34	8.238	5.156	3.082	8.982	7.996	0.986
35–39	7.712	5.006	2.706	8.007	7.528	0.479
40–44	6.148	4.070	2.078	6.658	6.509	0.149
45–49	4.637	3.282	1.355	5.571	5.418	0.153
50–54	3.404	2.882	0.522	4.137	3.967	0.170
55–59	2.727	2.307	0.420	3.586	3.407	0.179
60–64	2.100	1.705	0.395	3.046	2.848	0.198
65+	3.767	2.689	1.078	6.207	5.375	0.832
Total	94.414	74.365	20.049	102.746	96.185	6.561

[a]Totals may not add up because of rounding.

would not be 195.392 million people but 194.821 million people. Therefore, human losses among those born before the war were 24.74 million and total losses were 26.0 million people.

Of the total volume of human losses (*Table 23.4*) over 76 percent, or 20.0 million, were male. The cohorts of males born between 1901 and 1931 suffered the most losses; of the total losses their share is more than half or 10 million men. Using model life tables one can estimate the share of human losses as a result of total rise of mortality level caused by worsening life conditions. During the war, mortality level rose all over the country; thus, according to current estimates, negative natural increase took place not only in the Ukraine (Perkovskiy and Pirozhkov, 1990) and Byelorussia (Rakov, 1974), which were occupied, but also in Siberia, which was far from military actions (Isupov, 1990). According to our calculations, excess mortality may account for 9 to 10 million deaths that occurred during the war.

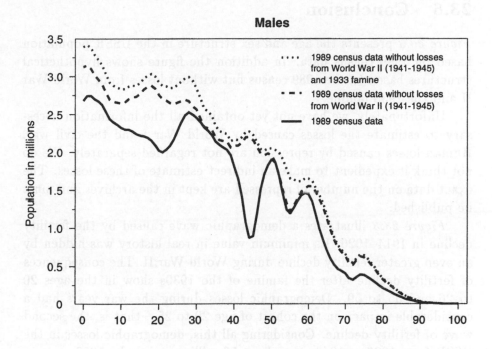

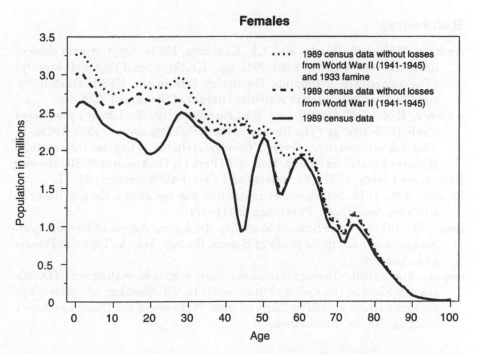

Figure 23.3. Demographic losses of the USSR population.

23.6   Conclusion

Figure 23.3 presents the age and sex structure in the USSR population based on 1989 census data. In addition the figure shows hypothetical structures based on the 1989 census but without losses from World War II and the 1933 famine.

Unfortunately, we have not yet obtained all the information necessary to estimate the losses caused by World War I and the civil war. Human losses caused by repression are not regarded separately. We do not think it expedient to make an indirect estimate of these losses. The exact data on the number of repressed are kept in the archives and must be published.

Figure 23.3 illustrates a demographic wave caused by the fertility decline in 1914–1920. Its minimum value in real history was hidden by an even greater fertility decline during World War II. The consequences of fertility decline after the famine of the 1930s show in the ages 20 to 26 and 55 to 59. Demographic losses during the war years had a considerable impact on the cohort of age 20 to 25 – that is, the second wave of fertility decline. Considering all this, demographic losses in the USSR from 1927 to 1946 were about 56 million people by 1959.

References

Andreev, E.M., L.E. Darsky, and T.L. Kharkova, 1990a. Opyt otsenki chislennosti naseleniya SSSR 1926–1941 gg: Kratkiye rezul'taty issledovaniya (The Experience Estimating Population Size in the USSR: Brief Investigation Results). *Vestnik statistiki* (Herald of Statistics) **7**:34–46.

Andreev, E.M., L.E. Darsky, and T.L. Kharkova, 1990b. Istoriya naseleniya SSSR 1920–1959 gg (The History of the USSR Population, 1920–1959). In *Ekspress-informatsiya, Seriya "Istoriya statistiki"* (Express-Information, History of Statistics Series). Issue 3–5 (Part 1). Goskomstat SSSR, Moscow.

Argumenty i fakty, 1989. (Arguments and Facts) **45**(November):11–17.

Biraben, J.N., 1976. Naissances et répartition par âge dans L'Empire Russe et en Union Sovietique. *Population* **2**:441–471.

Brass, W., 1971. On the Scale of Mortality: Biological Aspect of Demography. *Symposia of Society for Study of Human Biology*, Vol. X. Taylor & Francis Ltd., London.

Isupov, V.A., 1990. Demograficheskaya sfera v epokhu stalinizma. (Demographic Field in the Epoch of Stalinism). In V.I. Shishkin, ed. *Aktual'nye problemy istorii sovetskoy Sibiri* (Actual Problems of the History of Soviet Siberia). Nauka, Novosibirsk.

Iz arkhivov Goskomstata SSSR (The USSR State Committee on Statistics Archives), 1990. Itogi Vsesoyuznoy perepisi naseleniya 1937 g (The 1937 Population Census). *Vestnik statistiki* (Herald of Statistics) **7**:65–79; **8**:66–79.

Korchak-Chepurkovskiy, Y.A., 1970. O nekotorykh voprosakh istorii i metodiki perspektivnykh raschetov naseleniya (Some Questions of the History and Methods of Population Projections). In Y.A. Korchak-Chepurkovskiy, *Izbrannye demograficheskiye issledovaniya* (Selected Demographic Studies). Moscow.

Kozlov V.I., 1989. O lyudskikh poteryakh Sovetskogo Soyuza v Velikoy Otechestvennoy voyne 1941-1945 godov (On Human Losses in the USSR during the Great Patriotic War 1941–1945). *Istoriya SSSR* (History of the USSR) **2**:132–139.

Kurman, M.V., 1937. TsGANKh, f. 1562, op. 329 yed. khr. 132.

Kvasha, A.Y., 1990. Tsena pobed (The Cost of Victories). In V.I. Mukomel, ed., *SSSR: Demograficheskiy diagnoz* (USSR: Demographic Diagnosis). Progress, Moscow.

Livshits, F.D., 1990. Perepis' naseleniya 1937 goda (The 1937 Population Census). In A.G. Volkov, ed. *Demograficheskiye protsessy v SSSR* (Demographic Processes in the USSR). Moscow.

Lorimer, F., 1946. *The Population of the Soviet Union: History and Prospects*. League of Nations, Geneva.

Maksudov, S., 1989. *Poteri naseleniya SSSR* (Population Losses in the USSR). Chalidze Publications, Benson, VT.

Maryański, A., 1966. *Wspo'lczesne wedzówki ludów: Zarys geografii migracji* (Modern Human Migrations; Geography of Migration Essays). Zaklad Narodowy Imienia Ossolinskich-Wydawnictwo, Wroclaw, Warsaw, Krakow. Ossolin'skich-Wydawnictwo.

Perkovskiy, A.L., and S.I. Pirozhkov, 1990. Iz istorii demograficheskogo razvitiya 30 – 40kh godov: na primere Ukrainskoy SSR (From the History of Demographic Development of the 1930–1940s: The Ukraine as an Example). In *Ekonomika, Idemografiya, Statistika: Issledovaniya i problemy*. (Economics, Demography, Statistics: Studies and Problems). Nauka, Moscow.

Ptoukha, M.V., 1987. Naseleniye Ukrainy do 1960 goda (The Ukraine Population up to 1960). In T.V. Ryabuskin, ed. *Sovetskaya demografiya za 70 let*. (Soviet Demography over the Past 70 years). Moscow.

Rakov, A.A., 1974. *Belorussiya v demograficheskom izmerenii* (Byelorussia in Demographics Measuring). Minsk.

Rybakovskiy L.L., 1989. Dvadtsat' millionov ili bol'she? (Twenty Million or More?). *Politicheskoye obrazovaniye* (Political Education) **10**:96–98.

Sifman, R.I., 1974. *Dinamika rozhdaemosti v SSSR* (Fertility Dynamics in the USSR). Moscow.

Smertnost' i prodolzhitel'nost' zhizni naseleniya SSSR 1926–1927 (Mortality and the Expectation of Life in the USSR, 1926–1927), 1930. Moscow and Leningrad.

Stalin, I.V., 1952. *Voprosy leninizma* (The Problems of Leninism), 11th edition. Moscow.

Tatimov, M.B., 1989. *Soeial'naya obuslovlennost' demograficheskikh protsessov* (Social Conditionality of Demographic Processes). Alma-Ata.

Tolts, M.S., 1991. Perepis', prigovorennaya k zabveniyu (Census Sentenced to Oblivion). In A.G. Vishnevsky, ed., *Demografija i sociologija: Sem'ia i semeynaya politika* (Demography and Sociology: Family and Family Policy). Institute for Socioeconomic Studies of the Population, Moscow.

TsGANKh, f. 1562, op. 329, yed. khr. 256, 277; op. 336, yed. khr. 95.

TsGANKh, f. 1562, op. 329, yed. khr. 256, l. 38.

TsGANKh, f. 1562, op. 329, yed. khr. 132.

USSR State Security Committee, 1990. *Izvestiya* (News) February 14 1990.

Volkov, A.G., 1990. Iz istorii perepisi naseleniya 1937 goda (From the History of the 1937 Population Census). *Vestnik statistiki* (Herald of Statistics) **8**:45–56.

Vsesoyuznaya perepis' naseleniya 1926 goda (All-Union Population Census, 1926), 1929. Gosudarstvennoye sotsialno-ekonomicheskoye izdatelstvo, Moscow and Leningrad.

Chapter 24

Demographic Regularities and Irregularities: The Population Age Structure

Sergei Pirozhkov and Gaiané Safarova

Age–sex distribution is the most important population characteristic, even more significant than the population size. Researchers, however, have always devoted special attention to the latter. This is mainly due to the structural factor that affects the processes inherent in population, leaving its general quantitative characteristics unchanged.

Analysis of population age composition provides a thorough insight into separate processes of natural population movement and, hence, a clear picture of the characteristics of the population's reproduction. Sometimes it is impossible to explain changes in population size ignoring its age distribution, since the latter is the result of the evolution of the population's reproduction regime. Moreover, the age distribution makes a definite contribution to future population growth.

Changes in age distribution depend on two basic groups of factors: the regular or evolutionary factors, characterizing a smooth process of demographic development excluding any perturbations, and the crisis factors, characterizing disturbances in the normal course of demographic history such as wars, famines, and mass repressions. If the group of evolutionary factors determines the general regularities of change in the age composition, then the second group, the irregularities or crisis factors, interrupts the normal development of age distributions causing short-term breaks and discontinuities. Keyfitz (1987) introduced the concept of *demographic discontinuity* to describe these disruptions. Demographic discontinuity is caused by sudden changes in fertility and mortality levels of different cohorts that are affected by profound social perturbations. It generates long-term waves in age-structure changes.

This chapter attempts to illustrate to what extent changes in the age distribution of the USSR population were conditioned by the regular evolutionary factors and by the irregular crisis factors.

24.1 Evolution of Age Distribution during Demographic Transition

The demographic transition, the global tendency of change in reproduction regimes from high fertility and mortality levels to low ones, developed initially in Western European countries in the late 18th century and mid-19th century. In the late 19th century the demographic transition extended to the European region of the territory that later became the Soviet Union. For the period from 1896 to 1900 the crude birth rate for 50 provinces of Russia was 49.5 per 1,000 inhabitants, while the death rate was on average 32.1 per 1,000 (Rashin, 1956).

The fertility and mortality declines were more rapid in the Baltic states and the major cities of Russia. According to Novoselskiy and Payevskiy (*Smertnost' i prodolzhitel'nost zhizni*, 1930), life expectancy for the population of European Russia in 1896–1897 was 31.43 years for males and 33.36 years for females. These values were close to those in Western Europe in the late 18th and early 19th centuries.

Some Soviet demographers (Belova *et al.*, 1988) have determined that fertility limitation started in Russia with female cohorts born in 1870. At this time the total fertility rate was 7.0. For cohorts born between 1870 and 1900, it declined from 6.32 to 4.47. During the following period, TFR steadily declined, and it stabilized at the level of 2.3–2.2 only for cohorts born after 1925. Thus, it can be assumed that the active phase of demographic transition to the modern type of fertility took place in the period between 1920 and 1960 in the USSR. For synthetic cohorts, the total fertility rate stabilized at 2.5.

The beginning of demographic transition in European Russia was characterized by a mortality decline that was slower than the mortality decline in Western European countries. The demographic transition was conditioned by cataclysms such as the World War I (1914–1918), the civil war (1918–1920), and famines and epidemics in the 1920s. The total direct and indirect losses of the USSR population from 1914 to 1920 have been estimated at more than 25 million (Kvasha, 1990). It was against this background that the pre-transition population distribution was formed, which can be evaluated from the 1897 and 1926 censuses.

The analysis of these age distributions shows that their population profiles, on the whole, belong to the classic type of progressive structure represented by a triangle with broad base and narrow vertex (*Figure 24.1*). This type of structure takes place in populations with high fertility and mortality levels, low life expectancy, and the absence of population aging. In the age distribution obtained in the 1926 census a marked gap in the 5–9 age group was evident. Its total number was 1.8 million children less than that in the 10–14 age group. Evidently, the 5–9 age group was more affected by the consequences of World War I and the civil war and by other social crises of the 1920s than the 10–14 age group.

Changes in the Soviet Union's population age distribution following the 1926 census was considerably influenced by the evolution of a population reproduction regime in demographic transition (a factor that significantly affected the transformation of age composition) and by profound social perturbations of the 1930s and 1940s (famine in 1932–1933, mass repressions in the 1930s, and World War II). The latter influences have dramatically disturbed the natural course of USSR demographic history. Therefore both evolutionary and the crisis factors have had a decisive impact on the population age composition after 1926. This resulted, on the one hand, in a considerable decrease in the children's age-group proportions and in an increase in the proportions of the older population and, on the other hand, in significant deformations in the age distribution. The age compositions in 1926 and 1989 in the USSR are given in *Figure 24.2*.

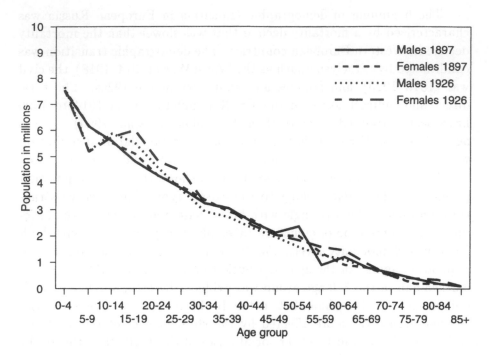

Figure 24.1. Age profiles of the populations of European Russia, 1897 and 1926.

24.2   The Procedure for Estimating the USSR Population in the 1930s and 1940s

Reliable data on the USSR's population reproduction and size were not available for the period from the late 1920s to the early 1950s. Some officially published summary indicators of the country's demographic development were falsified. The materials of the 1937 census were treated as "defective" and its organizers were repressed and called "the people's enemies." Modern analyses of USSR demographic development in the 1930s and qualified expert estimations of the 1937 census materials do not confirm the conclusions of the Soviet People's Commissar's resolution of 25 September 1937, which considered this census unsatisfactory (Volkov, 1990; Andreev *et al.*, 1990).

Lack of reliable statistical information on population trends in the 1930s and 1940s evoked estimates of USSR population losses due to the famine in 1932–1933, mass repressions, and so on, especially in the West.

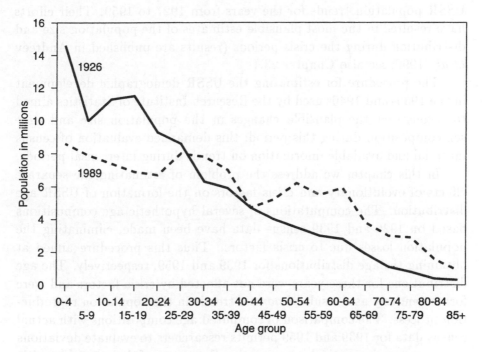

Figure 24.2. Age compositions in 1926 and 1989 in the USSR.

Most estimates did not take into account data on population trends from 1927 to 1938 and were not based on demographic accounts.

The figures of USSR population losses during the 1930s and 1940s cited in literature vary considerably. According to Anderson and Silver (1985) the USSR population lost from 3.2 to 5.5 million people; Maksudov (1989) estimates the USSR lost 9.8 million people. A figure of 9.9 million people (7.9 million losses from starvation or repression and 2 million losses from emigration) is given by Soviet historian Tzaplin (1989); this number is close to previous estimates.

Attempts at estimating the total population and its distribution in the 1930s were repeatedly made by USSR statistical offices (in 1944 by the committee under the guidance of Popov and in 1964 by Bekunova and Rodina), but these estimates were not published. In 1989–1990 scientists from the Department of Demography of the Research Institute of Statistics of the State Committee of the USSR on Statistics, after a thorough analysis of data from 1926, 1937, 1939, and 1959 censuses, using modern methods of demographic modeling, reconstructed data on

USSR population trends for the years from 1927 to 1959. Their efforts have resulted in the most plausible estimates of the population size and distribution during the crisis periods (results are published in Andreev *et al.*, 1990; see also Chapter 23.)

The procedure for estimating the USSR demographic development in the 1930s and 1940s used by the Research Institute of Statistics aimed to reconstruct the plausible changes in the population size and age–sex composition during this period; this demanded evaluation of census material and available information on trends during intercensal periods.

In this chapter we address the problem of estimating the separate effects of evolutionary and crisis factors on the formation of USSR age distribution. The computations of several hypothetic age compositions based on 1926 and 1939 census data have been made, eliminating the population losses due to crisis factors. Thus this procedure aimed at obtaining the age distributions for 1939 and 1959, respectively. The age distributions for these years were not affected by crisis factors and were formed entirely as a result of *normal* trends in the population reproduction process. The comparison of computed age compositions with actual census data for 1939 and 1959 permits researchers to evaluate deviations in age-group sizes, which characterize the scope of demographic crisis brought about by events such as famine in 1932–1933, mass repressions in the 1930s, and World War II.

At the same time, we suppose that the estimates of deviations of a hypothetic population's age-group sizes from the real sizes characterize not the net population losses due to crises but losses of the country's demographic potential as a result of social catastrophes of the 1930s and 1940s. The latter losses include both actual losses of the population and birth deficits due to the crisis (that is, children who were not born due to the parents' death, separation, and so on).

For the computation of hypothetic age distributions for 1939 and 1959 three data groups are required: age–sex composition in the initial period (1926 and 1939), total fertility rate (TFR), and life expectancy at birth (for males and females separately) in each transition step.

24.3 Estimates of Population Losses: 1926–1939

Computation of the USSR population age–sex distribution for 1939 has been carried out in two stages:

- The 1926 census data on age–sex structure, age-specific fertility rates, and the life table for 1926–1927 have been taken as initial data. Based on these data a one-step transition has been performed and population distribution for 1931 has been computed. The age–sex distribution for 1929 has been derived as the linear combination of age–sex compositions for 1926 and 1931.
- On the basis of the 1929 population distribution a five-year transition with two steps has been performed and as a result the age–sex composition for 1939 has been obtained.

Fertility and mortality indexes were available only for 1926–1927 and 1938–1939. For the intercensal period, 1927–1938, the values of TFR and life expectancy at birth were obtained by linear interpolation (*Table 24.1*).

The hypothetic age distribution for 1939 was computed based on previously published official age-structure data of the 1926 census and interpolated values of fertility and mortality levels (denoted as scenario A in Appendix *Table 24A.1*). Another hypothetic age composition was projected based on the reconstructed age distribution for 1927 (Andreev *et al.*, 1990) and the fertility and mortality levels as in scenario A; this age composition is denoted as scenario B (see Appendix *Table 24A.2*).

Comparing real (census or reconstructed) population age–sex composition for 1939 with scenarios A and B shows the USSR demographic losses due to the social cataclysms of the 1930s. The results of comparative analysis of real and hypothetic age distributions are given in *Table 24.2* and *Figure 24.3*. *Table 24.2* shows that the USSR total demographic potential losses (actual losses and birth deficits) due to the social cataclysms of the 1930s according to scenario A are equal to 22.1 million people, the age groups of children being the most affected (*Figure 24.3*). According to scenario B, the USSR demographic potential losses are equal to 22.7 million people. The estimates do not differ much; their divergence is caused by differences in initial age compositions.

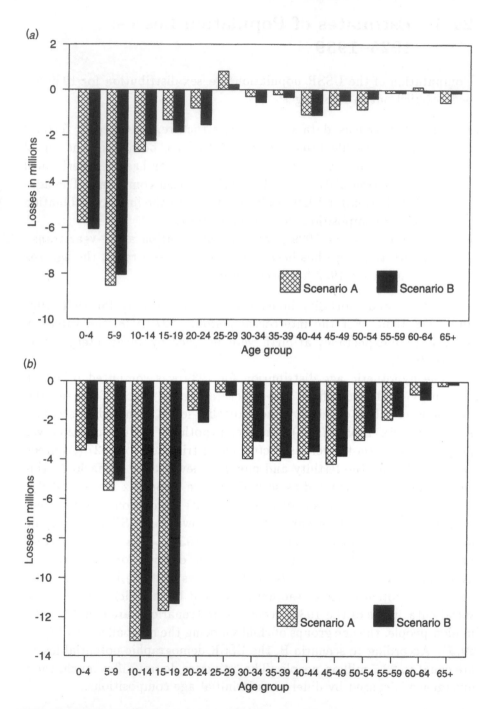

Figure 24.3. Losses of the USSR demographic potential in (*a*) 1926–1939 and (*b*) 1939–1959, in millions of people.

Table 24.1. Total fertility rate (TFR) and life expectancy at birth in 1928–1929 and 1933–1934.

Years	TFR	Life expectancy at birth	
		Males	Females
1928–1929	5.15	42.3	47.3
1933–1934	4.77	43.3	48.5

Table 24.2. Estimated losses of the USSR population in 1926–1939, in millions of people.

Scenario	Total losses	Male losses	Female losses
Scenario A	22.1	12.1	10.0
Scenario B	22.7	12.5	10.2

24.4 Estimates of Population Losses: 1939–1959

Computation of hypothetic age–sex composition for 1959 based on 1939 census data adjusted to modern borders has been carried out in four steps for two scenarios (A and B). Scenario A is based on published data of the 1939 and 1959 censuses and linear interpolated values of TFR and life expectancy at birth for intermediate points. In scenario B the reconstructed age–sex composition for 1939 (Andreev *et al.*, 1990) is taken as the initial composition and TFR and life expectancy at birth are the same as those in scenario A.

In the first step, the data available for 1938–1939 have been taken for TFR and life expectancy. For the second and third steps, TFR and life expectancy at birth have been obtained by linear interpolation (*Table 24.3*). The effects of World War II and famine in 1947 on fertility and mortality levels have been eliminated.

Table 24.3. Total fertility rate (TFR) and life expectancy at birth for 1943–1944 and 1948–1949.

Year	TFR	Life expectancy at birth	
		Males	Females
1943–1944	3.90	49.7	55.5
1948–1949	3.40	55.3	61.2

Table 24.4. Estimated losses of the USSR population in 1939–1959, in millions of people.

Scenario	Total losses	Male losses	Female losses
Scenario A	58.0	35.6	22.4
Scenario B	55.3	34.4	20.9

Table 24.5. Estimated losses of the USSR population in 1926–1959, in millions of people.

Scenario	Total losses	Male losses	Female losses
Scenario A	80.1	47.7	32.4
Scenario B	78.0	46.9	31.1

In the fourth step, the TFR for 1954–1955 has been used (this rate is less than the TFR in 1953–1954 and does not overstate the estimate). The life expectancy for males and females has been obtained from available data. The results of comparative analysis of real and projected age compositions are given in *Table 24.4* and *Figure 24.3*.

Scenario B is preferable as it is based on reconstructed age–sex distribution for 1939. According to scenario B, USSR population losses from World War II, the famine of 1947, and mass repressions of the 1940s and early 1950s are equal to 55.3 million people; males and children were most affected.

24.5   Estimates of Population Losses: 1926–1959

The estimates of the USSR's population losses due to crisis factors of the 1930s and 1940s are given in *Table 24.5*. The table shows that over the period of 30 years the USSR population could have grown by approximately 80 million people. Differences of estimates from scenarios A and B result from differences in the initial age structures for 1926 and 1939 taken for projections. As scenario B is based on corrected age compositions for the 1926 and 1939 censuses, it is preferable.

24.6   Instability of Age Distribution

Analysis of the age-distribution evolution in demographic transition and its deformations due to the crisis events in the 1930s and 1940s must be supplemented with a comparison of real age compositions and the corresponding stable age distributions computed for periods close to the censuses. Such analysis allows researchers to evaluate the accumulated age-structure deformations relative to the reproduction regime at a given time. For that the concept of instability suggested in Pirozhkov (1976), which characterizes the extent of structural shift of real population provided that its reproduction regime remains unchanged, will be used.

Previously the instability was considered on the basis of a deterministic stable population model. Later the system of instability indexes was enlarged on the basis of the Leslie matrix model (Rubinov and Safarova, 1988). Within the framework of the Leslie matrix model, the property of strong ergodicity – that is, convergence of age distribution of the population, with reproduction regime being invariable, to certain limiting age composition called the stable population – takes place.

In all cases where the real process considered tends to some limit, the problem of measuring the deviation from this limiting state arises. The deviation of the real population age distribution from corresponding stable age composition (the instability) is estimated below.

The concept of instability is of great practical significance in analyzing the real population age composition. It is an integral estimate of the effect of the reproduction regime proper and socioeconomic factors determining discontinuity of the *normal* process of demographic development. Thus, the evolutionary and crisis factors affecting the age composition can be estimated with the aid of instability indexes.

For the instability measurement the following indexes are used: σ_t is the standard deviation of real population proportions from corresponding stable population proportions; $\mathbf{I}_t^P$ is the total relative deviation of real population from the stable one, suggested in Pirozhkov (1976); and the index $\mathbf{I}_t^R$ is the total weighted deviation of real population from the stable one (Rubinov and Tchystyakova, 1986). These coefficients can be written as

$$\sigma_t = \sum_{t=0}^{\omega} \sqrt{\frac{(\tilde{x}_t^i - \tilde{\nu}^i)^2}{\omega + 1}} \ , \tag{24.1}$$

where $\tilde{x}_t^i$ is the proportion of the real population $\mathbf{X}_t$ in age interval indexed by i at time t, and

$$\tilde{x}_t^i = x_t^i \ / \sum_{\kappa=0}^{\omega} x_t^\kappa \ . \tag{24.2}$$

Similarly, $\tilde{\nu}^i$ is the fraction of i-th stable population $\mathbf{V} = (\nu^0, \ \nu^1, \ldots, \nu^\omega)$ age group:

$$\mathbf{I}_t^P = \sum_{i=0}^{\omega} \frac{|\tilde{x}_t^i - \tilde{\nu}^i|}{\tilde{\nu}^i} \tag{24.3}$$

$$\mathbf{I}_t^R(\mathbf{X}_0, \mathbf{V}) = \frac{1}{(\bar{\mathbf{g}}, \mathbf{X}_0)} \sum_{i=0}^{\omega} \bar{g}^i \ \left| \frac{x_t^i}{\lambda_0^t} \left(\frac{\bar{\mathbf{g}}, \mathbf{X}_0}{\bar{\mathbf{g}}, \mathbf{V}} \right) \nu^i \right| \ , \tag{24.4}$$

where $(,)$ denotes the scalar product.

Here, $\bar{\mathbf{g}} = (\bar{g}^0, \bar{g}^1, \ldots, \bar{g}^\omega)$ is the reproductive value (the left eigenvector of the Leslie matrix corresponding to positive eigenvalue λ_0 of the Leslie matrix; the vector $\bar{\mathbf{g}}$ components represent the values of the corresponding age group's contribution to population reproduction). It should be noted, however, that this index characterizes instability only in reproductive age groups since, by definition, the vector $\bar{\mathbf{g}}$ components are equal to zero beyond the limits of reproductive age.

These indexes essentially depend on the initial age structure. The following index (stabilization rate) characterizes the age-structure instability for populations in the process of stabilization, when there is no influence of initial age composition (Safarova, 1984). The stabilization rate, which is the speed of convergence of real population to the stable age distribution, is measured by the ratio

$$\gamma = |\lambda_1| \ / \ \lambda_0 \ , \tag{24.5}$$

where λ_0 is the stable population growth rate (the dominant eigenvalue of the Leslie matrix), λ_1 is the next largest eigenvalue of the Leslie matrix. Deviation of the real population from the stable population at time t represents the sum of the value of order γ^t and higher. Moreover, γ permits evaluation of stabilization period. If an accuracy ξ is chosen, then the value t, beginning from which γ^t becomes less than ξ, is just the length of stabilization period.

For populations in completed demographic transition, the values of γ (for models with five-year age groups and five-year transition step)

Table 24.6. Instability indexes of age composition for European Russia and the USSR populations.

Year	Stabili- zation rate	Standard deviation	Relative deviation $\mathbf{I}_t^P$	Weighted deviation $\mathbf{I}_t^R$
1896–1897	0.79	0.0101	4.29	0.06
1926–1927	0.78	0.0085	2.49	0.11
1938–1939	0.78	0.0105	3.44	0.13
1958–1959	0.80	0.0105	3.56	0.13
1969–1970	0.78	0.0112	3.44	0.14
1978–1979	0.78	0.0087	2.57	0.08
1988–1989	0.80	0.0070	2.19	0.07

are close to 0.8. The populations of 30 developed countries have been examined. For 20 countries we have inequality $0.78 \leq \gamma \leq 0.82$ (i.e., a stabilization period of 140–175 years with $\xi = 10^{-3}$), and only for two countries we have $\gamma < 0.75$ or $\gamma > 0.85$.

The fact that the stabilization rate varies in sufficiently small limits is analogous with behavior of the stable population growth rate, which is close to one for modern-type reproduction regimes. For European Russia and the USSR the stabilization period varies from 140 to 160 years (*Table 24.6*).

The analysis of the instability indexes considered shows that the highest values have been obtained for the 1960s and the 1970s. During this period the age-distribution deformations caused by perturbations of the 1930s and 1940s resulted in significant deviations of the real age structure from corresponding stable population. At the same time, the values of indexes σ and, in particular, of $\mathbf{I}_t^P$ in 1896–1897 are distinguished by the higher magnitudes. This phenomenon can be explained by considerable deviation of the most numerous fraction of the 0–4 age group in the real population (15 percent) versus the stable one (19 percent), due to high levels of infant mortality and incomplete registration of births. The higher values of the instability index from the 1930s to the 1970s can also be explained by the great territorial differences in reproduction regimes of the Soviet population; the European republics and the Central Asian republics are in different phases of transition. In the 1980s, the USSR population age-distribution instability diminished smoothly. It approached the stabilized analogue, which is solely

Table 24.7. Changes in instability indexes depending on reproduction
regime changes.

Year	Standard deviation	Relative deviation $\mathbf{I}_t^P$	Weighted deviation $\mathbf{I}_t^R$
1896–1897	0.0336	27.50	0.25
1926–1927	0.0264	15.04	0.22
1938–1939	0.0182	8.10	0.16
1958–1959	0.0078	2.63	0.08
1969–1970	0.0074	2.24	0.07
1978–1979	0.0082	2.46	0.07
1988–1989	0.0070	2.19	0.07

dependent on the evolution of the reproduction regime. This indicates
a greater homogeneity of population reproduction for the Soviet Union.

Nevertheless for the USSR as a whole, with its heterogeneous population, the instability is sufficiently complicated depending on both age
composition and reproduction regime. Excluding the age-distribution
influence, the instability indexes are considered to depend on the variation of the reproduction regime. For that we calculate the instability
indexes taking the age distribution for the USSR in 1989 and the reproduction regimes successively for 1896–1897, 1926–1927, . . . , 1978–1979.
Table 24.7 shows that all indexes considered respond strongly to the
reproduction-regime changes; $\mathbf{I}_t^P$ responds most strongly. Values of all
indexes rapidly decline up to the 1950s and then stabilize. This behavior
reflects the changes in the total fertility rate and life expectancy which
changed drastically up to the 1950s and then varied insignificantly. All
indexes considered reach their minimal values when the age composition
corresponds to the reproduction regime.

24.7   Contribution of Age Distribution
to Future Population Growth

Age distribution is an independent factor in future demographic development. The initial age distribution to future population growth can
be evaluated using two techniques. The first is the calculation of the
demographic growth potential based on both the stationary and stable population-growth models; the second is the decomposition of the
rate of natural increase into two components – one depending on the

Table 24.8. The USSR population growth potential accumulated in the age distribution based on the Bourgeois-Pichat technique.

Year	Gross potential	Net potential
1896–1897	1.73	1.26
1926–1927	1.74	1.44
1938–1939	1.59	1.34
1958–1959	1.35	1.24
1969–1970	1.29	1.17
1978–1979	1.24	1.16
1988–1989	1.21	1.14

reproduction-regime intensity and another depending on the age distribution. The procedures for computating the demographic growth potential have been developed by Bourgeois-Pichat (1971) and Andreev and Pirozhkov (1975), and the procedures for evaluating the population growth-rate components are described by Preston (1974).

The estimates of growth potential based on the stationary population model (*Table 24.8*) indicate a steady decline of the contribution of age composition to future population growth. The structure loses the portion of its potential accumulated at the expense of past high fertility levels, which indicates the end of demographic transition in most parts of the USSR territory by the early 1990s.

More detailed studies of the age-distribution contribution to the rate of natural increase can be done by decomposing the coefficient into two components: the first depending on the reproduction-regime intensity and the second depending on the age structure (*Table 24.9*). The growth rate has been decomposed using the following formulas (Kitagava, 1955):

$$r1 = [(1 + 1/Ro)/2](b - bsRo) - d + ds \qquad (24.6)$$

$$r2 = [(1 - 1/Ro)/2](b + bsRo) , \qquad (24.7)$$

where Ro is the net reproduction rate; b is the crude birth rate; bs and ds are the stationary population birth and death rates of equal value.

From *Table 24.9* it can clearly be seen that prior to demographic transition in the USSR the age structure was unessential in the growth rate of the population. Its effect markedly increases with rearrangement of the reproduction regime. Only in the last decade (1979–1989) does the contribution of the structure factor to population growth begin to

Table 24.9. Trends in components of the USSR population rate of natural increase, per thousand people.

		Components depending on	
	Total	Age structure	Reproduction regime
Year	r	$r1$	$r2$
1896–1897	17.7	−2.2	19.9
1926–1927	23.5	7.0	16.5
1938–1939	19.6	9.3	10.3
1958–1959	17.7	13.2	4.5
1969–1970	9.1	7.2	1.9
1978–1979	8.3	7.5	0.8
1988–1989	8.7	6.5	2.2

decrease. This appears to be associated with a decrease of age-structure instability and deformations due to crises.

24.8   Conclusion

Analysis of changes in the age–sex distribution of the USSR population over 90 years of demographic history demonstrates its considerable evolution resulting from changes in demographic processes. The essential nature of this evolution was expressed in both population aging and a reduction in the contribution of age structure to population growth. These trends are characteristic of countries that have passed through demographic transition and where the modern regime of population reproduction has been established.

At the same time, an important characteristic of change in the USSR population age distribution consisted in significant deformations of separate age cohorts due to crisis phenomena. The approximate estimates of population losses during the 1930s and 1940s described in this chapter differ from those cited in other sources. The estimates in this chapter include indirect losses expressed in decreases in the number of births of subsequent generations. These losses of the future, comparable with the direct population losses caused by physical destruction, serve as evidence of deformation of Soviet demographic development and are reflected for many decades in the lives of succeeding generations.

Appendix

Table 24A.1. Age–sex distributions of the USSR population according to 1939 census data and scenario A obtained from 1926 census data and trends of fertility and mortality not subject to crisis phenomena of the 1930s, in thousands of people.

Age group	Total Census	Scenario A	Males Census	Scenario A	Females Census	Scenario A
0–4	21,731	27,505	10,971	13,750	10,760	13,755
5–9	17,776	26,296	8,864	13,087	8,912	13,209
10–14	21,951	24,652	10,990	12,438	10,961	12,214
15–19	15,198	16,517	7,467	8,293	7,731	8,224
20–24	14,260	15,064	6,903	7,551	7,357	7,513
25–29	16,644	15,837	8,131	7,798	8,513	8,039
30–34	13,935	14,236	6,940	6,825	6,995	7,411
35–39	11,586	11,793	5,411	5,555	6,175	6,238
40–44	8,511	9,605	3,959	4,384	4,552	5,221
45–49	6,822	7,666	3,099	3,558	3,723	4,108
50–54	5,775	6,621	2,613	3,078	3,162	3,543
55–59	5,147	5,260	1,997	2,479	3,150	2,781
60–64	4,281	4,158	1,736	1,908	2,545	2,250
65–69	3,141	3,135	1,232	1,355	1,909	1,780
70–74	1,867	2,262	694	940	1,173	1,322
75–79	1,066	1,369	392	558	674	811
80+	866	723	295	279	571	444
Total	170,557	192,699	81,694	93,836	88,863	98,863

Table 24A.2. Reconstructed age–sex distribution of the USSR population according to 1939 census data and scenario B based on reconstructed 1926 census data and trends of fertility and mortality not subject to crisis phenomena of the 1930s, in thousands of people.

Age group	Total Census	Scenario B	Males Census	Scenario B	Females Census	Scenario B
0–4	21,749	27,805	10,980	13,900	10,769	13,905
5–9	18,091	26,133	9,026	13,003	9,065	13,130
10–14	21,035	23,266	10,525	11,678	10,510	11,588
15–19	15,479	17,333	7,630	8,646	7,849	8,687
20–24	14,033	15,574	6,730	7,610	7,303	7,964
25–29	16,325	16,101	7,917	7,918	8,408	8,183
30–34	13,699	14,252	6,811	7,047	6,888	7,205
35–39	11,398	11,725	5,343	5,608	6,055	6,117
40–44	8,543	9,652	3,977	4,817	4,566	4,835
45–49	6,800	7,274	3,098	3,402	3,702	3,872
50–54	5,735	6,114	2,553	2,788	3,182	3,326
55–59	5,055	5,169	2,003	2,252	3,052	2,917
60–64	4,161	4,250	1,696	1,818	2,465	2,432
65+	6,771	6,895	2,551	2,788	4,220	4,107
Total	168,874	191,543	80,840	93,275	88,034	98,268

Table 24A.3. Age–sex distributions of the USSR population according to 1959 census data and scenario A based on 1939 census data and trends of fertility and mortality not subject to crisis phenomena of the 1940s, in thousands of people.

Age group	Total		Males		Females	
	Census	Scenario A	Census	Scenario A	Census	Scenario A
0–4	24,333	27,878	12,405	14,099	11,928	13,779
5–9	22,029	27,600	11,203	13,894	10,826	13,709
10–14	15,337	28,568	7,808	14,303	7,529	14,265
15–19	16,472	28,159	8,259	14,002	8,213	14,157
20–24	20,343	21,827	10,056	10,696	10,287	11,131
25–29	18,190	18,741	8,917	9,291	9,273	9,450
30–34	18,999	22,949	8,611	11,415	10,388	11,534
35–39	11,590	15,636	4,528	7,620	7,062	8,016
40–44	10,408	14,352	3,998	6,861	6,410	7,491
45–49	12,264	16,474	4,706	7,902	7,558	8,572
50–54	10,447	13,437	4,010	6,512	6,437	6,925
55–59	8,699	10,662	2,906	4,775	5,793	5,887
60–64	6,696	7,343	2,347	3,207	4,349	4,136
65+	13,019	13,228	4,296	5,102	8,723	8,126
Total	208,826	266,854	94,050	129,679	114,776	137,178

Table 24A.4. Reconstructed age–sex distribution of the USSR population according to 1959 census data and scenario B based on reconstructed 1939 census data and trends of fertility and mortality not subject to crisis phenomena of the 1940s, in thousands of people.

Age group	Total		Males		Females	
	Census	Scenario B	Census	Scenario B	Census	Scenario B
0–4	24,347	27,547	12,413	13,932	11,934	13,615
5–9	22,276	27,336	11,328	13,761	10,948	13,575
10–14	15,201	28,333	7,660	14,185	7,541	14,148
15–19	16,716	28,037	8,291	13,941	8,425	14,096
20–24	19,334	21,451	9,573	10,716	9,761	10,735
25–29	18,395	19,121	8,991	9,459	9,404	9,662
30–34	18,691	21,766	8,449	10,781	10,242	10,985
35–39	12,233	16,120	4,829	7,897	7,404	8,223
40–44	10,654	14,224	4,098	6,708	6,556	7,516
45–49	12,455	16,253	4,771	7,756	7,684	8,497
50–54	10,616	13,209	4,048	6,377	6,568	6,832
55–59	8,640	10,411	3,006	4,705	5,634	5,706
60–64	6,537	7,442	2,366	3,253	4,171	4,189
65+	12,940	13,092	4,335	5,084	8,605	8,008
Total	209,035	264,342	94,158	128,555	114,877	135,787

References

Anderson, B., and B. Silver, 1985. Demographic Analysis and Population Catastrophes in the USSR. *Slavic Review* **244**(3):517-536.

Andreev, E.M., and S.I. Pirozhkov, 1975. O potentsiale demograficheskogo rosta (On Demographic Growth Potential). In *Naseleniye i okruzayuschaya sreda* (Population and Environment). Statistika, Moscow.

Andreev, E.M., L.E. Darsky, and T.L. Kharkova, 1990. Istoriya naseleniya SSSR, 1920-1959 gg (The History of the USSR Population, 1920-1959). In *Ekspress-informatsiya, Seriya "Istoriya statistiki"* (Express-Information, History of Statistics Series), Issue 3-5 (Part I). Goskomstat SSSR, Moscow.

Belova, V.A., G.A. Bondarskaya, and L.E. Darsky, 1988. Sovzemennye problemy i perepektivy rozhdayemosti (Current Problems and Prospects of Fertility). In A.G. Volkov, ed., *Methodologiya demograficheskogo prognoza* (Backgrounds of Population Forecasting). Nauka, Moscow.

Bourgeois-Pichat, J., 1971. Stable, Semi-stable Populations and Growth Potential. *Population Studies* **225**(2):235-254.

Keyfitz, N., 1987, *The Demographic Discontinuity of the 1940s*. WP-87-92. IIASA, Laxenburg.

Kitagava, E., 1955. Components of a Difference between Two Rates. *Journal of the American Statistical Association* **2500**:1168-1194.

Kvasha, A.J., 1990. Tsena Pobed (Cost of Victories). In V.I. Mukomel, ed., *SSSR: Demograficheskiy diagnoz* (USSR: Demographic Diagnosis). Progress, Moscow.

Maksudov, S., 1989. *Poteri naselniya SSSR* (Population Losses in the USSR). Chalidze Publications, Benson, VT (in Russian).

Pirozhkov, S.I., 1976. *Demograficheskiye protsessy i vozrastnaya structura naseleniya* (Demographic Processes and Population Age Distribution). Statistika, Moscow.

Preston, S., 1974. Empirical Analysis of the Contribution of Age Composition to Population Growth. *Demography* **270**(4):417-432.

Rashin, A.G., 1956. *Naseleniya Rossii za 100 let (1811-1913): Statististicheskiye otcherki* (Population of Russia over a Hundred Years (1811-1913): Statistical Reports). Gosstatizdat, Moscow.

Rubinov, A.M., and N.E. Tchystyakova, 1986. Vozrastnaya structura i potentsial rosta naseleniya (Age Composition and Population Growth Potential). In A.G. Volkov, ed., *Demograficheskiye protsessy i ikh zakonomernosti* (Demographic Processes and Their Regularities). Mysl, Moscow.

Rubinov, A.M., and G.L. Safarova, 1988. Pokazateli instabil'nosti vozrastnoy structury naseleniya (Population Age Distribution Instability Indices). In *Matematicheskoye modelirovaniye socialno-ekonomicheskikh i demograficheskikh protsessov* (Mathematical Simulation of Socioeconomic and Demographic Processes). Theses of All-Union Scientific Conference, Erevan.

Safarova, G.L., 1984. Otsenka skozosti stabilizatsii i tempa rosta stabil'nogo naseleniya v matrichnoy modeli vosproizvodstva naseleniya (Estimate of Stabilization Rate and Stable Population Growth Rate in Matrix Model of Population Reproduction). *Matematicheskiye metody v socialnikh naukakh* (Mathematical Methods in Social Sciences) (17):66–85.

Smertnost' i prodolzhitel'nost zhizni naseleniya SSSR, 1926–1927: Tablitsy smertnosti 1930 (Mortality and the Expectation of Life in the USSR, 1926–1927: Life Tables), 1930. Plankhozgiz, Moscow-Leningrad.

Tzaplin, V.V., 1989. Statistika zhertv stalinizma v 30-ye gody (Statistics of Stalin's Victims in the 1930s). Vopzosy istorii (*Problems of History*) (4):175–181.

Volkov, A.G., 1990. Perepis' naseleniya 1937: Vymysli i pravadal (The Census of 1937: Fabrication and Truth). In *Express-informatsiya Seriya, Istoriya Statistiki* (Express-information, History of Statistics Series), Issue 3–5 (Part II). Goskomstat, Moscow.

Chapter 25

Beyond Stable Theory: Intercohort Changes in USSR, USA, and Europe

Nathan Keyfitz

How large the upcoming generation will be is the most important single question of demography. Births statistics provide an indication of this – at least each year's births measure the size of the new cohort as it starts through life. Because it is subject to mortality as it gets older, measuring the sizes of the generations over time is more informative if one displays not exactly the births, but the births less the deaths of those dying young. We would want to subtract at least the infant deaths from the births, they having no influence on the size of the current generation when it reaches adulthood, nor the size of the one that will descend from it, and similarly to subtract the deaths between infancy and the age when reproduction starts, plus indeed some of the deaths that take place during the ages of reproduction.

And for many purposes we may find it less important to know about the absolute number of individuals in successive cohorts, than

The present paper is an application of theory developed in "The Profile of Intercohort Increase," *Mathematical Population Studies*, 1990 **2**(2):105–117.

the changes in their numbers from year to year. That is the quantity that here and elsewhere is called "intercohort change."

Not only is the number a desirable one for assessing the movement of a population, but it happens to be calculable with a minimum of data – nothing but a series of censuses is needed. The intercohort change can be calculated by a method that in no way depends on birth and death statistics. A civil registration system of sufficient accuracy for demographic purposes can come into existence only after development is well advanced. Complete and usable birth statistics for the United States and Canada were complete only in the 1920s; it had taken more than a century for those countries to put their civil registration in place.

The method of calculation of the intercohort change is straightforward, consisting as it does only of noting and averaging the difference between persons of given age in successive censuses. If the number of persons aged 30–34, say, in 1950 was $_{30-34}P_{1950}$ and the number of the same age in 1955 was $_{30-34}P_{1955}$, then one estimate of the intercohort change in question is the difference of these two numbers:

$$_{30-34}P_{1955} - {}_{30-34}P_{1950} .$$

As long as an appreciable number of individuals in each of the two cohorts remains alive every pair of censuses will give not one estimate of the increase, but an estimate for each separate age.

To compare these estimates as obtained from the several ages verifies the accuracy, or at least the consistency, of the censuses in question. It also checks the appropriateness of the method in the face of possible large variations in mortality. With typical mortality and good censuses the numbers fall in a narrow range. Using the young and middle ages of life typically gives estimates of the intercohort change varying by less than 5 percent and often only 1 or 2 percent. Where variation is larger than this there have been either unusual fluctuations in death rates, or else simple errors in the data. Thus we have here one of the few demographic measures where the calculation of error is built into the data.

The present paper contrasts the approach here presented with that of stable theory; gives a detailed example of the calculation and exhibits the error of the method; shows results for the USSR, and compares them with those for Europe and the USA, as well as the Less Developed Countries (LDCs) and the More Developed Countries (MDCs).

25.1   Contrast with Stable Population Theory

Stable population has served many purposes in its time. Over periods when the rates of birth, age by age of mother, and rates of death, again age by age of the person, are relatively constant, then the age distribution of the population is determinate. But in a time such as the present, when everything is changing, the representation of actual populations by the stable models is very approximate.

In fact at no time does the stable model calculate age distributions in this way with great accuracy. It has been useful in the opposite direction, in times of relative stability, and when there were no figures at all on births and deaths. Various devices were used and they all have their place in the demographic armamentarium.[1] Once mortality started to decline as the first phase of the demographic transition the stable theory seems less applicable, but there a device due to Ansley Coale has rescued the general approach (Coale, 1963).

Though stable theory may be far from an adequate representation of real populations, yet it does serve for thinking about them. It answers the question, "If everything else remains unchanged, and the birth rate falls by so much, what is the effect on the proportion of people over age 65?" or "If the birth rate falls to bare replacement, so that ultimately the population will stabilize, how much higher will it be than at the time when the birth rate fell?" or "Which is more important for the proportion of people, a small fall in the birth rate, or a small fall in the death rate at the older ages only?" It can be shown that a fall in the death rate equally at all ages has no effect on the age distribution.

This paper is concerned with the superposition on relative stability of a sharp discontinuity in the number of births surviving to maturity. After World War II mortality, especially infant mortality, fell sharply, and as a result of that (and possibly some increase in fertility as well) the population started sharply on an upward course. This was conspicuously shown in the post–World War II census age distributions for certain countries of Southeast Asia and elsewhere. It is well illustrated by the chart of age distributions for the world as a whole shown as *Figure 25.1*.

The device for analyzing age distributions was discussed earlier (Keyfitz, 1990) and applied to certain less developed countries of Southeast Asia. Here it will be tested on two countries for which data are available on mortality and fertility.

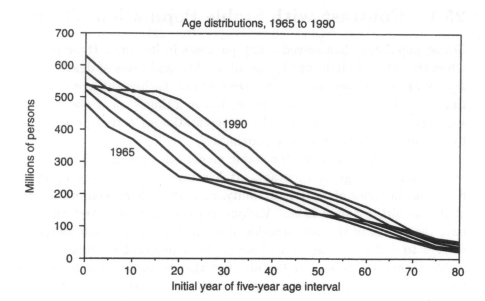

Figure 25.1. World population age distributions, 1965–1990, United Nations data. Note discontinuities for those born just after World War I and World War II.

The number of children 0–4 at last birthday in the USSR in 1950 is estimated at 18,823,000. Here as throughout we use UN estimates, and imagine them to have been derived from censuses taken in the years ending in 0 and 5 (*Table 25.1*). For convenience the entire range of ages is not shown; what appears in this and subsequent tables up to *Table 25.8* is a window covering ages up to 30, the last age group being 25–29 at last birthday. These six age groups will indicate the method used, that of course is applicable to the whole range of ages, and the whole period from 1950 to 2025, using UN estimates for the years after 1985.

The 18,823,000 figure represents births during the period 1945–1950, less the deaths among those births by the time the oldest of them had reached age 5. We can talk of them as the 1945–1950 cohort, and we can follow them down through life from *Table 25.1*. Thus five years later, in 1955 they would have numbered 18,552,000; by 1960 they would have declined to 18,511,000, and so on.

The decline is of course due mostly to mortality. The pure mortality comparison is disturbed by migration, either in or out, but in respect of so large a base population the amount of migration could not be a large

Table 25.1. Original age by year as estimated for the USSR by the United Nations, in thousands of persons, counted in hypothetical censuses at five-year intervals.

Age	1950	1955	1960	1965	1970	1975
0–4	18,823	22,896	24,512	23,487	21,072	21,606
5–9	13,296	18,552	22,736	24,374	24,005	20,906
10–14	22,037	13,166	18,511	22,707	25,073	23,864
15–19	17,530	21,824	13,135	18,478	22,469	24,797
20–24	19,669	17,323	21,736	13,083	18,109	22,199
25–29	13,960	19,395	17,223	21,611	13,859	16,969

proportion. We find, in fact, that the decline in the size of the cohort as we go forward in time is not strictly regular, and the irregularity must be due to varying levels of mortality, as well as to varying amounts of migration.

We may contrast the 1945–1950 cohort with that born in the following five years, the 1950–1955 cohort. The number of these latter caught in a census that might have been taken in the year 1955 would have been 22,896,000 the number caught in a 1960 census 22,736,000, etc. Again a fall in most five-year periods, as one would expect.

We could apply a life table backward to find the number of births that would have given rise to these individuals, after we had made suitable allowance for migration. To do so would have been to define the size of the cohort by the number of its births. But there are some advantages in not going through that, aside from the additional element of arbitrariness that it introduces.

Instead define the size of the cohort as the number of individuals in later life, say as the average in some sense of the persons alive at successive ages, at least up to the ages at which mortality becomes heavy. The cohort of course, 1000 years after it is born, ends up at size zero, and we do not average in the older ages, but think of its size during the years when mortality is moderate, say up to about age 50 or 60. The cohort size will be used in a way that will not be disturbed by the vagueness, and the reader is asked to bear with it for the moment.

The same data are rearranged in *Table 25.2*, so that cohorts fall in columns. Now the 18,552,000 supposed to be counted in 1955 are placed under the 18,823,000 imagined to be counted in 1950, and so forth.

Table 25.2. Data from *Table 25.1* shifted to place cohorts in columns, in thousands of persons.

Year in which the cohort is caught by a (hypothetical) census

Age	1925	1930	1935	1940	1945	1950	1955	1960	1965	1970	1975
0–4						18,823	22,896	24,512	23,487	21,072	21,606
5–9					13,296	18,552	22,736	24,374	24,005	20,906	
10–14				22,037	13,166	18,511	22,707	25,073	23,864		
15–19			17,530	21,824	13,135	18,478	22,469	24,797			
20–24		19,669	17,323	21,736	13,083	18,109	22,199				
25–29	13,960	19,395	17,223	21,611	13,859	16,969					

Table 25.3. Five-year increases with cohorts as columns, in thousands.

Age	1925	1930	1935	1940	1945	1950	1955	1960	1965	1970	1975
0–4						4,073	1,616	−1,025	−2,415	534	
5–9					5,256	4,184	1,638	−369	−3,099		
10–14				−8,871	5,345	4,196	2,366	−1,209			
15–19			4,294	−8,689	5,343	3,991	2,328				
20–24		−2,346	4,413	−8,653	5,026	4,090					
25–29	5,435	−2,172	4,388	−7,752	3,110						

Table 25.3 shows the difference of size of the several successive cohorts, as these are estimated afresh from each age and each pair of time periods. Thus the number in the 1945–1950 cohort as counted at age 0–4 is 18,823,000, and the number in the 1950–1955 cohort estimated at the same age is 22,896,000, so the intercohort difference is 4,073,000. And we can obtain a second estimate of the same intercohort difference by using ages 5–9, and taking 22,736,000–18,552,000, a difference that amounts to 4,184,000. And then when we go on to the difference obtained from ages 10–14, on the third line of *Table 25.2*, and obtain 4,196,000, and so on forward (*Table 25.3*).

Now the remarkable feature of *Table 25.3* is the constancy within columns. It is this constancy, the fact that the intercohort difference comes out very nearly the same no matter what age and time one calculates it from, that we will take advantage of in this paper. Most of the columns show some decline in the last row, due to the effect of mortality, but it is remarkable how little mortality interferes on the whole. All of the items under 1940 are negative corresponding to the lower fertility and greater infant mortality in all cohorts of 1940–1945 compared with 1935–1940; all of the items in the column for 1945 are positive and about 5 million in size (except for the last) due to the improvement from 1940–1945 to 1945–1950. In short the variation between columns is vastly greater than the variation between rows.

Table 25.4 summarizes *Table 25.3*, showing the lowest value in each column, the highest value, and the average. This is the result that we are looking for, especially the last column. The number opposite 1935, for example, is the difference between the 1930–1935 cohort and the 1935–1939 cohort; the next number below that one is the difference between the 1935–1939 and the 1940–1945 cohort, the large negative quantity being children not born during the war or dying in infancy.

The numbers here are only intended to exhibit the arithmetic, and for convenience they are confined to a 6 × 6 window on the age–time table, whose full extent is 16 × 16. Somewhat different numbers will appear in *Table 25.10*, which uses the whole range of data provided by the United Nations.

We have now reached the end of the work except for checking. If the method is at all accurate it ought to serve to reconstruct the original age–time data, *Table 25.1*. It ought to be possible to do this with the intercohort differences, provided we can take any one cohort from the

Table 25.4. Five-year intercohort increase.

Middle Year	Low	High	Mean
1925	5,435	5,435	5,435
1930	−2,346	−2,100	−2,259
1935	4,294	4,413	4,365
1940	−8,871	−8,000	−8,491
1945	3,110	5,345	4,816
1950	3,991	4,196	4,107
1955	1,616	2,366	1,987
1960	−1,209	−369	−868
1965	−3,099	−2,400	−2,757
1970	534	534	534

Table 25.5. Central cohort, born 1945–1950, in thousands, USSR.

Age	Number of people
0–4	18,823
5–9	18,552
10–14	18,511
15–19	18,478
20–24	18,109
25–29	16,969

observations. Let us take that born in 1945–1950 (*Table 25.5*) as the central cohort and add and subtract the intercohort differences shown as the right-hand column of *Table 25.4*.

That gives us *Table 25.6*, which comes close to reproducing the original *Table 25.1*. Insofar as it achieves this we have in effect shown that *Table 25.1* is redundant; that rather than containing $6 \times 6 = 36$ items of data it contains only $6 + 10 = 16$ items. In general, if we had a time showing n ages for n equally spaced points of time it would contain not $n \times n$ units of information, but only $3 \times n - 2$ items.

Table 25.7 shows the departure of *Table 25.6* from *Table 25.1* in thousands of persons, and *Table 25.8* shows the same departure in parts per thousand. At most ages these are comparable with the error of the census.

Because of the large swings in mortality in the USSR the discrepancy of *Table 25.8* is larger than the result of a similar calculation for the

Table 25.6. Reconstituted age–time table, in thousands, USSR.

Age	1950	1955	1960	1965	1970	1975
0–4	18,823	22,930	24,917	24,049	21,292	21,826
5–9	13,736	18,552	22,659	24,646	23,778	21,021
10–14	22,186	13,695	18,511	22,618	24,605	23,737
15–19	17,788	22,153	13,662	18,478	22,585	24,572
20–24	19,678	17,419	21,784	13,293	18,109	22,216
25–29	13,103	18,538	16,279	20,644	12,153	16,969

Table 25.7. Discrepancy of reconstituted table from original table, in thousands.

Age	1950	1955	1960	1965	1970	1975
0–4	0	34	405	562	220	220
5–9	440	0	−77	272	−227	115
10–14	149	529	0	−89	−468	−127
15–19	258	329	527	0	116	−225
20–24	9	96	48	210	0	17
25–29	−857	−857	−944	−967	−1,706	0

United States and Europe. For them, corresponding to *Table 25.8*, we have *Table 25.9*. For Europe and the United States four–fifths of the discrepancies are less than 1 percent, most of them much less.

I felt it necessary also to verify the GAUSS program that produced these results. This was easily done by feeding in as though they were data the reconstituted *Table 25.6*, which ought to produce discrepancies of zero in *Table 25.8*. That is exactly what turns up – there is no need to show here the table of zeros that emerges in place of *Tables 25.7, 25.8*, and *25.9* when one uses *Table 25.6* as input to the same program.

Table 25.8. Discrepancy in parts per thousand.

Age	1950	1955	1960	1965	1970	1975
0–4	0	1	17	24	10	10
5–9	33	0	−3	11	−9	6
10–14	7	40	0	−4	−19	−5
15–19	15	15	40	0	5	−9
20–24		6	2	16	0	1
25–29	−61	−44	−55	−45	−123	0

Table 25.9. Discrepancy in parts per thousand.

Age	1950	1955	1960	1965	1970	1975
Europe						
0–4	0	8	13	13	15	16
5–9	–5	0	4	4	3	1
10–14	–9	–2	0	1	–1	0
15–19	1	6	8	0	3	0
20–24	–4	–6	1	–10	0	–17
25–29	18	18	20	8	9	0
USA						
0–4	0	2	2	10	20	21
5–9	–5	0	7	17	20	14
10–14	–18	–11	0	4	2	–8
15–19	4	6	9	0	-6	–14
20–24	13	18	15	0	0	–6
25–29	6	6	4	3	7	0

Table 25.10 is not based on the above small window on the age–time
distribution, nor on the full distribution. It is desirable to use as much
of the data as possible, of course, but also to recognize that the oldest
ages are given with much less certainty. Hence for the analysis proper,
exhibited in *Table 25.10*, the work was confined to a 14 × 14 table,
which is to say using the ages from 0–4 to 65–69, and population figures
for the years from 1950 to 2020.

The numbers are shown for the USSR along with two other areas,
Europe as a whole and the United States. Again we think of the year
marked on the left as standing between the two five-year cohorts being
compared. At the start of the table all three areas showed rising cohort
sizes; then Europe's births started to fall and the fall was accentuated by
World War I. The big decline in cohort size from 1910–1915 to 1915–1920
in Europe was succeeded by an even larger rise five years later. Rises
continued through the 1920s in Europe and the USSR, but not in the
USA, whose cohort size started to fall in the late 1920s, and accelerated
in the 1930s.

Very different experience in World War II is shown by Europe and
the USSR, on the one hand, and the United States, on the other. The
USSR cohort of 1940–1945 was short by 7,354,000 compared with the
preceding cohort, while the corresponding US cohort was higher by
2,412,000 and Europe was intermediate between these two conditions.

Table 25.10. Five-year intercohort increase of period after the given date over the period before the given date.

Year	Europe	USSR	USA	World	LDCs	MDCs
1895	2,720	1,136	495	5,552	1,028	4,524
1900	2,575	1,450	552	11,626	6,635	4,991
1905	1,586	1,601	989	11,664	6,941	4,723
1910	−494	650	701	8,357	6,990	1,366
1915	−6,147	−2,316	595	224	7,331	−7,108
1920	7,841	3,291	795	24,120	11,498	12,622
1925	430	5,135	−414	20,087	13,665	6,422
1930	−194	−2,102	−826	13,192	15,329	−2,136
1935	506	3,694	474	15,254	10,490	4,764
1940	−780	−7,354	2,412	8,592	12,589	−3,997
1945	4,201	3,457	3,526	44,303	31,328	12,975
1950	1,364	4,883	2,404	58,843	51,338	7,505
1955	1,156	2,232	1,794	33,037	28,734	4,303
1960	1,322	−901	−363	51,340	50,775	565
1965	−228	−2,909	−2,433	46,055	51,134	−5,079
1970	−2,679	713	−1,180	23,163	25,371	−2,209
1975	−2,241	1,615	33	−3,417	−1,181	−2,236
1980	−1,512	1,886	1,407	41,941	41,004	936
1985	−626	118	359	47,782	48,453	−671
1990	−150	−1,274	−536	44,308	45,662	−1,354
1995	−416	−289	−527	22,093	22,917	−823
2000	−911	−115	479	6,236	6,923	−686

After the war there was a baby boom in all three areas, slightly earlier in Europe than in the United States. It tapered off in all three cases around 1960.

For further comparison the numbers for the LDCs and the MDCs are also shown in *Table 25.10*. The LDCs plus the MDCs add up to the world total, an additivity due to all of the steps in the calculation being linear. Note the negative value for the MDCs in the early 1930s, while the LDCs remained positive, though at a lower rate than earlier; the sharp rise in both after the War, that in the MDCs being short-lived, while in the LDCs the intercohort has kept increasing. The evolution of the LDCs includes increases of about 7 million per five-year cohort per five-year period up to about the time of World War I, and after that a jump to something like 12 million; following World War II another jump in a rise to about 50 million, all numbers referring to increases per five-year period for five-year cohorts. Working from census data alone we have thus arrived at the point from which the population explosion can be defined: the 1945–1950 change from five years earlier was more than double that of 1940–1945, and that of 1950–1955 in turn nearly double that of 1945–1950.

To follow the dating of the table, think of each of the years in the stub as a pivot, with the numbers of the table proper showing what happened just after compared with what happened just before the corresponding year in the stub. Referring to a different source, what birth statistics and (not entirely independent) estimates of births and deaths are available confirm the general magnitudes of the numbers in *Table 25.10.*[2]

25.2   Political Effects of the Discontinuity

We know that large cohorts are likely to have a hard time in a competitive economy; there is plenty of evidence for developed countries that their wages will be lower, their unemployment rates higher, and their career advancement slower, all compared with a small cohort that follows or precedes them. A member of a small cohort has fewer competitors for education and jobs. Of course the jobs can expand and contract, but there is some cost and delay in such adaptation of the economy to the personnel available and it is that cost and delay that disadvantages the large cohort.

All this is reversed when we move from the economic to the political realm. In a representative democracy the larger the group the more votes it will have and hence the more influence in the administration and the legislature. If there is no democracy the larger group will still have more weight, exercised perhaps through peaceful political pressure, perhaps through violent action in the streets. It may not be possible to vote the dictatorship out of power, but it is possible to make trouble for it in many ways, and the degree of troublemaking of a dissident group must be something like proportional to its numbers. Those disadvantaged economically seek political action as an offset.

The interplay of the economic and political is to be seen in many countries. Shortage of jobs they consider suitable for the entering members of the labor force causes many of them to become dissidents; when the situation becomes bad enough to drive people to violence, as in Indonesia in 1965 and in Chile today, the dissident youth forms the spearhead of the attack on the established power – Sukarno and Pinochet, respectively. Burmese conditions are shaping up in what has become a classical cycle of dictatorship, worsening economy, protest, and repression.

Within the disadvantaged cohort there is one group that is especially dissatisfied, and toward which all governments act with circumspection. This is the body of students, especially university students. There are several reasons why they can be dangerous to established authority:

1. They are likely to suffer more than the mass of the population from their large numbers and inadequate policies. Peasants too can be adversely affected by their numbers, in that they face land shortages, but it seems to take more deprivation to bring peasants to violence than educated city people. Perhaps the greater ease of communication and assembly in the city help the protest movement.

2. It is not so much deprivation that gets people out onto the streets, but their position relative to some reference group, and relative to expectations that can no longer be fulfilled. With the advent of independence 40 or so years ago there were in most developing countries very few educated people; those that there were could rise quickly to the heights of power, influence, and wealth. In many countries it is these fortunate individuals, members of very small cohorts of the schooled, who are the reference point, and relative to whom the subsequent very large number of the schooled feel themselves deprived.

3. Not yet being established in life the students have no job, house, or other possession at stake. They have no family to support, always a hostage for good behavior.

4. Education has developed their analytical skills. It enables them to understand their own interests, to forecast the consequences for their own future of existing policies and trends, and to draw conclusions on corruption in high places, to see through the inadequacies of official policies.

5. Being articulate they are able to interpret the economic problem for others, and cause the dissatisfaction to spread. And this interpretation (be it expressed in the language of Marxism or otherwise) they

can disseminate in writing and orally, sometimes using officially permitted means of communication, sometimes underground. In such
matters interpretation is everything; when peasants suffer hardship
that they think is due to bad weather they will be patient; if the
hardship is seen as due to bad management and downright exploita-
tion, then they are responsive to a call for action. The means to
disseminate a particular interpretation of events is a major source
of power, and not only in a democracy, and the fluent educated
young can often have access to radio and press that they use to
counter official interpretation.

6. For all these reasons the educated tend to have power, and it is the
young who are educated. In a stationary condition there may be no
difference in education between the generations, but the presently
developing countries are not stationary. A typical member of the
older generation may be illiterate or have primary schooling only,
and to him the college (or even secondary school) students have the
mystique of knowledge. Perhaps they are too young for actual rule,
and will have to mobilize behind a more established figure, but the
prestige of the knowledge that they are deemed to carry is great.

The authorities' fears of a thwarted, underemployed, perhaps over-
educated, youth cohort are not always imaginary. Examples of trou-
blesome groups include the Naxalities in India, the People's Liberation
Front in Ceylon, the Kabataang Makabayan in the Philippines, all heavily weighted with youth. On the other hand, the worried authorities can
easily exaggerate the degree of radicalism of student protest. At the
present time in Indonesia, for example, student opposition is mostly reformist within the framework of the existing military government, but
we cannot say how long it would remain purely reformist without the
supervision of the military regime.

We are faced in all this with a problem of imputation. All we know
is that there are (1) high proportions of young people and (2) much
unrest among young people. How far (1) *causes* (2), and what are the
conditioning circumstances under which (1) causes (2), scholars have yet
to say.

Beyond the problem of imputation is changing conditions. At the
time of the Tbilisi Conference young people were in revolt, in many
countries with violence, and almost always concerned with employment.
Now there is still violence but youth is not the main actor, which is rather

separatist movements. On one day recently international dispatches carried word of 11 different struggles for independence around the world, some by constitutional means but the majority violent. Of course this also may be related to the youth cohort, in that it is the young who are doing the fighting, and they hope that with independence their life chances will be better, but their object is not in the main material.

Notes

[1] One particularly ingenious device was due to the late Jean Bourgeois-Pichat (1957).
[2] As provided by the United Nations and other sources, for instance, excerpted in Keyfitz and Flieger (1990).

References

Bourgeois-Pichat, J., 1957. Utilisation de la notion de population stable pour mesurer la mortalité et la fécondité des populations sous-développés. *Bulletin de l'Institut International de Statistique* (Actes de la 30e Session).

Coale, A., 1963. Estimates of Various Demographic Measures through the Quasi-stable Age Distribution. *Emerging Techniques in Population Research: Proceedings of the 1962 Annual Conference of the Milbank Memorial Fund.* 175–193.

Keyfitz, N., 1990. The Profile of Intercohort Increase. *Mathematical Population Studies* 2(2):105–117.

Keyfitz, N., and W. Flieger, 1990. *World Population Growth and Aging: Demographic Trends in the Late Twentieth Century.* University of Chicago Press. Chicago, IL.

United Nations, 1988. *World Demographic Estimates and Projections, 1950–2025*, p. 385, New York, NY.

Index

Nationalities are listed under the names of the Republics. All references are to USSR unless stated otherwise.

477